T0039113

Methods in Molecular Biology™

Series Editor
John M. Walker
School of Life Sciences
University of Hertfordshire
Hatfield, Hertfordshire, AL10 9AB, UK

For further volumes:
http://www.springer.com/series/7651

Computer-Aided Tissue Engineering

Edited by

Michael A.K. Liebschner

Department of Neurosurgery, Baylor College of Medicine, Houston, TX, USA and Research Service Line, Michael E. DeBakey VA Medical Center, Houston, TX, USA

Editor
Michael A.K. Liebschner, Ph.D.
Department of Neurosurgery
Baylor College of Medicine
Houston, TX, USA
and
Research Service Line
Michael E. DeBakey VA Medical Center
Houston, TX, USA

ISSN 1064-3745 ISSN 1940-6029 (electronic)
ISBN 978-1-61779-763-7 ISBN 978-1-61779-764-4 (eBook)
DOI 10.1007/978-1-61779-764-4
Springer New York Dordrecht Heidelberg London

Library of Congress Control Number: 2012935698

Printed on acid-free paper

Humana Press is part of Springer Science+Business Media (www.springer.com)

Preface

Revolution in biological sciences and bioengineering has created an environment in which advances in life sciences are not only amendable to, but also require, the active participation of engineering design and manufacturing in order to achieve solutions for complex biological problems. This revolution, along with the advancements of modern design and manufacturing, biomaterials, biology, and biomedicine, has brought about the new field of computer-aided tissue engineering. Advances in computer-aided tissue engineering encompass broad applications in large-scale tissue engineering fabrication, artificial organs, orthopaedics implants, and biological chips. This book highlights the interdisciplinary nature of this topic and reviews the current state of computer-aided three-dimensional tissue modeling, tissue classification, and tissue fabrication and implantation. Particular focus is placed on rapid prototyping and direct digital fabrication for cell and organs, construction of tissue analogs, and precursors to 3D tissue scaffolds.

The objective of this book is to serve the scientific community and professionals by highlighting research issues and providing directions in the use of computer-aided tissue engineering for medical applications as well as innovative and advanced drug discovery. The book embraces current advances and successful implementations. It provides a coherent framework for researchers who are interested in these technologies and for clinicians who plan to implement them.

Houston, TX, USA *Michael A.K. Liebschner, Ph.D.*

Contents

Contributors

ARTI AHLUWALIA • *Centro "E. Piaggio", Faculty of Engineering, University of Pisa, Pisa, Italy*

HENRIQUE A. ALMEIDA • *Department of Mechanical Engineering, Institute for Polymers and Composites, School of Technology and Management, Polytechnic Institute of Leiria, Leiria, Portugal*

MD. SHAMSUL AREFIN • *Biomechanics and Tissue Engineering Group, Faculty of Engineering and Industrial Sciences, Swinburne University of Technology, Melbourne, Australia*

GILDA A. BARABINO • *Wallace H. Coulter Department of Biomedical Engineering, Georgia Institute of Technology, Atlanta, GA, USA*

PAULO J. BÁRTOLO • *Centre for Rapid and Sustainable Product Development, Department of Mechanical Engineering, Institute for Polymers and Composites, School of Technology and Management, Polytechnic Institute of Leiria, Leiria, Portugal*

BAHAR BILGEN • *Wallace H. Coulter Department of Biomedical Engineering, Georgia Institute of Technology, Atlanta, GA, USA*

SHENGYONG CAI • *School of Mechanical Engineering, Shanghai Jiao Tong University, Shanghai, China*

SHAOCHEN CHEN • *Pearlie D. Henderson Centennial Endowed Faculty Fellow, Department of Mechanical Engineering, The University of Texas at Austin, Austin, TX, USA; Department of NanoEngineering, Institute of Engineering in Medicine, University of California, San Diego, La Jolla, CA, USA*

BORIS N. CHICHKOV • *Nanotechnology Department, Laser Zentrum Hannover e.V., Hannover, Germany*

DONG-WOO CHO • *Department of Mechanical Engineering, Pohang Institute of Intelligent Robotics, Pohang University of Science and Technology (POSTECH), Pohang, South Korea; Division of Integrative Biosciences and Biotechnology, Pohang Institute of Intelligent Robotics, Pohang University of Science and Technology (POSTECH), Pohang, South Korea*

CHEE KAI CHUA • *Division of Systems and Engineering Management, School of Mechanical and Aerospace Engineering, Nanyang Technological University, Singapore, Singapore*

HUNG DO • *Biomechanics and Tissue Engineering Group, Faculty of Engineering and Industrial Sciences, Swinburne University of Technology, Melbourne, Australia*

VILLE ELLÄ • *Department of Biomedical Engineering, Tampere University of Technology, Tampere, Finland*

ANATH FISCHER • *Laboratory for CAD & LCE, Department of Mechanical Engineering, Technion—Israel Institute of Technology, Haifa, Israel*

ROBERT E. GULDBERG • *Parker H. Petit Institute for Bioengineering and Bioscience, George W. Woodruff School of Mechanical Engineering, Georgia Institute of Technology, Atlanta, GA, USA*

Dietmar W. Hutmacher • *Regenerative Medicine, Institute of Health and Biomedical Innovation, Queensland University of Technology, Kelvin Grove, Australia*
Yaron Holdstein • *Laboratory for CAD & LCE, Department of Mechanical Engineering, Technion—Israel Institute of Technology, Haifa, Israel*
Hyun-Wook Kang • *Department of Mechanical Engineering, Pohang University of Science and Technology (POSTECH), Pohang, South Korea*
Minna Kellomäki • *Department of Biomedical Engineering, Institute of Biomaterials, Tampere University of Technology, Tampere, Finland*
Krzysztof J. Kurzydlowski • *Faculty of Materials Science and Engineering, Warsaw University of Technology, Warszawa, Poland*
Min Lee • *Division of Advanced Prosthodontics, Biomaterials and Hospital Dentistry, University of California Los Angeles, Los Angeles, CA, USA*
Kah Fai Leong • *Division of Systems and Engineering Management, School of Mechanical and Aerospace Engineering, Nanyang Technological University, Singapore, Singapore*
Michael A.K. Liebschner • *Department of Neurosurgery, Baylor College of Medicine, Houston, TX, USA; Research Service Line, Michael E. DeBakey VA Medical Center, Houston, TX, USA*
Stefan Lohfeld • *National Centre for Biomedical Engineering Science, National University of Ireland Galway, Galway, Ireland*
Yi Lu • *Department of Mechanical Engineering, The University of Texas at Austin, Austin, TX, USA*
Peter X. Ma • *Department of Biologic and Materials Sciences, Department of Biomedical Engineering, Macromolecular Science and Engineering Center, The University of Michigan, Ann Arbor, MI, USA*
Peter E. McHugh • *Department of Mechanical and Biomedical Engineering, National University of Ireland Galway, Galway, Ireland; National Centre for Biomedical Engineering Science, National University of Ireland Galway, Galway, Ireland*
Ulrich Meyer • *Clinic for Maxillofacial and Plastic Facial Surgery, University of Düsseldorf, Münster, Germany*
Yos S. Morsi • *Biomechanics and Tissue Engineering Group, Faculty of Engineering and Industrial Sciences, Swinburne University of Technology, Melbourne, Australia*
Jörg Neunzehn • *Clinic for Cranio, Maxillofacial Surgery, University of Münster, Münster, Germany*
Pedro Yoshito Noritomi • *Three-Dimensional Technologies, Renato Archer Information Technology Center—CTI, Campinas, Brazil*
Tatiana Al-Chueyr Pereira Martins • *Three-Dimensional Technologies, Renato, Archer Information Technology Center—CTI, Campinas, Brazil*
Aleksandr Ovsianikov • *Institut für Werkstoffwissenschaft und Werkstofftechnologie (E308), TU Wien, Wien, Austria*
Amal Ahmed Owida • *Biomechanics and Tissue Engineering Group, Faculty of Engineering and Industrial Sciences, Swinburne University of Technology, Melbourne, Australia*
Anne Rajala • *Department of Biomedical Engineering, Institute of Biomaterials, Tampere University of Technology, Tampere, Finland*

KEVIN SHAKESHEFF • *School of Pharmacy, Wolfson Centre for Stem Cells, Tissue Engineering, and Modelling (STEM), Centre for Biomolecular Sciences (CBS), University of Nottingham, Nottingham, UK*
JORGE VICENTE LOPES DA SILVA • *Three-Dimensional Technologies, Renato Archer Information Technology Center—CTI, Campinas, Brazil*
LAURA A. SMITH • *Department of Biologic and Materials Sciences, The University of Michigan, Ann Arbor, MI, USA*
NOVELLA SUDARMADJI • *Division of Systems and Engineering Management, Rapid Prototyping Research Laboratory, School of Mechanical and Aerospace Engineering, Nanyang Technological University, Singapore, Singapore*
WOJCIECH SWIESZKOWSKI • *Division of Materials Design, Faculty of Materials Science and Engineering, Warsaw University of Technology, Warszawa, Poland*
AHMAD M. TARAWNEH • *Department of Neurosurgery, Baylor College of Medicine, Houston, TX, USA*
ANNALISA TIRELLA • *Centro "E. Piaggio", Faculty of Engineering, University of Pisa, Pisa, Italy*
MIKKO TUKIAINEN • *Department of Biomedical Engineering, Institute of Biomaterials, Tampere University of Technology, Tampere, Finland*
GIOVANNI VOZZI • *Centro "E. Piaggio", Faculty of Engineering, University of Pisa, Pisa, Italy; C.N.R. Institute for Composite and Biomedical Materials, Pisa, Italy*
XUNGAI WANG • *Centre for Material and Fibre Innovation, ITRI, Deakin University, Melbourne, Australia*
MATTHEW WETTERGREEN • *Department of Bioengineering, Caroline Collective, Houston, TX, USA*
HANS PETER WIESMANN • *Clinic for Cranio, Maxillofacial Surgery, University of Münster, Münster, Germany*
MARIA ANN WOODRUFF • *Biomaterials and Tissue Morphology Group, Institute of Health and Biomedical Innovation, Queensland University of Technology, Kelvin Grove, Australia*
BENJAMIN M. WU • *Department of Bioengineering, University of California Los Angeles, Los Angeles, CA, USA; Division of Advanced Prosthodontics, Biomaterials and Hospital Dentistry, University of California Los Angeles, Los Angeles, CA, USA*
JUNTONG XI • *School of Mechanical Engineering, Shanghai Jiao Tong University, Shanghai, China*

Chapter 1

Computer-Aided Tissue Engineering: Benefiting from the Control Over Scaffold Micro-Architecture

Ahmad M. Tarawneh, Matthew Wettergreen, and Michael A.K. Liebschner

Abstract

Minimization schema in nature affects the material arrangements of most objects, independent of scale. The field of cellular solids has focused on the generalization of these natural architectures (bone, wood, coral, cork, honeycombs) for material improvement and elucidation into natural growth mechanisms. We applied this approach for the comparison of a set of complex three-dimensional (3D) architectures containing the same material volume but dissimilar architectural arrangements. Ball and stick representations of these architectures at varied material volumes were characterized according to geometric properties, such as beam length, beam diameter, surface area, space filling efficiency, and pore volume. Modulus, deformation properties, and stress distributions as contributed solely by architectural arrangements was revealed through finite element simulations. We demonstrated that while density is the greatest factor in controlling modulus, optimal material arrangement could result in equal modulus values even with volumetric discrepancies of up to 10%. We showed that at low porosities, loss of architectural complexity allows these architectures to be modeled as closed celled solids. At these lower porosities, the smaller pores do not greatly contribute to the overall modulus of the architectures and that a stress backbone is responsible for the modulus. Our results further indicated that when considering a deposition-based growth pattern, such as occurs in nature, surface area plays a large role in the resulting strength of these architectures, specifically for systems like bone. This completed study represents the first step towards the development of mathematical algorithms to describe the mechanical properties of regular and symmetric architectures used for tissue regenerative applications. The eventual goal is to create logical set of rules that can explain the structural properties of an architecture based solely upon its geometry. The information could then be used in an automatic fashion to generate patient-specific scaffolds for the treatment of tissue defects.

Key words: Computer-aided tissue engineering, Bio-additive fabrication, Scaffold micro-architecture, Structure optimization, Mathematical algorithms, Micro-architecture control

1. Goal of Tissue Regeneration

1.1. Tissue Engineering Approach

Tissue engineering is an important emerging branch in biomedical engineering. It is a gateway into a future where few will suffer because of lack of organs or due to organ dysfunction and failure.

Michael A.K. Liebschner (ed.), *Computer-Aided Tissue Engineering*, Methods in Molecular Biology, vol. 868, DOI 10.1007/978-1-61779-764-4_1,

In the USA in year 2000 alone, approximately 72,000 people were on the waiting list for an organ transplant due to end-stage organ disease, but only 23,000 transplants were performed. At the end of 2007, there were 97,248 people registered on the organ waiting list. Only 27,578 organs were transplanted in the US that year (1).

The creation of tissues for use in the medical practice has been well reported. These tissues involve fabricated skin, liver, pancreas, kidney, esophagus, nerves, heart valves leaflets, cartilage, tendon, and bone. In the early 1990s, the first artificial skin product for burn treatment was introduced followed after that by cartilage constructs (2–9).

The underlying concept of tissue engineering aims to restore tissue structure and function through incorporation of different biological materials, such as cells, growth factors, and biopolymers to create environments, which promote the development of new tissues whose properties more closely match their native counterparts.

The tissue engineering approach differs from current treatments, which focus mainly on drugs as the main stay modality of the treatment to encourage the body to fight diseases on its own. Instead, a highly porous artificial extracellular matrix (ECM), or scaffold, is thought to be needed to accommodate cellular growth and tissue regeneration (10).

In tissue engineering, donor tissue is harvested and degraded into individual cells, which are then seeded into a carrier or scaffold. The resulting tissue construct is then grafted into the desired site of intact tissue of the patient. After implantation, the body takes over blood vessels that attach themselves to this construct followed by the degradation of the scaffold that allows the newly seeded tissue to mix up with its surroundings (11).

There is a wide variety of cell sources that can be used for tissue engineering. Fully differentiated cells, progenitors, and stem cells have shown a great promise in the creation of tissue constructs to be used in the medical field. Cells like stem cells have the ability to differentiate into a wide variety of cell types. So, in order to create a particular tissue construct with the desired functional as well as structural properties such cells must be differentiated under a controlled environment.

Scaffolds have to be designed with particular requirements in mind. Mechanical, architectural, transport and biological requirements should be met in the scaffold design to direct appropriate cellular attachment, migration, proliferation, and differentiation. Tissue engineering using 3D scaffolds has been found to be less than ideal in creation tissue constructs. The lack of mechanical strength of the 3D scaffolds in addition to the absence of interconnectivity between the porous structures in the scaffold itself were found to be the main cause behind 3D scaffold design drawbacks.

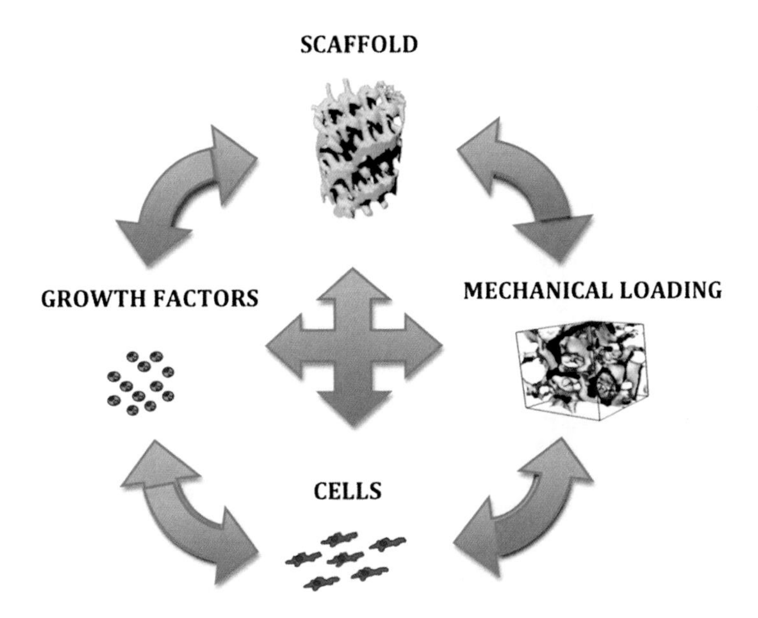

Fig. 1. Four paradigms of tissue engineering.

1.2. The Fourth Paradigm in Tissue Engineering (12)

The main paradigm of tissue engineering is the combination of cells, growth factors, and substrates. In practice, the three are melded into an implantable tissue replacement. Mechanically induced tissue formation may be considered the fourth paradigm of tissue engineering (Fig. 1).

Mechanical aspects, such as the appropriate mechanical cues as well as the frequency and magnitude of these cues, remain elusive, yet arguments that tissue is accentuated under biomechanical culture and has superior cell distributions can scarcely be refuted.

From a mechanical point of view, one of the important paradigms in tissue engineering is that a scaffold should mimic the biomechanical properties of the organ or tissue to be replaced. Mechanical factors have great impact on the cellular level architecture and function in scaffolds. Since there is an intimate relationship between mechanical properties of a scaffold and the porosity of its porous structures, it might be that the more flexible and porous the scaffold is the more it will allow cells to attach and proliferate in a more efficient way. In contrast, if the scaffold is stiffer but with less porosity cells will have difficulty attaching to the scaffold, which will be reflected on the final function of the scaffold.

2. Functional Tissue Engineering

2.1. Computer-Aided Tissue Engineering

The emergence of interactive computer-aided technologies in the field of tissue engineering research has led to the development of computer-aided tissue engineering (CATE). CATE can be defined

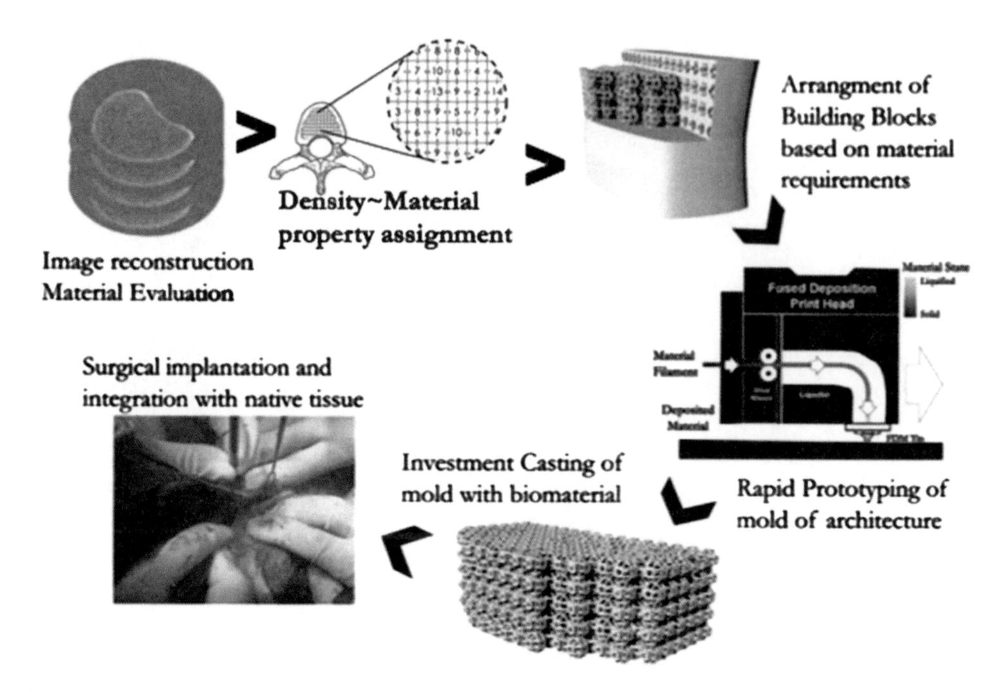

Fig. 2. Application concept of CATE where a patient's anatomy or defect area is acquired, translated into a boundary condition and then filled with engineered building blocks. Techniques such as direct fabrication (formerly known as rapid prototyping) fabricate the scaffolds, which are then surgically implanted.

as "the technology concerned with the use of computer systems to assist in the creation, modification, analysis and optimization of a design." The concept is derived from computer-aided design (CAD) and computer-aided manufacturing, where an automated process assists in the fabrication of engineering materials (Fig. 2).

In the success of tissue engineering, three-dimensional (3D) scaffolds play an important role as an extracellular material onto which cells attach, grow, and proliferate under certain requirements to form a new tissue.

Scaffolds bio-modeling, informatics, and bio-manufacturing are the three main concepts encompassing the use of CATE (13, 14). Tissue scaffolds modeling is the first step in this process. It involves obtaining a 3D image of the tissue, either by extraction from imaging modalities or with CAD generation of a 3D tissue model. The primary imaging techniques used in scaffolds bio-modeling are CT, MRI, and optical microscopy, by which a series of 2D images of the scanned area are obtained (Table 1). Combining these 2D images together creates a more meaningful 3D view, which can be used in developing a 3D anatomical model. This anatomical model can be used for contour-based generation and 3D image processing to create CAD models.

The second step in CATE is tissue informatics. In its broadest term, tissue informatics defines compiling information about each tissue from organ to subcellular level, but is most specifically referred to when analyzing the type and interaction of genes and

Table 1
Comparison of imaging modality for CATE application

	CT	MRI	Optical microscopy
Advantages	High resolution images Less exposure time during scanning High contrast to calcified hard tissue	Does not use ionizing radiation Provides images for both soft and hard tissues Able to differentiate tissues of similar densities	Able to differentiate tissues of similar densities
Disadvantages	Using ionizing radiation Unable to differentiate soft tissues of similar densities Useful in hard tissue modeling	Resolution not as good as CT or optical microscopy More exposure time during scanning	Limited applications due to small field of view

proteins within tissues. The third and final step in this process is the design of a scaffold based upon the location as well as the biological and biophysical requirements.

The concept behind CATE for biological replacements of bone is that the complex architecture of bone is too difficult to be replaced with a scaffold/implant construct which takes into consideration the small feature size, mechanical properties of the hierarchical scale in which they exist, fluid flow properties (permeability) and pore connectivity, and matching biomechanical properties. Therefore, as an alternative, the replacement of a bone defect is depicted as the assembly of smaller sub-volumes of simplified unit-cells that have characteristics that are discretely known and able to be manufactured. In this way, the creation of an implant consists of the optimization problem of selecting and matching primitive shapes, which match the local properties within the sub-volume they are meant to replace yet fit together. This is a complicated problem which brings into play several engineering tools, such as (1) CAD for the design of unit-cells, (2) finite element analysis, FEA, for the analysis of unit-cell mechanical properties, (3) optimization, for the fluid flow and topological mechanical properties, and (4) bio-additive fabrication, BAF, for the fabrication of complex and functional biomaterials that may find their application in tissue regenerative medicine.

In general, CATE is a useful and inexpensive way to explore design strategies and for providing personalized engineered solutions. It holds the promise of being able to very specifically and repeatedly offer tissue substitutes of almost unlimited quantity and maybe one day allow the fabrication of whole artificial organs and large tissue sections.

2.2. Bio-Additive Fabrication

The continuous revolution in different aspects of biological sciences, in general, and in functional tissue engineering, in particular, has led to the need of involving tissue engineering in finding the solutions for complex biological problems. This revolution in tissue engineering in addition to the development of modern design techniques and biomaterials has moved additive manufacturing to a whole new level of broad applications in biomedical engineering (15).

Additive manufacturing technology has been found to be a very promising technology allowing the versatility of using biomaterials and fabrication of scaffolds with complex micro-architecture (15).

The introduction of additive fabrication in combination with CAD has led to a new era, where models can be fabricated for the representation of patient-specific anatomical geometry. Additive manufacturing follow the principal of a layer-by-layer manufacturing, in which a computer tomography image can be reproduced similar to the actual physical model known as bio-modeling. These physical models can be held and felt, allowing us to have a better understanding of the most complex 3D geometry which cannot be offered by any other imaging modality (15).

There is a role of additive manufacturing in tissue engineering as they contribute to freeform fabrication of tissue scaffolds. 3D scaffolds are extracellular materials, where cells can attach, grow, and form new tissues. Different methods of indirect scaffold fabrication are known (i.e., porogen leaching, melt molding, fiber bonding, and gas foaming), which have some limitations regarding their interaction, reproducibility, shape, usage of porogens (i.e., wax and salt) as well as toxicity concerns due to using organic solvents. However, the major limitation of such techniques is drawn back from the random spatial orientation of the porogens with the combination of requiring a high porosity to achieve an interconnected porous network. This yields unpredictable mechanical properties that are generally only a few percent in value of the mechanical properties of the solid material. These limitations increased our awareness to find alternative methods to fabricate the scaffolds (15, 16).

Direct scaffold fabrication methods have been found to be favorable alternatives to the indirect methods (15). They include stereolithography (17), fused deposition modeling (FDM), laminated object modeling (18), selective laser sintering (SLS), 3D printing (3DP), precision extruding deposition (PED), 3D phase change printing, and micro-syringe polymer deposition. When utilizing such methods, there are no limitations regarding reproducibility and interaction (Fig. 3).

Bio-additive manufacturing is a biofabrication technology employing different sets of biomaterials as building blocks to construct a wide range of therapeutic products under certain requirements. It has a great clinical applications including but not limited to, cell and organ printing, drug delivery studies, and in understanding diseases pathogenesis. Different techniques have been used for cell manipulation in BAF, including syringe-based cell deposition (Fig. 4), inkjet-based cell printing, laser printing, cell manipulation by mechanical, optical, electrical, magnetic, ultrasound, and ionic methods.

Fig. 3. Four different architectures printed at 80 and 90% porosity using a DTM laser sinter station.

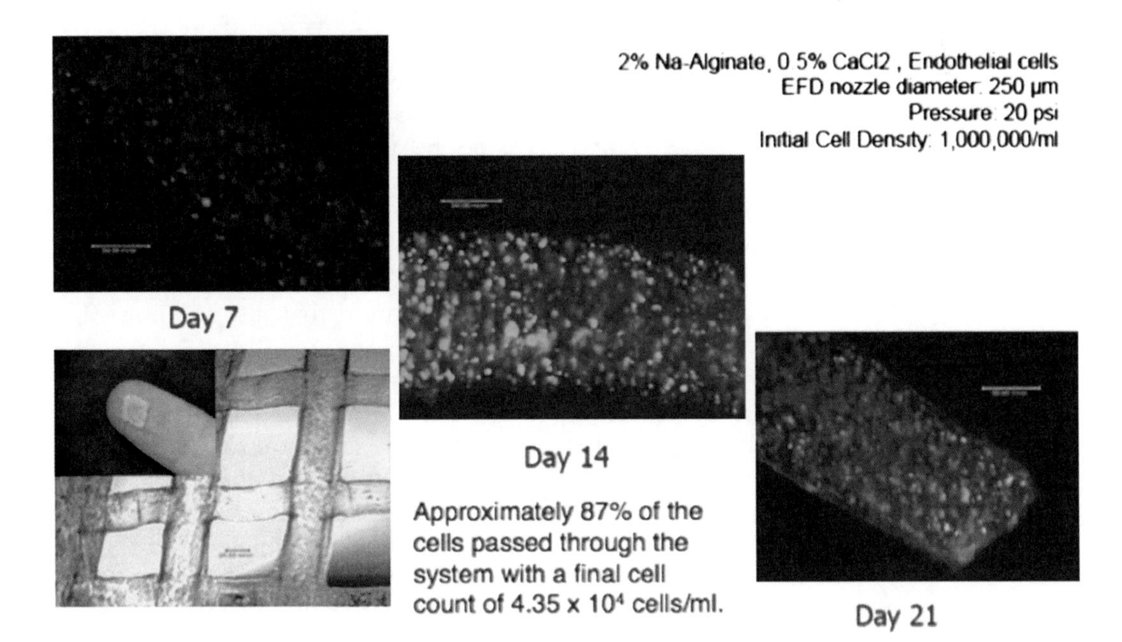

Fig. 4. Direct printing technique (syringe extrusion) of alginate material with embedded live endothelial cells and subsequent live/dead staining in culture.

Similar to any other new technology future promises and challenges are reported to bio-additive manufacturing. Future promises for BAF is to be able to construct different living parts or even a whole bio-functional living environment, create artificial implants that have similar structural and functional characteristics as the native tissues, and fabricate bio-mimetic micro-organs for drug evaluations. On the other hand, different challenges have been reported to this technology; (1) Inability of constructing scaffolds with feature size <100 μm; (2) Inability to model, design, and simulate biological organizations with time-dependent cellular properties; (3) Difficult integration of bio-additive manufacturing with nano systems because of their need of a clean environment; (4) Lack of viable fabrication processes to make heterogeneous scaffold materials; (5) Deficient knowledge of cell behavior within a 3D structural environment; and finally (6) Difficulty to maintain biological activity of cell-embedded materials before and after the fabrication process (15).

2.3. Biological Design in the Fourth Dimension

In early 1970s, engineers used 2D CAD as an electrical drafting tool to replace hand drawings. After that, a new generation of CAD, 3D CAD, emerged to provide a better design visualization and understanding of the designs architecture.

Fourth dimensional (4D) CAD model is an innovative way in representing 3D biological designs in a time-based manner (3D + time). They can be designed using the following steps: (1) 3D CAD model should be built from the 2D CAD model; (2) arranging 3D models and linking them through a third party technology (linking keys); (3) and finally creation of the simulation design.

4D CAD models allow us to have a better understanding of the design activities of the building block in a time sequence. In other words, the CAD models will not be stationary models but incorporate a time sequence within it. For example, 4D CAD models of bone replacement scaffolds will enable us to predict how the scaffold in integrated into the tissue defect and how the biomaterial of the scaffold is degraded and replaced by de novo tissue in a timely manner. This time sequence required more than just the knowledge of the scaffold micro-architecture.

Bone is a metabolically active living tissue that undergoes continuous remodeling throughout life. Bone remodeling is a series of complex sequential steps in which there is removal of old bone (resorption) and replacement with newly formed bone (formation). This remodeling process is necessary to maintain the structural integrity of the bone as well as to help maintaining the bone function as a reservoir of different types of minerals.

Two types of remodeling processes exist (i.e., external, internal). External remodeling or macromodeling is a biological process in which cortical surfaces of the bone will be reshaped (19). On the other hand, internal remodeling is a tissue-based level process involving mainly the trabecular cortices as well as the haversian systems.

The signal that initiates bone remodeling has not been identified (20), but evidence shows that microdamages or changes in the mechanical forces affecting the bone will trigger bone remodeling (21). Also, hormonal responses to changes in calcium and phosphorous concentration in the blood trigger bone remodeling.

In the adult population, 25% of trabecular bone and 5% of cortical bone is replaced every year by remodeling (22). The bone remodeling cycle involves a series of sequential steps that depends mainly on the interaction between two different cell lineages, namely the mesenchymal osteoblastic lineage and the hematopoietic osteoclastic lineage. This cycle occurs in small packets of cells called basic multicellular units (BMUs).

The activation phase of bone remodeling is the first phase in bone remodeling in which the resting bone surface converts into remodeling surface. It involves the interaction of osteoblast and osteoclast precursor cell, which leads to the differentiation, migration, and fusion of the multinucleated osteoclasts. These cells can now interact with mineralized bone matrix and initiate resorption by secreting particular lysosomal enzymes leaving irregular cavities on the trabecular bone surface, called Howship's lacunae. The next phase in bone remodeling after the "resorption" phase is the "reversal" phase in which there is a switch from bone matrix destruction to repair. In this phase, osteoblast precursor cells locally proliferate and differentiate into osteoblasts. The final phase is the "formation phase" where the cavities that have been formed by resorption can be completely filled in by successive layers of osteoblast.

3. Scaffold Design Requirements

3.1. Mechanical Requirements

Each tissue in the body has a unique function, and therefore a different set of mechanical requirements. Bone, for example, has an exceptional list of mechanical requirements, which include the capability of providing structural rigidity to protect soft tissue organs, acting as a reservoir for minerals, and providing a framework for the transfer of muscle forces to locomotion (12, 23). Therefore, the success of bone scaffold as measured in vivo is determined by its ability to participate efficiently in repairing bone defect and maintain its mechanical stability under different loading pressures.

Mechanical characteristics of the scaffold are critical either individually or in concert. Scaffold regional stiffness/strength, micro-architectural level of mechanical surface strain, and void fraction amount and orientation have an important role in scaffolds tissue engineering. Fine control of these factors has a great influence in the overall mechanical characteristics of the scaffold.

The reason that mechanical characteristics are important measures in scaffold tissue engineering is that growing cells (i.e., fibroblast) may generate a huge amount of contractile force that needs a strong scaffold to contain it. Weak scaffolds, in contrast, will collapse under this amount of substantial force leading to the distortion of the overall scaffold architecture (24, 25).

The loading conditions of bone stay within a specific force range during the normal daily activities. This indicates the need of a dual function implant. After implantation the implant must be able to maintain mechanical stability by serving as a load-bearing device. During the course of its integration and degeneration, the scaffold must completely transfer the mechanical load to the bone. In essence, the scaffold adopts a therapeutic role rather than a weight-bearing role. Allowing ingrowth by high progression of scaffold degradation and subsequent strength loss, the load is transferred gradually to the bone stimulating bone healing. Therefore, scaffolds must be tailored to incorporate more than just the adequate stress concentrations in order to sustain the load and remodeling of a particular area of the bone (26, 27) but must account for the fourth dimension in scaffold design.

3.2. Biological Requirements

One of the most important factors in tissue engineering is the choice of suitable material for the scaffold. The desirable characteristics of this material in terms of biocompatibility as well as biodegradability have a significant role in overall outcome of the tissue. Material biocompatibility means that the material does not provoke unwanted tissue response to the implant, and at the same time it has the ability to possess the right surface chemistry to promote cell attachment and function. On the other hand, biodegradability means the ability of the material to be degraded into nontoxic products, leaving behind the desired living tissue (28).

Other important biological requirements are cell loading, distribution, and attachment. Cell attachment represents the ability of the cells to bind to the surrounding structures. The binding of cells to the surrounding structures is a receptor mediated interaction. This interaction takes place in the natural extracellular matrix (ECM) through different adhesive peptides (i.e., integrin). In contrast, the connection between scaffolds and the environment depends on the number of characteristics. Such characteristics include the contact angle, mechanical stiffness, and microtopography (24). It has been clearly shown that surface characteristics can influence the degree of cell adhesion (29, 30).

Another required biological component in scaffold design is its capacity for fictionalization by adding biochemical markers that enhance cell interaction (24, 31).

It is well known that biological activity of any living tissue is regulated by mechanical signals. As each tissue in the body has a unique biological activity, it has different mechanical requirements. Bone, for example, has a set of mechanical requirements that enable it to maintain its function that differ from other tissues (12, 23).

3.3. Transport Requirements (24)

Transport, in the context of tissue engineering, means the process by which necessary materials (e.g., nutritions) for cell function move through a scaffold to reach the cells within that scaffold. These materials help maintain healthy cell function. Transport of these materials should be allowed to take place in both directions, inside and outside the scaffold, depending on the nature of these materials. Molecules like oxygen, glucose, and various chemical factors should be able to reach the cells inside the scaffold easily. Insufficient supply or the inability to provide the appropriate amount of nutrients to the cells inside the scaffold will result in cell death. On the other hand, metabolic wastes like carbon dioxide and other metabolic byproducts should be transferred outside the scaffold. Inability to do so will result in increase in the acidity in the environment surrounding the scaffold, again leading to cell death. Both conditions aid the "shell" effect observed in scaffolds that have been cultured in either static culture or non-perfusing cultures (e.g., spinner flasks).

Nutrient and waste transfer are coupled together by the transport phenomenon of diffusion and flow. Diffusion is a process of random motion of particles causing the distribution of these particles homogenously in the environment. Adding flow to the desired medium will increase the rate of the directed diffusion within that medium.

Transport of nutrients within the body is controlled by both flow and diffusion. Blood, for example, travels through the arteries until reaching the capillaries under the flow principle. From the capillaries, the blood moves towards the interstitial space through diffusion. Transport within resting cells in vitro is mainly controlled by diffusion not flow.

Maintaining the transfer of nutrients and waste products in and out the scaffold is a critical factor in scaffold tissue engineering that can be successfully dealt with by making connections between the pores of the scaffold. Other factor that may affect diffusion over time is cell growth. As cells proliferate, they may occlude the pores of the scaffold affecting the ability to transport nutrients and waste products in and out the scaffold, respectively, as well as leading to tissue necrosis.

A diffusion network is a new approach where we can deliver nutrients towards the center of the scaffold. This can be achieved by using a material that is not conductive to cell attachment and occlusion. Hydrogels have been used to create a vertical path within the scaffold through which nutrients can reach the interior part of the scaffold, thereby bypassing blockage caused by other materials.

3.4. Geometric Requirements

Structural considerations of scaffold design include anatomical compatibility, geometrical fitting, porosity, pore size, and surface area as well as the interconnection between pores. Control over these factors affects the geometrical properties of the scaffold (24, 32, 33).

Porous structures play an important role in scaffold design either in vitro or in vivo. In vitro, pores can be used to seed cells prior to implantation, whereas in vivo they can be used as a conductive structure. Desirable porosities in the scaffolds are around 90% with pore size ranges from 100 to 200 Mm (16). This porosity value stems not only from the porosity of natural bone tissue, but also from the ability to fabricate an interconnected porous network with indirect fabrication techniques (see Subheading 2.2 for details). The pore surface area is important in the attachment and the viability of the scaffold in the desired place of implantation. Another factor is the communication between the pores. These communications were found to allow more cellular migration and circulation of the extracellular nutrition and cell signals.

Common techniques for creating porous structures in scaffolds is through the casting of the scaffold in paraffin spheres to create a uniform density of spherical pores or expanding pores size and interconnectivity through dissolving salt particles (23, 34, 35).

4. Modulation of (+) Space

4.1. Building Block Approach

Scaffold design is critically important as it affects the ultimate in vivo functional requirements of the scaffold. Before the development of computer-generated 3D models, biomaterials science used to adopt a "trial-and-error" approach to modify existing designs based on the in vivo or in vitro results. Improvement in cell culture techniques and conditions as well as the development of advanced bioreactors has greatly improved the dependence on the in vitro results. CAD has also played a role in decreasing the amount of experimental tests needed and the overall scaffold design process. CAD can be used to generate and modify 3D models of scaffold designs for tissue engineering.

Simple shapes can be used and assembled together in a more complex architecture based on computer-generated 3D models (Fig. 5). Different medical imaging techniques in addition to CAD processes have been combined to a generate scaffold geometry that has the exact same shape as the void volume it needs to fill.

CAD can be used in conjunction with additive fabrication techniques. Additive fabrication is a manufacturing technique used to construct a structure layer by layer to end up with a totally controllable structure. It includes selective laser sintering, FDM, stereolithography, phase change printing, electron beam melting, and 3D printing.

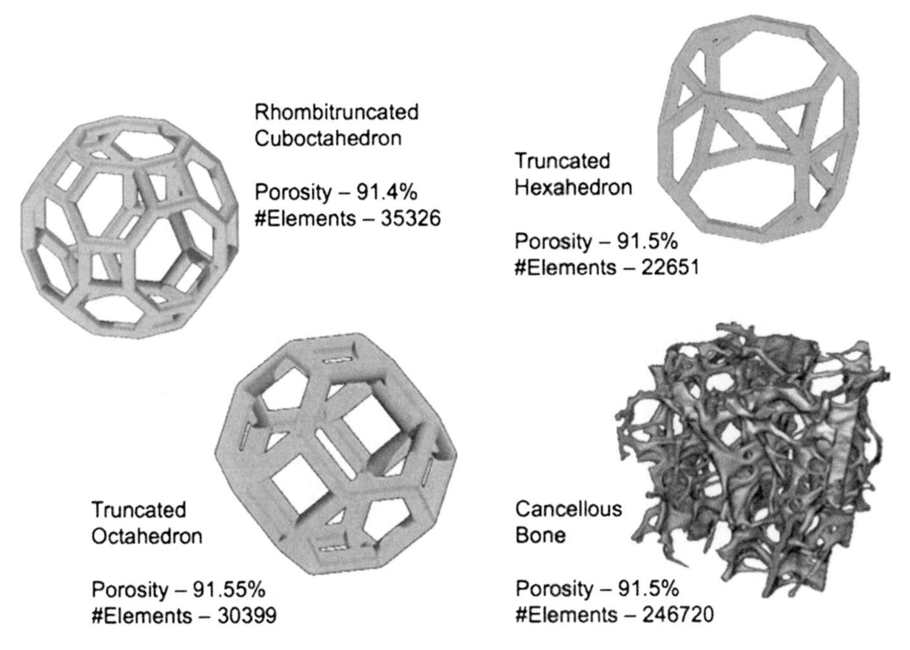

Fig. 5. Simplification of the seemingly random bone micro-architecture (*lower right corner*) into architectures derived from platonic and Archimedian solids.

Traditionally, scanning electron microscopy or transmission electron microscopy has been used with a great success to provide us with a detailed picture of the surface of the scaffold. However, with the recent advances in the use of 3D scaffolds with pore structures new nondestructive imaging techniques have been favored. These techniques (e.g., microcomputed tomography or synchrotron tomography) are mainly based on scanning the biomaterial through X-rays as the material rotates within the X-ray beam. These imaging techniques have allowed us to study and calculate the overall porosity as well as its distribution in a nondestructive manner and also it has been used to monitor the 3D mineralization over time.

The effect of the architecture of simple polyhedral shapes on modulus and stiffness has been explored in 3D. Based on the results of previous studies, we proposed the creation of a library of different shapes of building blocks that can be used to generate replacement materials through the different structural organization of the tissue. Determination of the general characteristics and properties of the building blocks does provide a library with comprehensive knowledge about each one of its building blocks in terms of porosity, material characteristics, and deformation patterns. Then, the building blocks can be assembled together to create a 3D representation of the desired tissue.

Generation of building blocks library with different block shapes requires the usage of CAD file databases as well as the use of FEA for cellular solid property characterization. Based on the smallest feature size able to be fabricated with direct additive fabrication, we proposed the smallest microstructures generated to be confined within a 27.0 mm^3 cubic volume, a product of the superposition of three 1.0 mm resolution QCT layers in three orthogonal directions. As needed, different elementary shapes (i.e., beams, cylinders, and spheres) may be arranged within these confines producing well-defined isotropy or anisotropy.

Combining different building blocks into one structure requires merging them with some preventive measures to reduce edge effects. The efficacy of generating building blocks based upon mechanical demands has been shown. However, lack of a common interface will distort finite element results. We introduced a common interface between the building blocks in the form of a torus (Fig. 6). The torus is being sliced into two halves through its vertical axis. Each part of two adjacent building blocks contained half of the sliced torus. By matching the two building blocks the torus halves are joined together again. Using torus as a common interface has been associated with many advantages for example, reducing the stress concentrations within the building blocks due to its rounded edges as well as preventing material mismatch between dissimilar architectures. The disadvantage is that the individual micro-architecture of the building blocks has to account for the connection to and subsequent load transfer through this torus. A selection scheme of various interface modalities for combinations of building blocks might be beneficial.

Fig. 6. A half torus functions as common building block interface.

4.2. Bio-Mimetic Micro-Architecture Optimization

It is important that the scaffold intended to replace bone tissue to resemble it in terms of structure and function. While an exact replication is not required, it is important that certain "bone-like" features to be preserved. In doing so, many challenges are facing us like (1) providing adequate mechanical support; (2) expose de novo tissue to physiological loading ranges; and (3) to allow for a smooth transition of functionality during degradation.

The mechanical properties of bone tissue at the tissue level have an elastic modulus between 10 and 20 GPa, exceeding mechanical properties of existing biomedical materials by a factor of 10 or more. On the other hand, bone tissue at the architectural level has an elastic modulus between 50 and 300 MPa, values that can be achieved with biomaterials, however, not at the same porosity levels. Using these values as a guideline, structural organization of biomaterials may lead to the design of synthetic scaffolds with material properties exceeding the required values for trabecular bone.

There are, however, two major challenges. The first one is that although the structural organization of bone tissue seems to follow mechanical loading principles (i.e., planes of material symmetry), its architectural organization still seems random and cannot be mathematically explained. The second challenge is that the elastic modulus mismatch between the scaffold biomaterial and bone tissue causes the scaffold to deform more compared to bone tissue for the same magnitude of load. Following Perren's interfragmentary strain theory "a tissue which ruptures or fails at a certain strain level cannot be formed in a region of precursor tissue which is experiencing strain levels greater than this level." The load-dependent deformation of the scaffold needs to be within a certain strain range in order for bone tissue to form and to eventually transfer its functionality over to the de novo tissue. As most scaffold materials follow bulk erosion over time, a lack of de novo tissue formation in a high loading region may lead a catastrophic failure of the whole construct during the degradation process. Scaffold micro-architecture optimization then aims at reducing such peak strain values and promoting a uniform tissue mechanical environment (36).

4.3. Modulation of Porosity and Architecture

Minimization schema in nature affects the material arrangement of objects. Geometries intending minimal energy expenditure or structural integrity give rise to highly ordered systems with defined, repeated architecture (37–39). Sometimes, the focus of the minimization is on the structural arrangement itself where in the case of cancellous bone and honeycombs, high strength is obtained even with the arrangement of low material volumes (40, 41).

The arc of cellular solids field has moved from the full characterization of 2D, regular as well as irregular symmetric architectures to the characterization of irregular 3D architectures.

Fig. 7. Wireframe renderings of selected polyhedra. *Left to right*: Hexahedron (H), truncated hexahedron (TH), rhombitruncated cuboctahedron (RC), truncated octahedron (TO). Polyhedra in this figure are displayed at 80% porosity.

A systemic decomposition of the geometric properties of the 3D architectures executed similarly to previous 2D analyses combined with deformation mechanisms could lead to material-independent geometrically derived equations governing mechanical properties for complex 3D architectures, such as bone. In order to achieve this goal, systemic exploration of regular architecture with slight geometric variation across a range of material volumes should be considered. The Platonic and Archimedian solids are architectures which fulfill this requirements; they are the simplest geometric shapes which exhibit symmetry and regularity. The symmetry, regularity, and similarities across the set of polyhedral lend themselves to straightforward modeling and geometric characterization (42).

Four different shapes were modeled for characterization from both the platonic and Archimedian sets. A hexahedron (H), truncated hexahedron (TH), rhombitruncated cuboctahedron (RC), and truncated octahedran (TO) (Fig. 7). All polyhedra are open celled and were created at five volumetric porosities (50, 60, 70, 80, and 90) with a constant material envelope allowing for a direct comparison between the specific arrangements of material in each architecture. Each architecture was classified according to porosity, beam length, beam diameter, surface area, space filling, and pore size ratio. The H has most of its material oriented in the loading direction. The TH has slightly less material oriented out of the loading direction due to the truncation of the corners. The TO and the TH have the same number of vertices and beams, but differ in material arrangement and beam spacing distance. Finally, the RC has twice the number of beams as the TO and TH. All polyhedra have the same connectivity index.

Polyhedra were subjected to a linear finite element simulation using unconfined compression. Architectures were meshed with a greater than 18,000 tetrahedral elements, as determined via a convergence study. Space filling efficiency (outer volume/inner volume) was calculated across the porosity range for each polyhedra. The outer volume, which amounts to the space exterior to the polyhedral, was defined as the Boolean subtraction of the space filling form of the polyhedral and the ball and stick form from the bounding box volume (43). The interior volume, which amounts to the space interior to the polyhedral, was defined as Boolean subtraction of the ball and stick form of the polyhedral from the space filling form of the polyhedra.

The geometric characteristics of the surveyed polyhedra are summarized in the Table 2.

Each polyhedra has a unique material architecture, meaning that the space filling pattern is also unique. The space filling pattern is a ratio of the volume external to the polyhedra versus the volume interior to it. Below 60% porosity all the polyhedra exhibit a significant drop in the inner volume consistent with a power law change, with the hexahedron having the lowest rate of change in shape conservation and the RC having the highest rate of change.

As shown in the Fig. 8; more complex shapes tend to shrink inward towards the center, which can be assessed by the pore

Table 2
Polyhedra geometric characteristics

	Hexahedron	Truncated hexahedron	Rhombitruncated cuboctahedron	Truncated octahedron
Struts	12	36	72	36
Vertices	8	24	48	24
Connectivity index	3	3	3	3

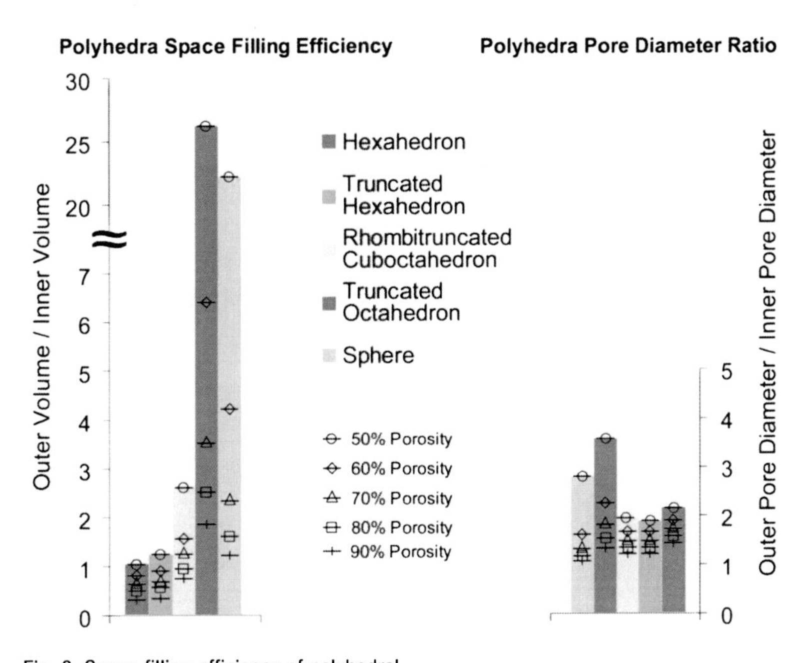

Fig. 8. Space filling efficiency of polyhedral.

diameter ratio. Excluding the TO, the pore diameter ratio varies linearly for all polyhedra across porosities and the slope are near equal. The pore ratio for the truncated octahedron is nearly twice that of the polyhedra at 50% porosity but is nearly equal to the RC at >70% porosity.

Mechanical testing of fabricated polyhedra building blocks illustrated that architectures sharing the same material volume but differing in material arrangement result in different mechanical properties (Fig. 9). All four architectures were fabricated at 80 and 90% porosity using elective laser sintering and mechanically compressed beyond failure in a commercial loading apparatus under quasi static loading conditions. The loading apparatus was set in displacement mode.

Differences in the mechanical loading profile were seen in the architecture simply by adjusting their porosity, as shown in Fig. 9. The H still had the highest modulus and yield point and the TH still has the lowest of both, when increasing the porosity from 80 to 90%. At the 90% porosity, all of the architectures demonstrated a much better strain response with fracture strain occurring at higher strains than the same architecture at a lower porosity. The H and the TO again had similar fracture strains, which at both porosities were the lowest of the four architectures. The TO began to show the signs of a densification region, with a strain hardening response that increased the stress immediately following the ultimate strength point. The RC again had densification regions, this time resulting in an ultimate strength, which occurred during a secondary peak for several of the samples. This secondary peak gives the RC a secondary modulus and strength, which was higher than any of the other architectures at the same porosity. Obviously, because of this the RC has the highest ultimate strain and fracture strain while the H has the lowest.

Interestingly enough, we tend to only pay attention to the elastic modulus and strength values when evaluating scaffold material properties. However, if we look from a practical point of view, the H scaffolds, while stronger, disintegrate almost in an explosive fashion. These architectures are very brittle. The RC architecture, while having only half the strength, continues to support load after failure. This sustained loading capacity may determine the faith of the scaffold during the degradation period. The ability of the de novo bone tissue to heal localized defects within a fractured scaffold may preserve functionality. On the other hand, if a scaffold completely disintegrates when overloaded, it will be rendered useless.

An overview of the surface mechanical environment depending on architecture and porosity can be shown by the elemental stress distribution of the architectures. Figure 10 illustrates a histogram of elemental stress distribution of the hexahedron architecture for five different porosities. The ratio of tensile to compressive stress

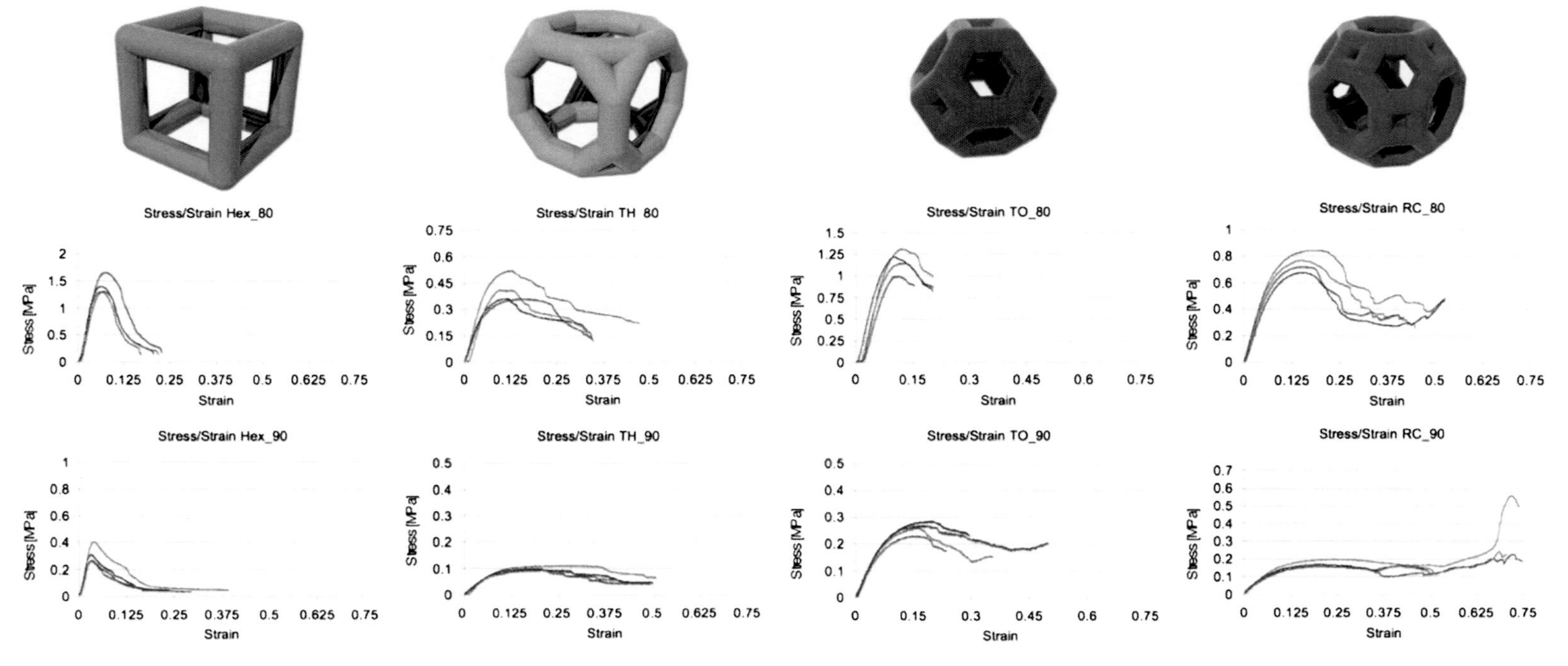

Fig. 9. Comparison of the biomechanical properties of four different micro-architectures when built at 80 and 90% porosity.

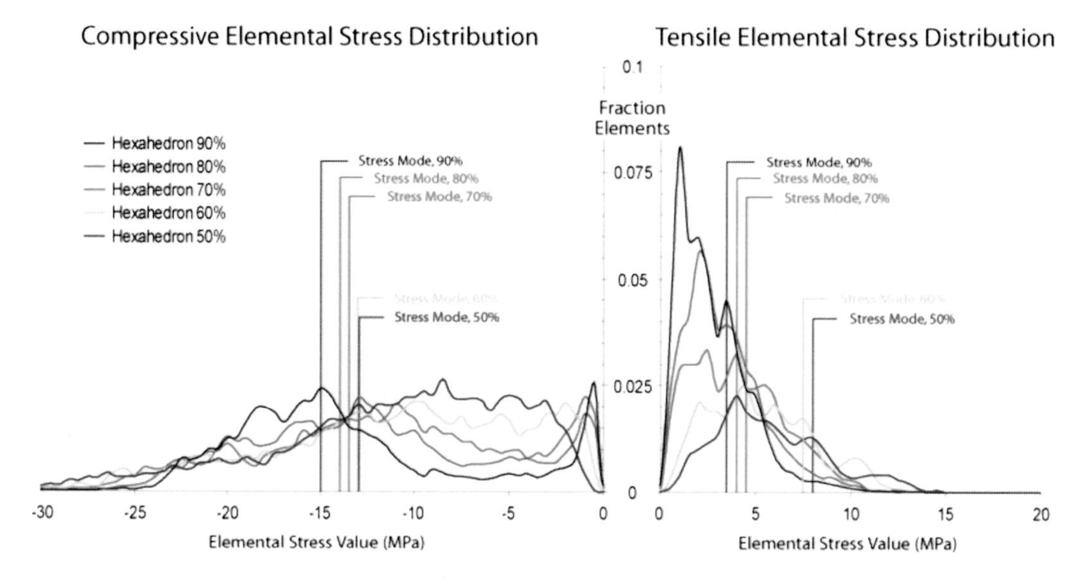

Fig. 10. Elemental stress distribution of the same architecture (hexahedron) at varying material volumes. Observational mode (highest number of occurrences) is displayed as the peaks of each data set. Stress value means are indicated by vertical ticks.

over the entire dynamic range of the porosities is always relatively the same, with about 16.5% of the stress values lying in the tensile region while the overall building block is loaded in compression.

Bone tissue architecture has a great impact on its structural properties as well as its mechanobiology. The spatial orientation of trabecular bone tissue within the spine indicates a mainly longitudinal loading of the spine. On the other hand, the almost symmetric distribution of material properties within the femoral head indicates a diverse loading axis during gait. By mimicking the micro-architectures of our building blocks to the same material volumes (porosity) of bone tissue, we can directly compare the effect of material spatial organization and architectures. The number of beams and the complexity of their arrangement affect the rate of change of beam length/diameter with respect to porosity. Low beam lengths indicate high beam numbers contained within a building block. The range of beam diameters provides information on the way by which the architecture changes shape during porosity adjustment. The TO has the widest range of the building blocks.

The complexity and arrangement of beams within a building block affects the resulting surface area of that architecture. The simplest shapes (H, TH) have a maximum surface area at 50% porosity and decrease with subsequent material removal. The more complex shapes (TO, RC) have tightly arranged beams with diameters, which encroach upon the beam spacing distance. Independent of beam arrangement complexity, all architectures approach the same surface area limit at porosities greater than 90%.

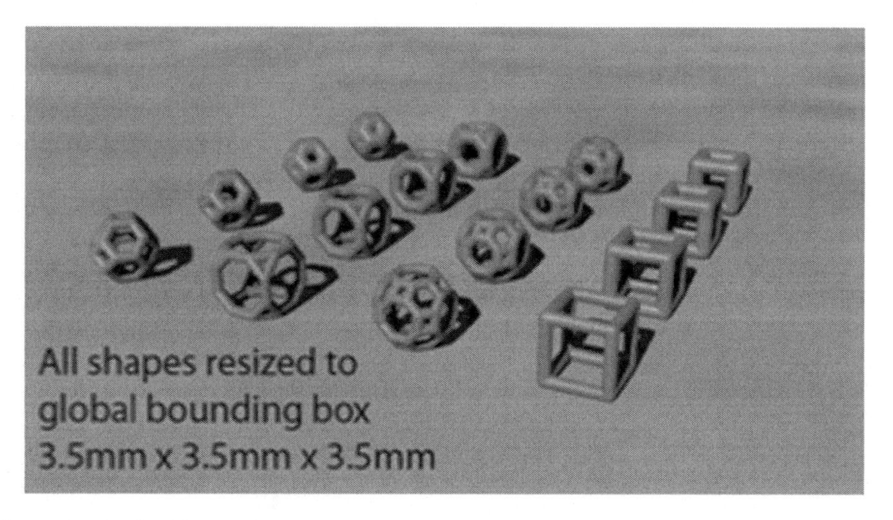

Fig. 11. Shape change of polyhedra with porosity.

Space filling is a measure of how polyhedra fill their bounding box. Architectures with a high connectivity index have the ability to equally distribute material in comparison to architectures of lower connectivity, which do not afford that possibility. At high porosities all architectures act the same, sharing space equally on the exterior as well as the interior. On the other hand, at low porosities, architectures with high beam numbers or those who are architecturally complex eventually overlap themselves, dependent on the topology of the vertices (Fig. 11). In complex architectures (RC, TO), the shape transition towards low porosities follows a power law, which indicates a greater modification to the overall geometry of the object, in contrast to the higher porosities where small topological differences result in large changes in mechanical properties and surface area. Therefore, the majority of the architectural changes when moving towards a higher porosity were associated with surface area expansion–contraction rather than volumetric arrangement, which promotes mechanical stability or architectural complexity.

4.4. Influence of Surface Mechanical Environment

The designers of scaffolds for tissue engineering have recognized the importance of material selection, biocompatibility, and open pore architecture with porous interconnectivity for nutrients and byproducts transfer; however, they have neglected the role of material organization on the final tissue produced.

Each tissue in the body has a unique function and therefore a different set of mechanical requirements. Bone, for example, should have certain mechanical requirements in order for it to function properly. Some of these requirements are the capability to provide structural rigidity, act as a reservoir for ions and calcium regulation, and provide a framework for the transfer of muscle forces. Two thirds of the composition of bone is mineralized tissue and the remaining one-third is organic material, cells, and water. Therefore, the mechanical properties of bone are determined by the interaction of the mineral and the organic contents as well as the orientation (woven, lamellar) in each scale size.

Mechanical environment plays an important role in designing scaffolds that can be used to replace load-bearing (i.e., bone) and non-load bearing (i.e., adipose) tissues. Bone is a viable tissue, in which, the way the mechanical signals are transuded is extremely important as it determines the biological activity of it. As previously mentioned, the type of mechanical loading (axial in the spine or diverse in the femoral head) will influence the spatial orientation and anisotropy of bone tissue. A bone replacement material will need to incorporate such information in order to sustain the functionality of the initial tissue.

In 1892, Wolff's law was the first to mention the relationship between bone micro-architecture orientation and loading direction. In his paper, he claimed that a mathematical law must exist that describes the spatial orientation of bone tissue with respect to mechanical loading while the overall bone shape is determined by genetics. Still till today, no one was able to establish such a mathematical law.

Almost 90 years later, Harold Frost introduced his "Mechanostat Theory," which is based on a mechanical usage window that explains the metabolic adaptation of bone to mechanical signals and an On/Off mechanism that helps to explain the apparent level biological machinery of the bone. Although this theory is more sophisticated than Wolff's law, it does not include any dynamic or time-dependent effects, which we know is required for the bone to be perfused properly.

The mechanical usage of the bone is defined as the voluntary load exerted on a skeleton during specific period of time (i.e., weeks) and is delineated by units of strain. It is divided into four main parts; disuse window, adapted window, mild overload window (MOW), and pathological overload window (POW). In the disuse window (DW), loss of bone mass will occur if strain values are below 50 $\mu\varepsilon$. This may be the case in patients who are bedridden for an extended period of time. The strain levels between 50 and 100 $\mu\varepsilon$ are considered physiological normal and are believed not to cause a change in bone mass. The next window is the adapted window (AW). In this window, strains between 100 and 1,000 $\mu\varepsilon$ trigger the remodeling of an old tissue to a new one preserving its original architecture and strength. In the MOW, there will be an increase in the bone mass due to lamellar remodeling that took place at a lower strain level of AW. Finally, at strains more than 3,000 MES, remodeling subsides and microdamage increases leading eventually to a decrease in the bone strength in what is known as the pathological overload window.

The "mechanostat theory" explains the mechanically induced bone formation without considering all the cellular details. The term mechanostat is similar to thermostat, where there should be a deviation from the set point to trigger the subsequent events to take place.

Finally, some limitations became evident to Wolff's law as well as the mechanostat theory by Frost. Although dynamic loading plays a vital role in the mechano transduction in load-bearing tissues like bone and cartilage, it has not been addressed properly in the previously mentioned theories neither in terms of magnitude nor in terms of loading type effects on the tissue. In bone, a small magnitude of dynamic loading increases bone formation by recruitment of surface cells, whereas longer loading periods seemed to be detrimental to a loaded implant within rat tibia when compared with static controls. Also, it was found that cells differ in their preference to the dynamic loading. Cartilage, for example, has greater preference to smaller magnitude dynamic loading compared to higher magnitude loading.

5. Outlook

5.1. Tissue Growth Optimization

One of the main goals in scaffold optimization is the ability to provide an idealized mechanical environment for the de novo tissue growth based on the mechanical loading stimulus provided by the body. Since the scaffold is supposed to provide functionality from day 1, it is considered that mechanical loading aspect persists throughout the life of the scaffold. In contrast, in vitro seeded cells tend to migrate and be effective for only days, maybe weeks. The same principal can be reflected on biochemical stimuli, such as growth factors that can be absorbed within minutes from the scaffold surface. When using certain encapsulation techniques, however, smaller dosages of growth factors can be released over a longer period of time.

5.2. Potential for CATE

The number of people needing a transplant continues to rise faster than the number of donors. According to the US department of health and human services, there are more than 106,000 people waiting for organ transplant (44). Several thousand transplant candidates are added each year to the national waiting list. Each day, 77 people receive organ transplants but unfortunately this is not met with the growing demand of transplanted organs. Eighteen people die each day while waiting for transplants that cannot take place because of the shortage of donated organs.

Doctors have successfully transplanted different organs like hearts, lungs, kidneys, livers, and other tissues for many years, but problems still exist. Many grafts do not survive permanently as we hope.

This has led to the development of a new era where tissues can be created with similar functional as well as structural characteristics of the desired organ, by the incorporation of different biological materials such as cells, growth factors, and biopolymers. Our vision for the future is that eventually artificial organs can be fabricated with standard additive fabrication techniques. Such procedures would make the replacement of failing or damages organs feasible for every patient and provide a virtually unlimited supply of replacement tissue.

References

1. US Department of Health & Human Services OPTN/SRTR Annual Report Trends in Organ Donation and Transplantation in the United States, 1998-2007 http://optn.transplant.hrsa.gov/ar2008/
2. Evans GRD, Brandt K, Widmer S et al (1999) In vivo evaluation of poly(L-lactic acid) porous conduits for peripheral nerve regeneration. Biomaterials 20:1109
3. Cooper ML, Hansbrough J, Spielvogel RL et al (1991) In vivo optimization of a living dermal substitute employing cultured fibroblast on a biodegradable polyglycolic acid or polyglactin mesh. Biomaterials 12:243
4. Cima LG, Ingber D, Vacanti JP et al (1991) Hepatocyte culture on biodegradable polymeric substrates. Biotechnol Bioeng 38:145
5. Sato M, Ando N, Ozawa S, Hiroyuki Miki, Hayashi K, Kitajima M (1997) Artificial esophagus. Mater Sci Forum 9250:105
6. Shinoka T, Mayer J (1997) New frontiers in tissue engineering: tissue engineered heart valves. In: Atala A, Mooney AJ (eds) Synthetic biodegradable polymers scaffolds. Birkhauser, Boston, pp 187–198
7. Ashiku SK, Randolph M, Vacanti CA (1997) Tissue engineering cartilage. Mater Sci Forum 250:129
8. Kim TH, Jannetta C, Vacanti JP, Upton J, Vacanti CA (1995) Engineered bone from polyglycolic acid polymer scaffold and periosteum. Spring Meeting of the Materials Research Society, San Francisco, CA, 17–21
9. Cao YL, Vacanti JP, Ma PX, Paige KT, Upton J, Langer R, Vacanti CA (1995) Tissue engineering of the tendon. Mater Res Soc Proc 394:83–89
10. Wettergreen MA, Bucklen B, Sun W, Liebschner MAK (2005) Computer-aided tissue engineering of a human vertebral body. Ann Biomed Eng 33(10):1333–1343
11. Sun W, Lal P (2002) Recent development on computer aided tissue engineering – a review. Comput Methods Programs Biomed 67(2):85–103
12. Liebschner M, Bucklen B, Wettergreen M (2005) Mechanical aspects of tissue engineering. Semin Plast Surg 19:217–228
13. Jacobs CR, Davis B, Reiger CJ, et al (1999) The impact of boundary conditions and mesh size on the accuracy of cancellous bone tissue modulus determination using large-scale finite-element modeling. In: North American congress on biomechanics, 1999
14. Wettergreen M, Bucklen B, Sun W et al (2005) Computer-aided tissue engineering in whole bone replacement treatment. In: ASME international mechanical engineering congress and exposition, 2005
15. Bourell DL, Leu M, Rosen DW (2009) Roadmap for additive manufacturing identifying the future of freeform processing. Laboratory for freeform fabrication, advance manufacturing center, p 22–27
16. Freyman TM, Yannas I, Gibson LJ (2001) Cellular materials as porous scaffolds for tissue engineering. Prog Mater Sci 46:273–282
17. Krams R, Wentzel JJ, Cespedes I, Vinke R, Carlier S, van der Steen AF, Lancee CT, Slager CJ (1999) Effect of catheter placement on 3-D velocity profiles in curved tubes resembling the human coronary system. Ultrasound Med Biol 25(5):803–810
18. Peltola SM, Melchels FP, Grijpma DW, Kellomaki M (2008) A review of rapid prototyping techniques for tissue engineering purposes. Ann Med 40(4):268–280
19. Frost HM (1987) The mechanostat: a proposed pathogenic mechanism of osteoporoses and the bone mass effects of mechanical and nonmechanical agents. Bone Miner 2(2):73–85

20. Mullender M, El Haj AJ, Yang Y, van Duin MA, Burger EH, Klein-Nulend J (2004) Mechanotransduction of bone cells in vitro: mechanobiology of bone tissue. Med Biol Eng Comput 42(1):14–21
21. Turner FB, Andreassi JL 2nd, Ferguson J, Titus S, Tse A, Taylor SM, Moran RG (1999) Tissue-specific expression of functional isoforms of mouse folypoly-gamma-glutamae synthetase: a basis for targeting folate antimetabolites. Cancer Res 59(24):6074–6079
22. Martin I, Wendt D, Heberer M (2004) The role of bioreactors in tissue engineering. Trends Biotechnol 22:80–86
23. Cowin SC (ed) (1989) Bone mechanics handbook, 2nd edn. CRC, Boca Raton, FL
24. Darling AL (2005) Functional design and fabrication of heterogeneous tissue engineering scaffolds. Dissertation, Drexel University
25. Ramtani S (2004) Mechanical modeling of cell/ECM and cell/cell interactions during the contraction of a fibroblast-populated collagen microsphere: theory and model simulation. J Biomech 37(11):1709–1718
26. Fritton SP, Rubin C (2001) In vivo measurement of bone deformations using strain gauges. In: Cowin SC (ed) Bone biomechanics handbook. CRC, Boca Raton, FL, pp 1–41
27. Liebschner MAK, Wettergreen MA (2003) Optimization of bone scaffold engineering for load bearing applications. In: Ashammakhi N, Ferretti P (eds) Topics in tissue engineering 22:1–39
28. Liebschner MAK (2009) Mechano-biology as an optimization principle for biological tissue engineering. Virtual and Physical Prototyping 4(4):183–193
29. Berry CC, Campbell G, Spadiccino A, Roberton M, Curtis ASG (2004) The influence of microscale topography of fibroblast attachment and motility. Biomaterials 25:5781–5788
30. Drotleff S, Lungwitz U, Breuniga M, Dennis A, Blunk T, Tessmar J, Gopferich A (2004) Biomimetic polymers in pharmaceutical and biomedical sciences. Eur J Pharm Biopharm 58:385–407
31. Wang L, Shelton R, Cooper PR, Lawson M, Trif JT, Barralet JE (2003) Evaluation of sodium alginate for bone marrow cell tissue engineering. Biomaterials (24):3475–3481
32. Ho ST, Hutmacher DW (2006) A comparison of micro CT with other techniques used in the characterization of scaffolds. Biomaterials 27:1362–1376
33. Hutmacher DW, Sittinger M, Risbud MV (2004) Scaffold-based tissue engineering: rationale for computer aided design and solid free-form fabrication systems. Trends Biotechnol 22(7):354–362
34. Murphy W, Dennis RG, Kileny JL, Mooney D (2002) Salt fusion: an approach to improve pore interconnectivity within tissue engineering scaffolds. Tissue Eng 8(1):43–52
35. Ma PX, Choi J (2001) Biodegradable polymer scaffolds with well-defined interconnected spherical pore network. Tissue Eng 7:23–33
36. Perren SM (1979) Physical and biological aspects of fracture healing with special reference to internal fixation. Clin Orthop Relat Res 138:175–196
37. Pearce P (1990) Structure in nature is a strategy for design. MIT Press, Cambridge, MA, p 245
38. Ball P (1999) The self-made tapestry: pattern formation in nature. Oxford University Press, Oxford, p 287
39. Gibson LJ (2005) Biomechanics of cellular solids. J Biomech 38(3):377–399
40. Mosekilde L (1993) Vertebral structure and strength in vivo and in vitro. Calcif Tissue Int 53(Suppl 1):S121–S125, discussion S125-126
41. Sander EA, Downs JC, Hart RT, Burgoyne CF, Nauman EA (2006) A cellular solid model of the lamina cribrosa: mechanical dependence on morphology. J Biomech Eng 128(6):879–889
42. Cromwell PR (1997) Polyhedra. Cambridge University Press, Cambridge, p 451
43. Sun W, Hu X (2002) Reasoning Boolean operation based CAD modeling for heterogeneous objects. Comput Aided Design 34:481–488
44. Rogers-Foy JM, Powers DL, Brosnan DA, Barefoot SF, Friedman RJ, LaBerge M (1999) Hydroxyapatite composites designed for antibiotic drug delivery and bone reconstruction: a caprine model. J Invest Surg 12(5):263–275

Chapter 2

Computer-Aided Approach for Customized Cell-Based Defect Reconstruction

Ulrich Meyer, Jörg Neunzehn, and Hans Peter Wiesmann

Abstract

Computer-aided technologies like computer-aided design (CAD), computer-aided manufacturing (CAM), and a lot of other features like finite element method (FEM) have been recently employed for use in medical ways like in extracorporeal bone tissue engineering strategies. Aim of this pilot experimental study was to test whether autologous osteoblast-like cells cultured in vitro on individualized scaffolds can be used to support bone regeneration in a clinical environment.

Mandibular bone defects were surgically introduced into the mandibles of Göttinger minipigs and the scaffold of the defect site was modelled by CAD/CAM techniques. From the minipigs harvested autologous bone cells from the porcine calvaria were cultivated in bioreactors. The cultured osteoblast-like cells were seeded on polylactic acid/polyglycolic acid (PLA/PGA) copolymer scaffolds being generated by rapid prototyping. The bone defects were then reconstructed by implanting these tissue-constructs into bone defects.

The postoperative computerized topographic scans as well as the intraoperative sites demonstrated the accurate fit in the defect sites. The individual created, implanted scaffold constructs enriched with the porcine osteoblast-like cells were well tolerated and appeared to support bone formation, as revealed by immunohistochemical and histological analyses.

The results of this investigations indicated that the in vitro expanded osteoblast-like cells spread on a resorbable individualized, computer-aided fabricated scaffold is capable of promoting the repair of bone tissue defects in vivo. The shown results warrant further attempts to combine computer modelling and tissue engineering for use in different ways in bone reconstructive surgery.

Key words: Animal model, Bioengineering, Bone, Computer-aided techniques, Individualized scaffolds, Osteoblasts, Tissue engineering

1. Introduction

The reparation of bone and bone defects is an issue of intensive investigation in a lot of medical discipline and also in maxillofacial surgery. The augmentation of insufficient bone volume and the replacement of lost or damaged bone structures is a recent challenge

Michael A.K. Liebschner (ed.), *Computer-Aided Tissue Engineering*, Methods in Molecular Biology, vol. 868,
DOI 10.1007/978-1-61779-764-4_2,

in modern preimplantologic strategies and a very important step concerning the healing process.

Current approaches in bone reconstructive surgery exploit biomaterials, autographs or allograft, although restrictions on all of these strategies exist. The fact, that tissue grafts contain living cells and tissue-inducing substances conferring biologic plasticity is one important advantage over biomaterials. Within the last 15 years, the emerging field of tissue engineering has reached a level of clinical applicability that now offers promising alternatives for maxillofacial bone reconstruction (1).

In all areas of the modern research, computer-aided techniques play an important role. In particular in all sciences dealing with materials and their properties, features like computer-aided design (CAD), computer-aided manufacturing (CAM), and finite element method (FEM) are still of high importance.

First, the concept of CAD/CAM was mentioned decades ago in the field of dental technology and dentistry. The technology is used to produce dental restorations rapidly in prosthodontics and oral implantology. Until now, there are several commercial CAD/CAM systems on market. Nowadays, a great number of material blocks are made from different materials/biomaterials like feldspar ceramics, glass-ceramics, alumina oxide, zirconium oxide, titanium, composite materials, etc. (2).

This is also the case in various partial disciplines of medicine.

In addition, the FEM also plays an important role in modern dental research and craniofacial research. The FEM not only serves to investigate the properties of materials, it is also used to explore the behaviour of implants in bone (Fig. 1) or the charge of tooth enamel, dentine, and the pulp under the effect of mastification (Fig. 2).

This combination of all these technical disciplines adds into a new application called computer-aided tissue engineering (CATE) which involves the application of enabling computer-aided

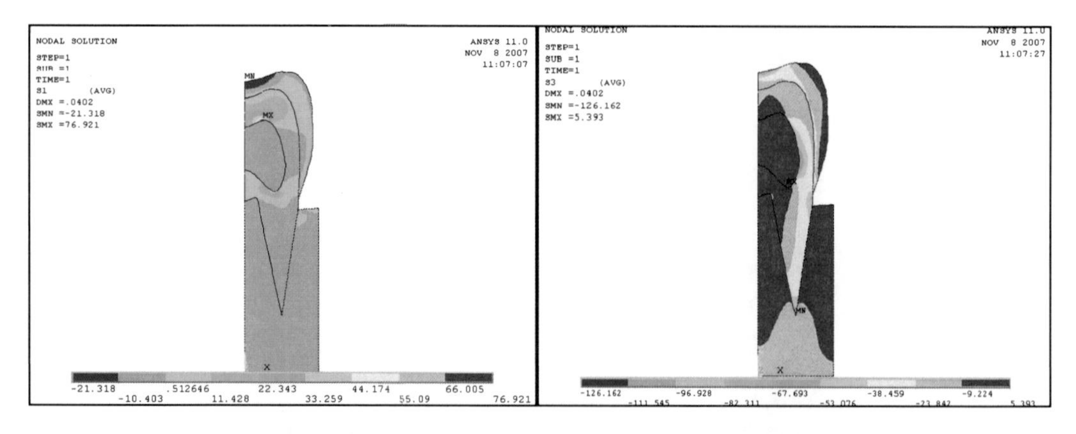

Fig. 1. FEM-construction of different matters (enamel, dentine, pulp) under force effect.

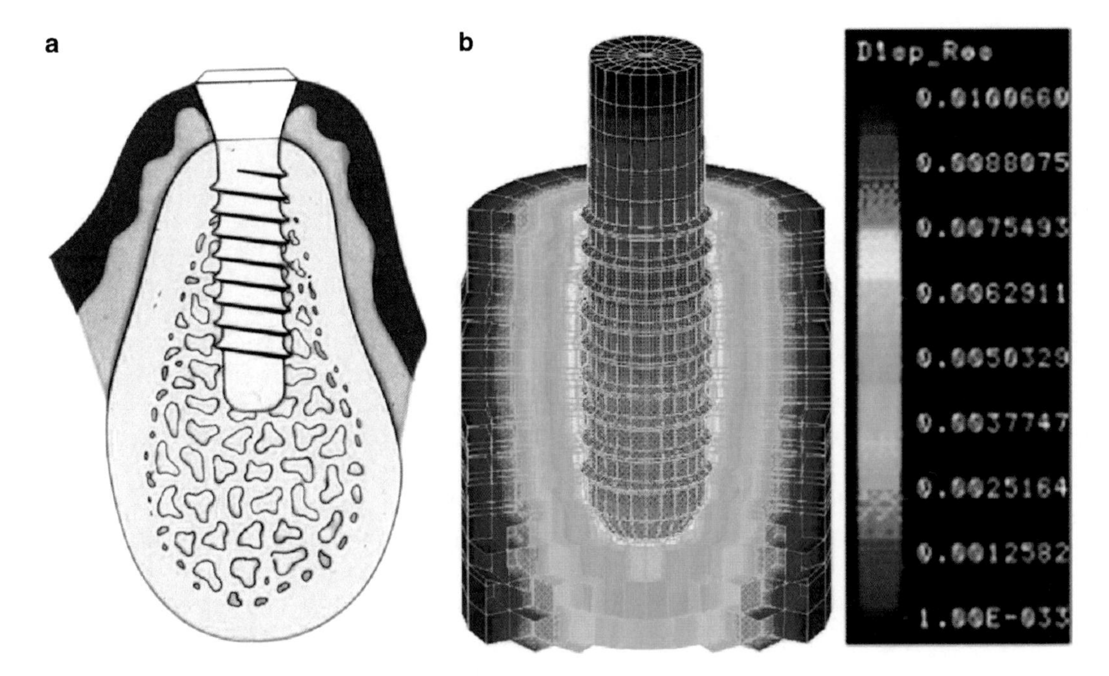

Fig. 2. (**a**) Scheme of implant/bone relation (**b**) FEM-construction of an inserted implant under force effect.

technologies, including beside the mentioned CAD, CAM, FEM image processing, rapid prototyping (RP) for modelling, designing, simulation, and manufacturing of biological tissue and organ substitutes. Another used method in CATE is solid freeform fabrication (SFF).

Unlike conventional machining, involving constant removal of materials, SFF builds parts by selectively adding materials, layer by layer, as specified by a computer program. Every layer represents the shape of the cross-section of the model at a specific level. Nowadays, this technique is viewed as a well-organized way of reproducibly generating scaffolds of desired properties on a large scale (3).

For extracorporeal tissue engineering, a complete three-dimensional tissue consisting of autologous or donor cells within a scaffold is grown in vitro and implanted once it has reached "maturity" (4). The engineered bone scaffold should in principle resemble in an idealized fashion the three-dimensional gross-, micro-, and nanomorphology of the bone to be replaced. When the anatomic structure is of major concern for the functional outcome, individualized tissue constructs, according to the external geometry, are desirable. A resent overview of the state of bone tissue engineering has given by Alsberg and associates (1) and Goldstein (5). Bone as a load-bearing organ is very susceptible towards stress transduction. A perfect congruency between the defect site and the artificial substrate is, therefore, important. Almost all tissue engineering concepts require to use some form of biocompatible scaffold, which serves as a template for cell proliferation and ultimately for tissue formation.

Two very important and essential components of successful tissue engineering are optimized biocompatible and resorbable materials (4) organized into a complex morphology. Bone scaffolds, like virtually all other kinds of scaffolds used in tissue engineering, are intended to degrade slowly and be replaced by newly formed regenerative tissue when implanted at appropriate sites.

Advanced computer-aided methods and technologies, medical imaging, and also manufacturing had assisted in these advances and created new possibilities in the development and use of bone tissue substitutes (for review, *see* Sun and Lal (6)). This approach combines cell biologic methods with image-based techniques. With the aid of especially adapted CAD/CAM systems, it is now possible to build up individualized bone scaffolds out of the body.

The application of non-invasive imaging techniques, such as computerized tomography (CT) and magnetic resonance imaging in combination with minor surgical procedures for the explantation and isolation of bone cells allows extracorporeal tissue constructions with only minimal donor morbidity. To monitor the clinical success of bone and tissue reconstructive procedures imaging techniques and data processing are employed.

The development and creation of bone-like tissue structures in vitro has to take into account the complexity of and scale required for clinical use (4). The fate of the extra corporeally propagated cells in terms of their viability and differentiation during culture and after the implantation is one critical aspect of such a tissue engineering approach. Even though a lot of clinical studies and investigations have documented the long-term healing process of bone defects after restoration with hybrid bone substitutes only a few studies have focused on the cell-driven mode of bone regeneration in the early and later postoperative stages of craniofacial defect reconstruction.

The aim of the present study was to construct an individual external shaped hybrid bone substitute in vitro by computer-aided techniques and rapid prototyping and assess its clinical applicability in mandibular defect reconstruction.

Attention was placed on the features of defect regeneration in a clinical setting as well as on control of the external shape and size of the tissue construct.

2. Material

2.1. Cell Culture

1. High Growth Enhancement Medium (ICN Biomedicals GmbH, Eschwege, Germany), (storage: 4°C).
2. Foetal calf serum (Biochrom KG, Berlin, Germany), (storage: −20°C).

3. 250 μg/mL Amphotericin B (Biochrom KG, Berlin, Germany), (storage: −20°C).
4. 10,000 IU/mL Penicillin (Biochrom KG, Berlin, Germany), (storage: −20°C).
5. 10 mg/mL Streptomycin (Biochrom KG, Berlin, Germany), (storage: −20°C).
6. 200 mmol/L L-glutamine (Biochrom KG, Berlin, Germany), (storage: −20°C).
7. 10 mmol/L Glycerophosphate (Sigma-Aldrich, St Louis, MO), (storage: −20°C).
8. 25 μg/mL Ascorbic acid (Sigma-Aldrich, St Louis, MO), (storage: −20°C).
9. Tyrode's solution (Biochrom KG Seromed), (storage: −20°C).
10. Prefabricated polylactic acid/polyglycolic acid (PLA/PGA).
11. Phosphate buffer solution (PBS), (storage: 4°C).

2.2. Animal Model

1. Atropine (0.06 mg/kg), (storage: 4°C).
2. Azaperone (0.03 mL/kg), (storage: 4°C).
3. Ketamine (10 mg/kg), (storage: 4°C).
4. T61 (Hoechst, Germany), (storage: 4°C).

2.3. Scaffold Construction

1. CT (Philips, Eindhoven, Netherlands).
2. G3 Workstation (Apple, Cupertino, CA).
3. Macintosh operating system version 8.5 (Apple, Cupertino, CA).
4. NIH-Image 1.61 (US National Institute of Health, Bethesda, MD).
5. 3D modelling program Rhinoceros (Robert McNeel & Associates, Seattle, WA).

2.4. Histology and Immunohistochemistry

1. Anti-osteocalcin from clone OC4-30 (Takara, Shiga, Japan) diluted 1:500 (storage: −20°C).
2. Anti-osteonectin from clone OSN4-2 (Takara, Shiga, Japan) diluted 1:500 (storage: −20°C).
3. Dako Antibody Diluent (Dakopatts, Glostrup, Denmark). (storage: −20°C).
4. Eosin (storage: 4°C).

   1 g Eosin (Chroma, Münster, Germany).

   50 mL Alcohol 96%.

   50 mL Aqua. dest.

5. 4% Phosphate buffered paraformaldehyde solution (pH 7.2), (storage: 4°C).
6. Haematoxilin, (storage: 4°C).
   1,000 mL Aqua dest.
   1 g Haematoxylin (Merck, Darmstadt, Germany).
   0.2 g NaJ (Merck, Darmstadt, Germany).
   50 g Kalialaun (Merck, Darmstadt, Germany).
   50 g Chlorathydrat (Roth, Karlsruhe, Germany).
   1 g Citric acid (Merck, Darmstadt, Germany).
7. Tris-buffered saline (TBS), (storage: 4°C).
8. Xylene (Merck, Darmstadt, Germany), (storage: room temperature).

3. Methods

3.1. Cell Culture

The used osteoblast-like cells for this study were derived from cranial periosteum harvested from the minipigs.

1. The needed periosteum was removed from the pigs' calvaria under general anaesthesia (Fig. 3d).
2. After removal it was cut into 3- to 6-mm pieces and transferred to culture dishes with the osteogenic layer facing downward. Osteoprogenitor cells migrated from the tissue explants.
3. The osteoblast-like cells were cultured for 5 weeks in High Growth Enhancement Medium (ICN Biomedicals GmbH, Eschwege, Germany) supplemented with 10% foetal calf serum, 250 μg/mL amphotericin B, 10,000 IU/mL penicillin, 10 mg/mL streptomycin, 200 mmol/L L-glutamine (Biochrom KG, Berlin, Germany), 10 mmol/L glycerophosphate, and 25 μg/mL ascorbic acid (Sigma-Aldrich, St Louis, MO) at 37°C and 5% CO_2 in humidified air. The medium was replaced twice a week.
4. To control their osteoblast-like features (4), the differentiation state of cells was continuously monitored before seeding and during growth in the scaffold by immunocytochemical methods in an established fashion (8).

   There were only taken cells from the first passage for the construction of the hybrid complex.
5. Primary culture cells were harvested by incubation in collagenase-containing Tyrode's solution (Biochrom KG Seromed), collected and centrifuged to create pellets.

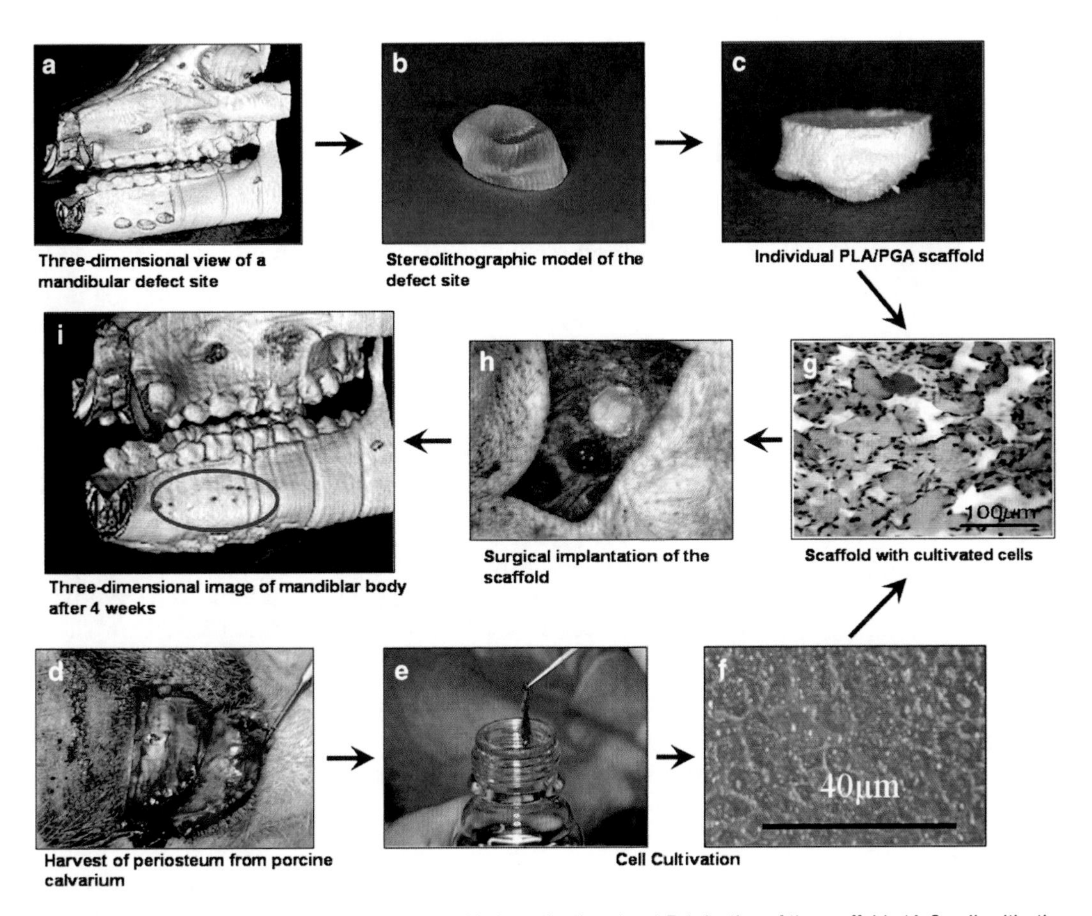

Fig. 3. Chart of the different working steps done in this investigation. (**a**–**c**) Fabrication of the scaffolds (**d**–**f**) cell cultivation (**g**–**i**) implantation of cell-loaded scaffolds and healing.

6. The resuspended cells were seeded into a bioreactor on the prefabricated PLA/PGA construct. The similar cell culture conditions like those for the outgrowth culture of the periosteum were used.
7. Cell viability was assessed at different times during cultivation. The hybrid materials were washed three times with PBS and prepared for implantation 3 days after seeding.

Control hybrid materials which were not used for the in vivo implantation were additionally cultured for up to a period of 3 weeks in the bioreactor and evaluated by immunohistochemical methods to assess the extracorporeal long-term outcome of cells in the scaffold (8).

3.2. Animal Model

The present experimental study was performed on six mature Göttinger minipigs that were in the age of 14–16 months, with an average body weight of 35 kg. The minipigs were chosen for this study to ensure a mandibular size comparable to the human mandible for bone substitute fabrication. In previous studies, this animal model has been successfully used for bone reconstruction procedures. All surgery was performed under sterile conditions in an animal operating theatre with an intramuscular injection of ketamine (10 mg/kg), atropine (0.06 mg/kg), and azaperone (0.03 mL/kg) to induce anaesthesia. The study was approved by the Animal Ethics Committee of the University of Münster under the reference number G 67/2202. All animals underwent the same operative procedure (9).

3.3. Scaffold Construction

1. Four mandibular defects were created in the mandible of each minipig 4 weeks after the removal of the periosteum. The mandible was exposed through a submandibular approach under general anaesthesia. There were created three defects with diameters of 1 cm by using a bur to drill subperiosteally in the mandibular body (Fig. 3a).
2. Computerized tomography scans of the mandible were performed after wound closure (Philips, Eindhoven, Netherlands). The scanning process was performed in 2-mm slices with 1-mm overlap technique.
3. The data storage was done on an optic disk using the Digital Imaging and Communications in Medicine (DICOM) format. If a three-dimensional visualization was desired, the DICOM data sets were transferred to a G3 Workstation (Apple, Cupertino, CA) for image reformation and viewing.
4. All hard- and software implications of the system were coordinated with the Macintosh operating system version 8.5 (Apple). NIH-Image 1.61 (US National Institute of Health, Bethesda, MD) was used for DICOM import, data analysis and volume-rendering visualization. On the basis of the DICOM data set, a surface reconstruction of the volume of each defect was performed using a special software tool which was developed at the University of Münster's Laboratory of Biophysics. With the aid of this program, written in the computer language PV-Wave (Visual Numerics, Houston, TX) the single CT slices were analyzed and the contours of the defect area were detected.
5. The detected defect outlines were then combined into a three-dimensional surface through a dense net of triangles (9).
6. The finishing of the scaffold design was completed by transforming this triangle net into a NURBS (non-uniform rational b-splines) surface witch was done using a 3D modelling program (Rhinoceros, Robert McNeel & Associates, Seattle, WA). Unlike triangle nets NURBS allow a smooth tooling of the surface for final adaptations. At last, the data set was transferred to a stereolithography (STL) format, which was used to create a model made of synthetic resin via rapid prototyping

(stereolithography) (Fig. 3b), after which the PLA/PGA construct was formed (Fig. 3c).

7. After the generation of a digital image representation of the external scaffold, the defect site was manufactured as an external form by rapid prototyping (H&H Engineering and Rapid Prototyping, Lemgo, Germany). Rapid prototyping is the name for several techniques, which read in data from CAD drawings and manufacture automatically three-dimensional objects layer by layer according to the virtual design. The utilization of RP in tissue engineering enables the production of three-dimensional scaffolds with complex geometries and very fine structures. All the parts of the internal structure-like material composition and pore size were controlled by a defined polymerization of PLA/PGA beads (50% PLA, 48% PGA, 2% carbon acid (H_2CO_3)) of various sizes. To allow the ingrowth of vessels and enhance nutrition and oxygen supply by diffusion, the pore size of the inner part of the scaffold was approximately 200 μm in diameter.

3.4. Tissue Engineering

1. The autologous porcine bone cells were spread on the scaffold and cultured in a bioreactor (Fig. 3g) for 3 days for the engineering of the viable bone tissue substitutes.
2. Then, the created tissue constructs were implanted into the mandibular defects 1 week after defect creation through the submandibular approach used previously (Fig. 3h). The periosteum was carefully sutured, and the wound was closed by multilayered sutures at the end of the operation.
3. The animals were inspected continuously for signs of wound dehiscence or infection for the first several days. After this period of daily observance, their general health was assessed weekly. The animals were feed a normal diet.
4. After the fixed periods of 3 and 30 days, the animals were sacrificed with an overdose of an excitation-free narcotic solution (T61, Hoechst, Germany) administered intravenously.
5. Following euthanasia, a CT scan was performed, and mandibular block specimens containing the bone substitute with surrounding tissue were removed from all the animals. To remove unnecessary portions of bone and soft tissue, the block samples were sectioned by a saw and finally prepared for histologic analysis.

3.5. Histology and Immunohistochemistry

To accomplish the immunohistochemical and histologic analysis,

1. The extra corporeally engineered constructs as well as the mandibular specimens were fixed in 4% phosphate buffered paraformaldehyde solution (pH 7.2) for 4 h by room temperature.
2. After decalcifying the samples in ethylenediaminetetraacetate acid (EDTA) they were embedded in paraffin. Consecutive

sections, 5 μm in thickness, were mounted on slides coated with L-lysine.

3. Having deparaffinized with xylene the sections subsequently were passed through a series of decreasing concentrations of alcohol into deionized water.
4. Shortly after rinsing in TBS, sections were stained with haematoxilin-eosin for conventional histology or alternatively incubated with primary antibodies in moisture chambers for 16 h at 4°C for immunohistochemistry.

   All the primary antibodies used in this study were diluted in Dako Antibody Diluent (Dakopatts, Glostrup, Denmark). Both monoclonal antibodies obtained from Takara (Shiga, Japan) anti-osteocalcin from clone OC4-30 and anti-osteonectin from clone OSN4-2 were diluted 1:500. The Dako EnVision technique (Dakapatts) was applied for the immunohistochemical staining.
5. The sections were rinsed three times in TBS after the incubation with primary antibodies.
6. For a short time, the sections were incubated with polymeric conjugates of soluble dextran polymers labelled with alkaline phosphatase enzymes.
7. Bound antibodies were detected using the Fast Red staining method with levamisole.
8. After 25 min the chromogenic reaction was stopped, and the treated sections were counterstained with Mayer's haematoxylin.
9. There were also performed positive controls using anti-actin as a primary antibody and negative controls with non-immune serum.

4. Results

4.1. Scaffold Fabrication

The use of rapid prototyping of mandibular body defects allowed the scaffold fabrication of individualized bone substitutes (Fig. 3c). Fabrication of defined porous PLA/PGA (50/50) copolymers resulted in a scaffold morphology with a spongiosal internal part. An individual site-specific insertion of the implant during surgery was allowed by the shape of the copolymers. After creating the defects in the pigs' mandible 4 days were needed to fabricate these individual fitting scaffolds. This interval is important for the clinical situation, where bone remodelling may alter the defect geometry after the surgical event.

4.2. Scaffold Culture

From the pigs taken osteoblast-like cells (Fig. 3f) commenced to attach approximately 4 h after seeding and actively colonized the surface of the PLA/PGA matrix within 3 days of culture. The attachment of the osteoblast-like cells at the PLA/PGA surface was demonstrated by scanning electron microscopy investigations. Careful review of the histological specimens obtained after 3 and 14 days in culture demonstrated evidence of osteoblast-like polygonal cells at the surface of the material. With the aid of microscopic methods, it was observed that the cells proliferated in the bulk material for at least up to 14 days. Frequently, the osteoblast-like cells filled up the space of the interconnecting pores between the neighbouring PLA/PGA particles. At higher magnification, the cells displayed firm attachment towards the material surface without signs of interfacial layer formation. Cells remained in their differentiated stage during the entire culture time. A lot of cells were cuboidal in appearance both at the material surface as well as in the macro pores within the bulk of the material. Immunohistochemical investigation revealed the synthesis of bone-specific marker proteins, as seen at the presence of newly deposited osteocalcin in the pericellular matrix. Signs of apoptotic cell death were not apparent as judged by conventional histology and immunohistology or as evaluated by a modified terminal deoxynucleotide transferase-mediated dUTP nick-end labelling (TUNEL) method.

4.3. Histologic Examination

On the PLA/PGA restorative material seeded porcine osteoblasts were viable under cell culture conditions. Furthermore, after implantation, the hybrid material displayed histological signs comparable to those found in the bioreactor. Viable and differentiated cells were observed throughout the implanted material. The cells built up a direct contact to the surface of the underlying PLA/PGA material (Fig. 4a). Comparison of the histology of the early (3 days post implantation) and later time points (4 weeks post implantation) revealed evidence of a progressive degradation of the scaffold

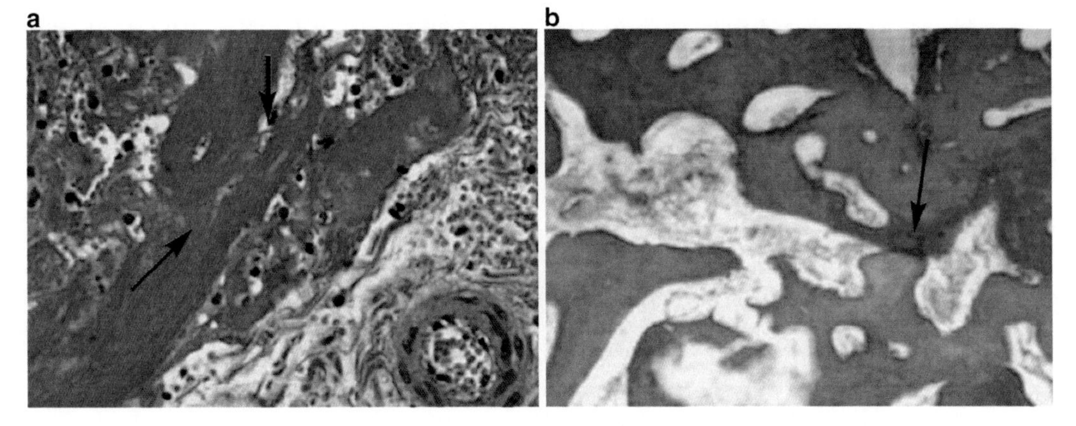

Fig. 4. (**a**) Histology of bone regeneration 3 days after implantation (*arrows* mark regions of mineralized matrix; original magnification ×10). (**b**) Defect site 30 days post implantation (*arrows* mark regions of mineralized matrix; original magnification ×10).

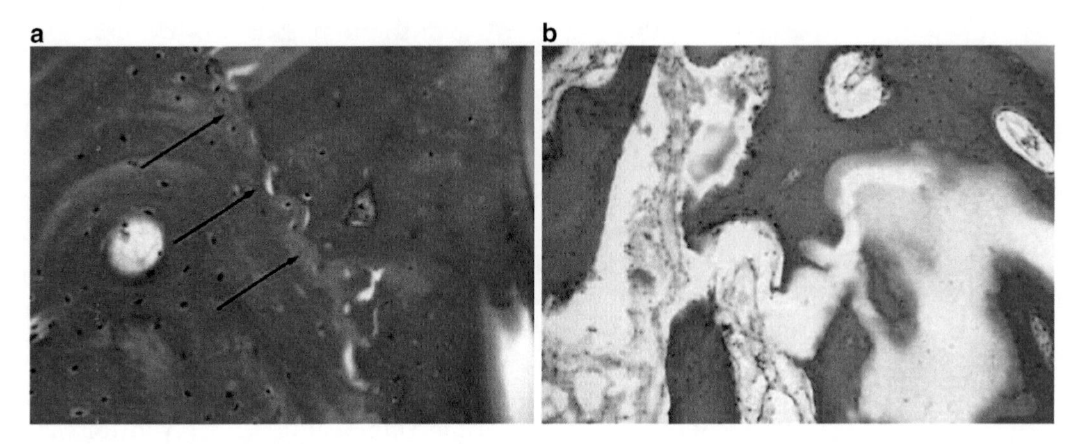

Fig. 5. (**a**) The *arrows* indicate new lamellar bone in some regions of the defect site separated from the original bone (original magnification ×40). (**b**) Four weeks after implantation: PLA/PGA remnants at late stage of degradation (original magnification ×40).

material during the 1-month post implantation period, as shown by an increase in the amount of newly formed bone. In some marginal areas, the bone tissue of the original bone defect site contacted the hybrid material. After 4 weeks, lamellar and spongiosal bone traversed the entire width of the defect and were integrated with the scaffold in all defect sites examined (Fig. 4b).

Cells lined the surface of the mineralized matrix and synthesized osteocalcin and osteonectin, which was indicative of their osteoblastoid phenotype. In some areas between the original bone and the regenerative tissue, small gaps were seen (Fig. 5a), which indicated the beginning of bone formation from the inner part to the defect site. Mature bone cells appeared to be attached to the remnants of the PLA/PGA particles (Fig. 5b). There was no interfacial layer observed around the particles at later stages of scaffold integration. For the most part of the implantation sites were filled by new bone tissue, with only some small amounts of PLA/PGA particles left in the defect area. First signs of vascularization were observed after 4 weeks as judged by the presence of endothelial cells and accompanying mesenchymal cells. Occasionally, a few multinucleated cells associated with erosin pits were seen at the scaffold surface.

4.4. Radiographic Analysis

A subsequent filling of the defect by an osseous-like material after 3 days and 4 weeks post implantation were demonstrated by CT scans (Fig. 3i). As shown in Fig. 5b the density of the implanted tissue substitutes at day 3 was lower than in areas of original cortical or spongiosal bone, but slightly higher than in the adjacent soft tissue. After a period of 4 weeks the density of the regeneration area was significantly higher than that of the early post implantation period, indicating a progressive osteoid formation during tissue repair. From the peripheral to the central areas of the defect site at both observation times a reduced density gradient was detected.

5. Discussion

This experimental study demonstrated that autologous osteoblast-like cells (in this case porcine ones) can be cultured extra corporeally and seeded on prefabricate individualized PLA/PGA scaffolds without obvious loss of viability and differentiation. The cells remained viable for at least 30 days after the implantation of the hybrid material into a mandibular defect sites. Thus, in vitro expanded osteoblast-like cells spread on the resorbable individualized scaffold are capable of promoting the repair of bony defects in vivo.

Effects like the biocompatibility of the material, the design of the internal and external structure of the scaffold like the porosity, the time period between the defect creation and the beginning of reconstruction, and the choice of cultured cells are critical requirements for the success of tissue engineering.

The following four basic steps were of importance for the success of the strategy to develop individualized hybrid substitutes:

- Periosteum was harvested by a minor surgical procedure, and osteoblast-like cells therefrom were isolated and multiplied in culture.
- The shape and the size of the created defect were defined by image processing and digitization of CT scan.
- The scaffold was created by polymerization of PLA/PGA beads of various sizes within the external scaffold designed by rapid prototyping.
- Finally, the construct was occupied by autologous bone cells in a bioreactor for later implantation at the defect site.

For a long time, it has been known that craniofacial bone has a vast capacity for regeneration when using osteoblast-like cells (7). Extracorporeal bone tissue engineering requires living osteoblast-like cells, and in a few cases that is all that may be needed. Fortunately, it is possible to maintain and propagate osteoblast-like cell types outside the human body for prolonged periods (4). It was possible to show that autologous cells are capable of forming new bone in animal models when a number of different vehicles are exploited (10–13). There are no legal concerns with the use of autologous cells, and there should be no problems resulting from immune rejection. The study at hand confirms that autologous osteoblast-like cells are easily harvested by a minor surgical procedure and can be successfully propagated in culture, making them a reliable cell source for tissue engineering in clinical settings.

The described method allows fabrication of resorbable and biocompatibility cell-containing scaffold with controlled external geometry in a reasonable time frame for use in mandibular surgery. In particular, the combined application of cell-based and tissue engineering CAD/CAM technology permits the tailoring of external scaffold geometries for bone tissue requirements. The fabricated bone scaffolds in this study seemed largely to resemble the external morphology of the bone to be replaced. The placing, orientation, spacing, and maintenance of autologous osteoblasts and their synthesized extracellular products in the construct, as revealed by immunohistochemical investigations was allowed by the scaffolds internal architecture, based on porogen leaching.

The beneficial biocompatibility may be attributed in part to the PLA/PGA matrices copolymerized in the presence of a carbonate buffer system. A large number of investigators have reported on the well-known favourable biocompatibility of PLA/PGA and found no impairment of cell function when using these polymers as a restorative material (14–16). As shown by Ishaug and associates the alkaline phosphatase activity expressed by osteoblasts cultured on PLA was comparable to that on tissue culture polystyrene (17).

Otto and colleagues indicated that the differentiation of bone cells is promoted PLA constructs (18). In contrast to the beneficial biocompatibility of bulk materials, degradable materials present problems of their own. The shift in pH related to hydrolysis is one major problem of PLA/PGA. The lowering of the pH by degradation threatens the viability of surrounding cells. The pH stability of the PLA/PGA-carbonate scaffold used seemed to contribute to the favourable results seen in the present study. The reduced gradient from peripheral to central areas of the defect site at both observation times indicates that bone repair was more pronounced at the defect borders. This might be based on the fact that in these areas both cell populations, the host cells and the transplanted cells, account for the repair process. In addition, the nutrition of transplanted cells is better at superficial transplant sites.

In tissue engineering used scaffolds are commonly fabricated from degradable materials (19, 20) and are shaped intraoperatively to fit into the defect site. The material composition, pore size, and chemistry were based on previous studies in extracorporeal bone tissue engineering studies (4). The conventional fabrication of the internal structure despite the use of CAD/CAM technique is an imitating factor of the present study. In contrast to conventional methods, the internal design of scaffolds can also be achieved individually on the basis of digital data formats (21, 22).

To design internal scaffold structures, including fibre bonding, solvent casting, particulate leaching, membrane lamination, melt moulding, temperature-induced phase separation, gas foaming, and three-dimensional printing, many different processing techniques have been developed (23). A large quantity of internal scaffold characteristics, such as porosity and pore size, has been modelled by such fabrication techniques. In most cases, the computer-driven techniques for scaffold fabrication have been evaluated on a material-specific level or controlled for their biologic behaviour in cell culture (24). However, techniques for computer-controlled modelling processes, which allow researchers to fabricate individualized cell/scaffold complexes for use in extracorporeal tissue engineering and, eventually, clinical settings, are seldom investigated in craniofacial reconstruction. With the aid of the presented strategy, it has been shown that it is possible to fabricate a cellular hybrid substitute with a desired external geometry.

- The applicability of individualized bone substitutes has been demonstrated by various studies (25). By various software tools, the modelling of the external and internal structure on the basis of digital data was performed (26). Whereas most studies used nonviable materials or were performed only on theoretical assumptions, the application of the cell-driven, image-based extracorporeal tissue engineering approach in the clinical setting has seldom been performed. Cells in combination with or without a scaffold were used in various studies addressing assessment of the clinical outcome of bone tissue engineering. The cells were either cultured on a suitable biomaterial prior to implantation or seeded on a substitute and injected into the defect. Although most authors have described the healing of bone defects by implantation of cellular hybrid materials in long-term experiments (27–31), it is unclear whether the finding can be attributed to the biomaterial alone or its cellular components. The addition of bone marrow-derived cells to various materials has been shown to enhance the performance of a wide variety of implanted bone substitutes (27). In a lot of studies, young animals or graft sites that have been surrounded by normal healthy tissue or wound sites that almost surely ossificate alone have been used (28, 29). It has been suggested from these studies that bone healing seems to be more rapid, complete, and reproducible when a cell-based therapy is employed (30, 31). However, some experimental studies have revealed that bone healing is accelerated when only extracts from cultured human osteosarcoma cells are injected (32), which indicates that the addition of nonviable cellular components to the matrix also supports bone regeneration.

6. Conclusion

In this preliminary study, it has been indicated that autologous osteoblast-like cells can be easily harvested and kept viable while proliferating in a bioreactor for further use in bone tissue engineering. The presented information confirms that it is feasible to harvest small amounts of autologous periosteum with minimal morbidity, expand the cell therein, and after it seed them on individualized scaffold in bioreactors (33) to repair mandibular bone defects (7). The combination of the two different techniques of CAD/CAM systems and tissue engineering methods allows a cell-based biomimetic approach. The advances in image-based design and bioreactor cell cultivation techniques make it possible to engineer patient-specific hybrid bone constructs that may improve bone defect reconstructions (34).

References

1. Alsberg E, Hill EE, Mooney DJ (2001) Craniofacial tissue engineering. Crit Rev Oral Biol Med 12(1):64–75
2. Liu XZ, Lv PJ, Wang Y (2008) J Peking Univ Health Sci 40(6):654–657
3. Hutmacher DW, Sittinger M, Risbud MV (2004) Scaffold-based tissue engineering: rationale for computer-aided design and solid free-form fabrication systems. Trends Biotechnol 22(7):354–362
4. Meyer U, Joos U, Wiesmann HP (2004) Biological and biophysical principles in extracorporal bone tissue engineering. Part I. Int J Oral Maxillofac Surg 33(4):325–332
5. Goldstein SA (2002) Tissue engineering: functional assessment and clinical outcome. Ann N Y Acad Sci 961:183–192
6. Sun W, Lal P (2002) Recent development on computer aided tissue engineering-a review. Comput Methods Programs Biomed 67 (2):85–103
7. Breitbart AS, Grande DA, Kessler R, Ryaby JT, Fitzsimmons R, Grant RT (1998) Tissue engineered bone repair of calvarial defects using cultured periosteal cells. Plast Reconst Surg 101:567–574
8. Meyer U, Meyer T, Jones DB (1998) Attachment kinetics, proliferation rates and vinculin assembly of bovine osteoblasts cultured on different pre-coated artificial substrates. J Mater Sci Mater Med 9(6):301–307
9. Meyer U, Büchter A, Hohoff A, Stoffels E, Szuwart T, Runte CH, Dirksen D, Wiesmann HP (2005) Image-based extracorporeal tissue engineering of individualized bone constructs. Int J Oral Maxillofac Implants 20(6):882–890
10. Owen M (1988) Marrow stromal stem cells. J Cell Sci Suppl 10:63–76
11. Gundle R, Joyner CJ, Triffitt JT (1995) Human bone tissue formation in diffusion chamber culture in vivo by bone-derived cells and marrow stromal fibroblastic cells. Bone 16 (6):597–601
12. Krebsbach PH, Kuznetsov SA, Bianco P, Robey PG (1999) Bone marrow stromal cells: characterization and clinical application. Crit Rev Oral Biol Med 10(2):165–181
13. Kuznetsov SA, Krebsbach PH, Satomura K, Kerr J, Riminucci M, Benayahu D, Robey PG (1997) Single-colony derived strains of human marrow stromal fibroblasts form bone after transplantation in vivo. J Bone Miner Res 12(9):1335–1347
14. Bos RR, Rozema FR, Boering G, Nijenhuis AJ, Pennings AJ, Verwey AB, Nieuwenhuis P, Jansen HW (1991) Degradation of and tissue reaction to biodegradable poly(L-lactide) for use as internal fixation of fractures: a study in rats. Biomaterials 12(1):32–36
15. Majola A, Vainionpaa S, Vihtonen K, Mero M, Vasenius J, Tormala P, Rokkanen P (1991) Absorption, biocompatibility, and fixation properties of polylactic acid in bone tissue: an experimental study in rats. Clin Orthop Relat Res 268:260–269
16. VanSliedregt A, Radder AM, de Groot K, Van Blietierswijk CA (1992) In vitro biocompatibility

testing of polylactides. Part 1. Proliferation of different cell types. J Mater Sci Mater Med 3:365–372

17. Ishaug SL, Yaszemski MJ, Bizios R, Mikos AG (1994) Osteoblast function on synthetic biodegradable polymers. J Biomed Mater Res 28 (12):1445–1453
18. Otto TE, Patka P, Haarman HJTHM, Klein CPAT, Vriesde R (1994) Intramedullary bone formation after polylactic acid wire implantation. J Mater Sci Mater Med 5:407–412
19. Wiesmann HP, Nazer N, Klatt C, Szuwart T, Meyer U (2003) Bone tissue engineering by primary osteoblast-like cells in a monolayer system and 3-dimensional collagen gel. J Oral Maxillofac Surg 61(12):1455–1462
20. Yang S, Leong KF, Du Z, Chua CK (2001) The design of scaffolds for use in tissue engineering. Part I. Traditional factors. Tissue Eng 7(6):679–689
21. Landers R, Hubner U, Schmelzeisen R, Mulhaupt R (2002) Rapid prototyping of scaffolds derived from thermoreversible hydrogels and tailored for applications in tissue engineering. Biomaterials 23(23):4437–4447
22. Runte C, Dirksen D, Delere H, Thomas C, Runte B, Meyer U, von Bally G, Bollmann F (2002) Optical data acquisition for computer-assisted design of facial prostheses. Int J Prosthodont 15(2):129–132
23. Widmer MS, Mikos AG (1998) Fabrication of biodegradable polymer scaffolds for tissue engineering. In: Patrick CW Jr, Mikos AG, McIntire LV (eds) Frontiers in tissue engineering. Elsevier, New York, NY, pp 107–117
24. Hutmacher DW, Schantz T, Zein I, Ng KW, Teoh SH, Tan KC (2001) Mechanical properties and cell cultural response of polycaprolactone scaffolds designed and fabricated via fused deposition modeling. J Biomed Mater Res 55 (2):203–216
25. Eufinger H, Wehmoller M (2002) Microsurgical tissue transfer and individual computer-aided designed and manufactured prefabricated titanium implants for complex craniofacial reconstruction. Scand J Plast Reconstr Surg Hand Surg 36(6):326–331
26. Eufinger H, Saylor B (2001) Computer-assisted prefabrication of individual craniofacial implants. AORN J 74(5):648–654, quiz 655–656, 658–662
27. Kadiyala S, Young RG, Thiede MA, Bruder SP (1997) Culture expanded canine mesenchymal stem cells possess osteochondrogenic potential in vivo and in vitro. Cell Transplant 6 (2):125–134
28. Piersma AH, Ploemacher RE, Brockbank KG (1983) Transplantation of bone marrow fibroblastoid stromal cells in mice via the intravenous route. Br J Haematol 54(2):285–290
29. Schliephake H, Knebel JW, Aufderheide M, Tauscher M (2001) Use of cultivated osteoprogenitor cells to increase bone formation in segmental mandibular defects: an experimental pilot study in sheep. Int J Oral Maxillofac Surg 30(6):531–537
30. Schmelzeisen R, Schimming R, Sittinger M (2003) Making bone: implant insertion into tissue-engineered bone for maxillary sinus floor augmentation-a preliminary report. J Craniomaxillofac Surg 31(1):34–39
31. Bruder SP, Fox BS (1999) Tissue engineering of bone. Cell based strategies. Clin Orthop Relat Res (367 Suppl):S68–S83
32. Hunt TR, Schwappach JR, Anderson HC (1996) Healing of a segmental defect in the rat femur with use of an extract from a cultured human osteosarcoma cell-line (Saos-2). A preliminary report. J Bone Joint Surg Am 78 (1):41–48
33. Wiesmann HP, Joos U, Meyer U (2004) Biological and biophysical principles in extracorporeal bone tissue engineering. Part II. Int J Oral Maxillofac Surg 33(6):523–530
34. Meyer U, Meyer T, Handschel J, Wiesmann HP (2009) Fundamentals of tissue engineering and regenerative medicine. Springer, Heidelberg, New York

Chapter 3

A Novel Bone Scaffold Design Approach Based on Shape Function and All-Hexahedral Mesh Refinement

Shengyong Cai, Juntong Xi, and Chee Kai Chua

Abstract

Tissue engineering is the application of interdisciplinary knowledge in the building and repairing of tissues. Generally, an engineered tissue is a combination of living cells and a support structure called a scaffold. The scaffold provides support for bone-producing cells and can be used to heal or replace a defective bone. In this chapter, a novel bone scaffold design approach based on shape function and an all-hexahedral mesh refinement method is presented. Based on the shape function in the finite element method, an all-hexahedral mesh is used to design a porous bone scaffold. First, the individual pore based on the subdivided individual element is modeled; then, the Boolean operation union among the pores is used to generate the whole pore model of TE bone scaffold; finally, the bone scaffold which contains various irregular pores can be modeled by the Boolean operation difference between the solid model and the whole pore model. From the SEM images, the pore size distribution in the native bone is not randomly distributed and there are gradients for pore size distribution. Therefore, a control approach for pore size distribution in the bone scaffold based on the hexahedral mesh refinement is also proposed in this chapter. A well-defined pore size distribution can be achieved based on the fact that a hexahedral element size distribution can be obtained through an all-hexahedral mesh refinement and the pore morphology and size are under the control of the hexahedral element. The designed bone scaffold can be converted to a universal 3D file format (such as STL or STEP) which could be used for rapid prototyping (RP). Finally, 3D printing (*Spectrum Z510*), a type of RP system, is adopted to fabricate these bone scaffolds. The successfully fabricated scaffolds validate the novel computer-aided design approach in this research.

Key words: Bone scaffold, Tissue engineering, Finite element method, Mesh refinement, Pore size distribution, Shape function

1. Introduction

Every year, millions of patients confront bone tissue loss due to surgery, disease, or trauma. For example, in the USA alone, as many as 20 million patients per year suffer from various organ- and tissue-related maladies, including burns, skin ulcers, diabetes, bone, cartilage, and connective tissue defects and diseases. While more than eight million surgical procedures are performed annually to treat

Michael A.K. Liebschner (ed.), *Computer-Aided Tissue Engineering*, Methods in Molecular Biology, vol. 868,
DOI 10.1007/978-1-61779-764-4_3,

these cases, over 7,000 people are on transplant waiting lists, with an additional 100,000 patients dying prior to meeting the waiting list qualifications (1). Tissue or organ transplantation is among the few options available for patients with excessive skin loss, heart or liver failure, and many common ailments, and the demand for replacement tissue greatly exceeds the supply, even before one considers the serious constraints of immunological tissue type matching to avoid immune rejection (2). Tissue engineering (TE) promises to help sidestep constraints on availability and overcome the scientific challenges, with huge medical benefits. As an alternative solution to these cases, tissue engineering involves the development of biological substitutes to restore or replace lost tissue function (3). One approach for engineering tissue involves seeding biodegradable scaffolds with donor cells and/or growth factors, and then culturing and implanting the scaffolds to induce and direct the growth of new healthy tissue (4). Examples of tissue-engineered substitutes that are currently being investigated throughout the world include skin, cartilage, bone, vasculature, heart, breast, and liver (5). Tissue engineering combines progenitor cells, biocompatible scaffolds, biological differentiation reagents, and growth factors to create the supportive environment for cell growth and differentiation (6). Studies have shown that the constitutive properties of cells are contact dependent (7, 8). For bone cells to make bone, they have to encounter a surface and secrete material that becomes mineralized and then ossified into bone. By implanting a scaffold, bone regeneration is prompted by mimicking conditions in nature that favor bone growth (9). The scaffold creates the necessary contact points that cue the cells to grow and provides the necessary support for cells to proliferate and maintain their differentiated function and its architecture defines the ultimate shape of the new bone and cartilage. In this method, the bone automatically replaces itself, allowing healing of injuries, such as fractures. But in some traumatic cases, where there is too much damage for the bone to heal on its own, bone grafting or moving bone from one body part to another is a common solution. Treatment concepts based on those techniques would eliminate problems of donor site scarcity, immune rejection, and pathogen transfer (10). Osteoblasts, chondrocytes, and mesenchymal stem cells obtained from the patient's hard and soft tissues can be expanded in culture and seeded onto a scaffold that will slowly degrade and resorb as the tissue structures grow *in vitro* and/or *in vivo* (3). However, the modeling, design, and fabrication of a tissue scaffold with intricate architecture, porosity, and pore size for desired tissue properties present a challenge in tissue engineering (11).

In this chapter, a novel bone scaffold design approach based on the shape function and all-hexahedral mesh refinement is presented. Based on the shape function in the finite element method

(FEM), various irregular pores are created in the bone scaffold after an all-hexahedral mesh generation. A defined pore size distribution in the bone scaffold can be achieved through all-hexahedral mesh refinement since pore size and morphology are controlled by the subdivided hexahedral elements. The details are shown in the published paper (12).

2. Methods

Suppose that regular pores (e.g., spheres) are modeled under the control of regular elements (e.g., cubes), regular pores would be deformed when regular elements are deformed. The shape function in the finite element method can establish the relationship between the regular pore and irregular pore. As a result, the porous bone scaffold, which contains various irregular pores, can be designed. Because the pore size and morphology are controlled by the hexahedral elements, the pore size distribution in the bone scaffold can be controlled through the control of the hexahedral element size distribution, that is, all-hexahedral mesh refinement.

2.1. Pore Modeling Based on Shape Function

There are many approaches for mapping the basic unit into arbitrary units while isoparametric transformation using the shape function is widely used in the FEM. The basic idea for isoparametric transformation is as follows: deform irregular lines, irregular plane polygons and irregular spatial polyhedrons in the Cartesian coordinates into the standard lines, standard plane polygons and standard spatial polyhedrons in the natural coordinates. The contrary is also true.

Two kinds of coordinate systems are often used in the isoparametric transformation method: Cartesian coordinate system and natural coordinate system. There are three vectors in the Cartesian coordinate system which can be denoted by the signs x, y, z; the natural coordinate system can be denoted by three vectors ε, η, and ζ of which the range is $[-1, 1]$. In Fig. 1, a standard cube whose side length is 2 is mapped into an arbitrary hexahedron. The mapping relation is (13):

$$\begin{pmatrix} x \\ y \\ z \end{pmatrix} = \sum_{i=1}^{8} N_i \begin{pmatrix} x_i \\ y_i \\ z_i \end{pmatrix} \qquad i = 1, \cdots, 8, \tag{1}$$

where x, y, and z denote the inner coordinates of the irregular hexahedron in the Cartesian coordinate system, x_i, y_i, and z_i denote the node coordinates of the irregular hexahedron in the Cartesian coordinate system, and N_i denotes the shape function at the node i of the hexahedron.

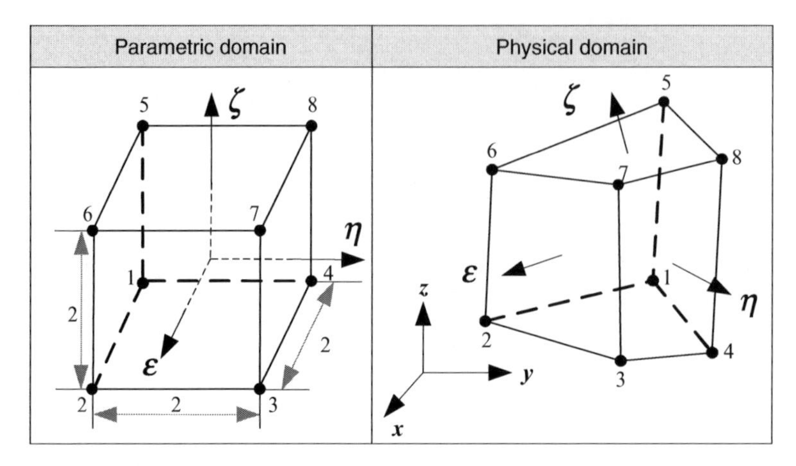

Fig. 1. Mapping relationship in the isoparametric element method.

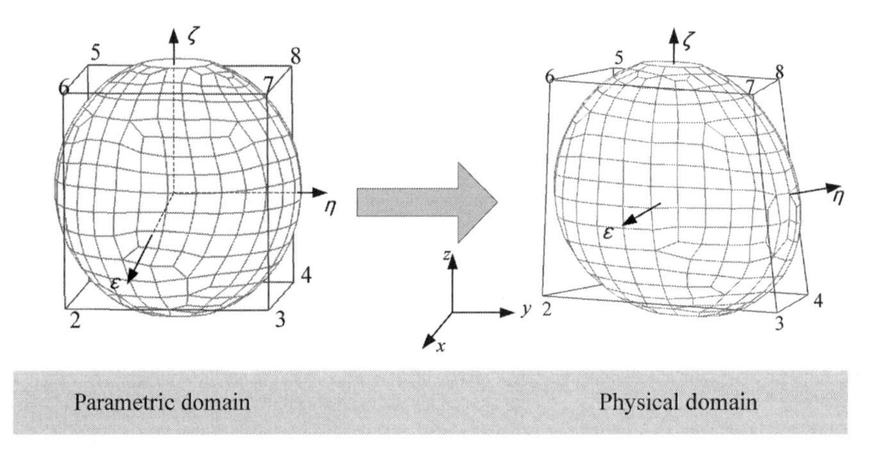

Fig. 2. Mapping relationship of pore-making element.

In Eq. 1, shape function N_i can be obtained by the Lagrange interpolation:

$$N_i = \frac{1}{8}(1 + \varepsilon_i\varepsilon)(1 + \eta_i\eta)(1 + \zeta_i\zeta) \quad i = 1, \cdots, 8, \tag{2}$$

where ε_i, η_i, and ζ_i denote the node coordinates of the standard cube in the natural coordinate system and ε, η, and ζ denote the inner coordinates of the standard cube in the natural coordinate system.

The basic pore-making element (the left one in Fig. 1) in the parametric domain is a standard sphere under the control of standard cube. All the hexahedral elements can be interpolated based on the eight-node hexahedral isoparametric shape function by accepting nodes' information as inputs. However, the inner pore-making element model cannot be interpolated just by the eight-node hexahedral isoparametric shape function. Therefore, a gridding sphere is used to represent and model the pore-making element. The details are shown in Fig. 2.

2.2. Bone Scaffold Modeling

In this section, the process and route of modeling a porous bone scaffold are introduced. The main route for designing a bone scaffold consists of three parts: 3D all-hexahedral mesh generation, irregular pore-making element modeling, and TE bone scaffold modeling by using Boolean modeling. See also Fig. 3 for the pore modeling method.

(a) The contour model is necessary before TE bone scaffold modeling. The profile model can be obtained by section scanning and measurement using CT/MRI on the bone specimen.

(b) After the contour model is constructed, an all-hexahedral mesh generation on the spongy bone with hexahedral elements is performed. Henceforth, a large number of various irregular hexahedral elements can be obtained.

(c) Accepting eight nodes on an arbitrary hexahedral element as inputs, the basic pore-making element in the parametric domain can be mapped into various irregular pore-making elements in the space domain based on an eight-node isoparametric shape function.

(d) The pore model of bone scaffold can be obtained by a Boolean operation, *union*, on various irregular pore-making elements.

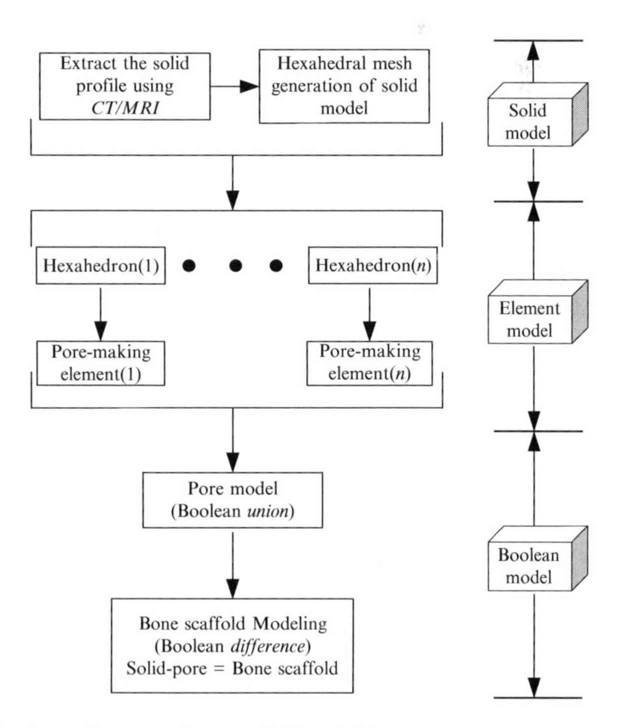

Fig. 3. Roadmap of a porous bone scaffold modeling.

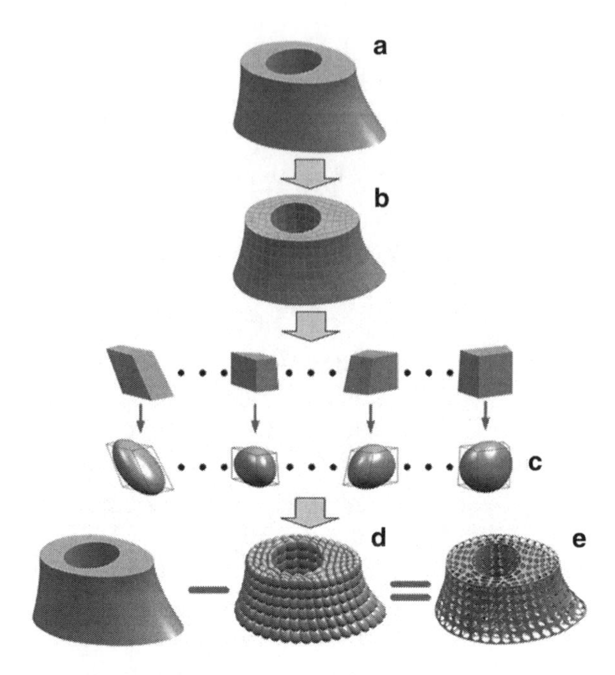

Fig. 4. Solid modeling for bone scaffold (Boolean Operation): (**a**) truncated bone model; (**b**) the subdivided truncated bone which is represented with various irregular hexahedral elements; (**c**) the various irregular pore elements; (**d**) the whole pore model after the Boolean operation *union* among the irregular pores; (**e**) the resulting bone scaffold which contains various irregular pores after the Boolean operation *difference* between the solid model and the whole pore model.

(e) The bone scaffold can be obtained by a Boolean operation, *difference,* between the contour model and pore model in the step d.

Details for the bone scaffold design procedure, interior pores, and pore shape are shown in Fig. 4.

2.3. Pore Size Distribution Control for a Porous Scaffold

Because the bone scaffold is modeled based on an all-hexahedral mesh, pore size distribution in the bone scaffold can be controlled through an all-hexahedral mesh refinement. 3D automated adaptive refinement of all-hexahedral mesh to achieve the defined pore size distribution in the bone scaffold is carried out. Before adaptive refinement of all-hexahedral mesh, first a 3D all-hexahedral mesh must be provided. Figure 5 shows details of the templates for adaptive refinement of an all-hexahedral mesh.

The templates (14, 15) shown in Fig. 5 can be applied based on eight nodes. Template *a0* is the original element. When eight nodes on the hexahedral element need to be refined, template *a* called an all-refinement template can be employed. This leads to the generation of 27 subelements. When four coplanar nodes on the hexahedral element need to be refined, template *b* called face-refinement template could be used. This would lead to the generation of 13 subelements.

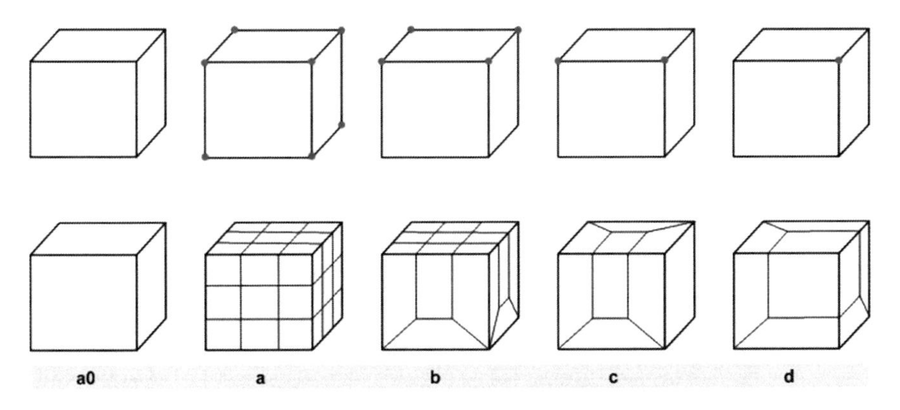

Fig. 5. Templates of all-hexahedral mesh refinement: (**a0**) original hexahedral element; (**a**) all-refinement template; (**b**) face-refinement template; (**c**) edge-refinement template; (**d**) point-refinement template.

When two nodes with common edge on the hexahedral element need to be refined, template *c* called the edge-refinement template could be adopted. As a result, five subelements are generated. When only a node on hexahedral element needs to be refined, template *d* called point-refinement template could be used. Similarly, this would lead to the generation of four subelements.

To perform all-hexahedral mesh refinement, first a target region must be specified. The target region is a proportion in the original 3D hexahedral mesh, where the density needs to be increased. It is known to us that the pore morphologies in the bone scaffold are controlled by various subdivided hexahedral elements. As the target region contains smaller pores than other regions, if an all-hexahedral mesh refinement on the target region is performed, there will be smaller hexahedral elements to be generated which leads to the generation of smaller pores in the target region. The target region can be specified by the curvature criterion, size criterion, or even user-defined criterion. All of these criteria identify the target regions, which possess smaller pores. This could be realized by users' interaction or programmer-designed interface. Because an all-hexahedral mesh refinement is performed on the original hexahedral mesh, the refinement region selection is reduced to the target hexahedrons' selection.

The steps of the algorithm can be described as follows.

(a) Select the target regions, which contain smaller pores than other regions, based on user-defined criteria correlated with the natural bone scaffold requirements biologically. Since the adaptive mesh refinement is performed on the original all-hexahedral mesh, the target region selection is then reduced to select the target hexahedrons.

(b) Subdivide the selected hexahedral elements based on the all-refinement template (template *a*).

(c) Check whether faces, edges, and points of the last-step selected elements need to be refined.

If *yes*, perform the corresponding refinement based on the templates listed in the Fig. 5. Then, go to step c.

If *no*, go to step d.

(d) End.

Figure 6 shows an example regarding the procedure for bone scaffold design based on shape function and adaptive refinement of all-hexahedral mesh. Firstly, a spongy bone model and a scaffold exterior model are extracted from the CT images. This step can be achieved by the software *Mimics 8.1*. Secondly, an all-hexahedral mesh generation on the spongy bone model based on the average

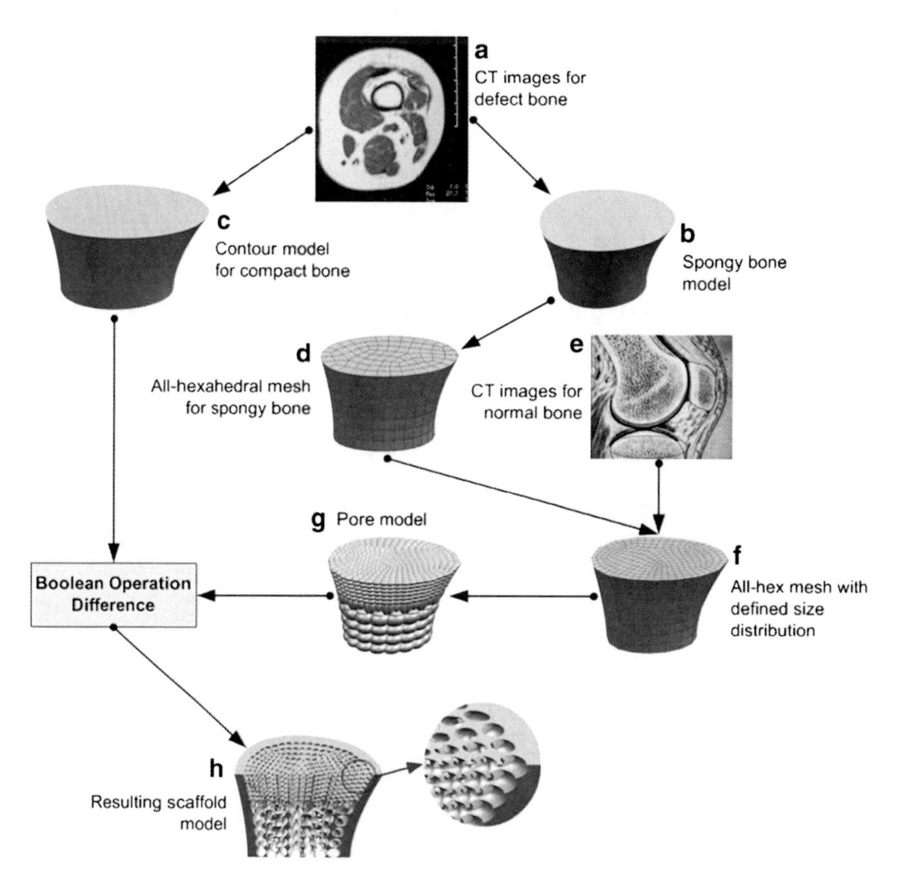

Fig. 6. Procedure for bone scaffold design based on shape function and conformal all-hexahedral mesh refinement: (**a**) CT (as shown here) or MRI images for defect bone serve as a starting point for designing spongy bone contour and scaffold exterior; (**b**) the spongy bone contour is created; (**c**) the scaffold exterior model is created, including the spongy bone segment and cortical bone segment; (**d**) 3D all-hexahedral mesh is generated based on grip mapping algorithms; (**e**) CT images for normal bone with defined pore size distribution; (**f**) refined all-hexahedral mesh is generated based on the conformal all-hexahedral mesh refinement algorithms and defined pore size distribution in the normal bone; (**g**) pores are modeled based on shape function in the finite element method; (**h**) half-sectional view for bone scaffold: Boolean *difference* operation between the scaffold exterior model and the whole pore model leads to the generation of porous scaffold with well-defined pore size distribution.

element size is performed, which is indirectly related with average pore size. Thirdly, the all-hexahedral mesh refinement on the target regions, where various smaller hexahedral elements are generated, is carried out. Fourthly, the results of last step are accepted as inputs, that is, various hexahedral elements, and various irregular pores based on the shape function, are generated. A Boolean operation *union* among the various irregular pores would lead to the generation of the whole pore model of bone scaffold. Finally, the Boolean operation *difference* between the scaffold exterior model and the whole pore model would generate the resulting bone scaffold model. The scaffold exterior model comprises the spongy bone segment and cortical bone segment. The spongy bone segment which contains various irregular pores is a concern, whereas the cortical bone segment is not. Therefore, the cortical bone segment just represents the solid model of the cortical bone which does not contain many Haversian canals and Volkmann's canals.

3. Materials

After the bone scaffolds have been designed, a suitable biomaterial must be chosen to manufacture them. Traditional tissue repair techniques, especially with synthetic materials, may lead to poor integration with the existing bone or tissue structure, potentially transferring dangerous pathogens into the body, and sometimes leading to a complete rejection of the graft. An ideal tissue scaffold should have hierarchical porous structures comparable to that of a human bone. Not only does this require 3D macroporous structures on the micro- or millimeter scale to facilitate transport of nutrients and tissue growth, but also surface features on the nanometer scale to improve the conformation of typical adhesive proteins and accelerate cell attachment and proliferation. Besides, the scaffold materials should be biocompatible and bioresorbable with a controllable degradation and resorption rate to match cell/tissue growth *in vitro/in vivo*. Mechanical properties for these scaffold materials should be strong enough to match those of the tissues at the site of implantation (16, 17). The effects of these materials on the tissue regeneration and cell growth will be the future research interests. The designed bone scaffold can be easily converted to the universal 3D file format (such as STL or STEP) which could be used for rapid prototyping. Finally, 3D printing (*Spectrum Z510*), a type of rapid prototyping (RP) systems, is adopted to fabricate these bone scaffolds. Figure 7 is the fabricated bone scaffold using $CaCO_3$. The successfully fabricated scaffolds validate the novel computer-aided design approach in this research.

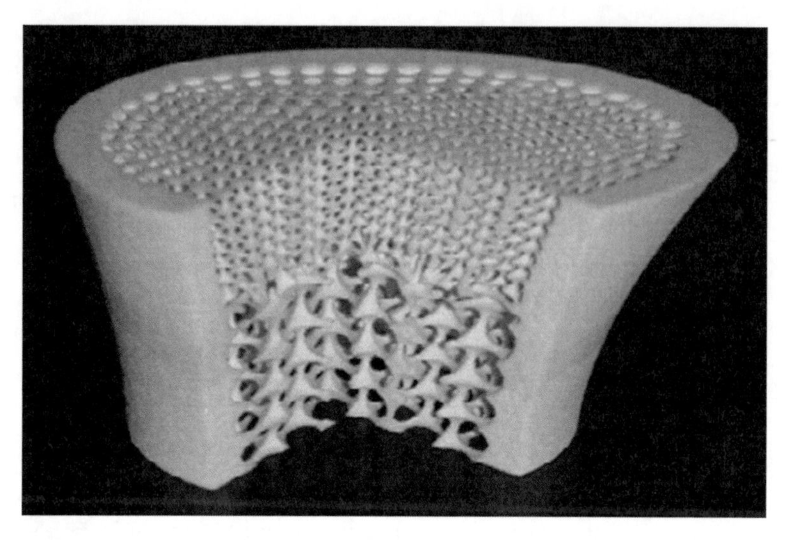

Fig. 7. Bone scaffold fabricated by 3D printing.

4. Discussion

One problem is how to fabricate these bone scaffolds with complex geometries. Inkjet-based deposition (18) provides a good solution to this case. Inkjet printing technology can dispense spheres of fluid with diameters of 15–200 μm (2 pl–5 nl) at rates of 0–25,000/s for single droplets on demand, and up to 1 MHz for continuous droplets. Another problem is that powders which are used to support the bone scaffold are usually trapped. The studies (19, 20) would provide a good solution to this problem when the bone scaffold is fabricated by using rapid prototyping.

Acknowledgment

We gratefully acknowledge the support from National Natural Science Foundation of China (NO.50575139) and the Programme of Introducing Talents of Discipline to Universities (NO B06012) for this chapter.

References

1. US scientific registry for organ transplantation and the organ procurement and transplant network. Annual report. Richmond, VA: UNOS, 1990
2. Saltzman WM (2004/07) Tissue engineering: engineering principles for the design of replacement organs and tissues [M]. Oxford University Press, New York, USA

3. Vacanti L (1993) Tissue engineering. Science 260(5110):920–6
4. Popov VK, Antonov EN, Bagratashvili VN et al (2006) Biodegradable scaffolds for tissue engineering fabricated by surface selective laser sintering. IFMBE proceedings, 3rd Kuala Lumpur international conference on biomedical engineering, 2006
5. Masood SH, Singh JP, Morsi Y (2005) The design and manufacturing of porous scaffolds for tissue engineering using rapid prototyping. International Journal of Advanced Manufacturing Technology 27:415–420
6. Dadsetan M, Heffeeran TE, Szatkowski JP et al (2008) Effect of hydrogel porosity on marrow stromal cell phenotypic expression. Biomaterials 29:2193–2202
7. Wang N, Butler JP, Ingber DE (1993) Mechanotransduction across the cell surface and through the cytoskeleton. Science 260: 1124–1127
8. Wang N, Ingber DE (1994) Control of cytoskeletal mechanics by extracellular matrix, cell shape, and mechanical tension. Biophys J 66: 2181–2189
9. Kraus KH (2006) Mesenchymal stem cells and bone regeneration. Veterinary Surgery 35: 232–242
10. Hutmacher DW (2000) Scaffold in tissue engineering bone and cartilage. Biomaterials 21: 2529–2543
11. Starly B, Lau W, Sun W et al (2006) Internal architecture design and freeform fabrication of tissue replacement structures. Computer-Aided Design 38:115–124
12. Cai S, Xi J (2008) A control approach for pore size distribution in the bone scaffold based on the hexahedral mesh refinement. Computer-Aided Design 40:1040–1050
13. Jiang Xiaoyu (1993/02) The basis for the finite element method [M], The Tsinghua Press, Beijing, P.R. China
14. Harris Nathan J, Benzley Steven E, Owen Steven J (2004) Conformal refinement of all-hexahedral element meshes based on multiple twist plane insertion. In: Proceedings of IMR' 2004, pp 157–168
15. Hong-mei Z, Guo-qun Z, Song Y (2005) 3D automated adaptive refinement of all-hexahedral element mesh. Chinese J Plast Eng 12(7): 195–199
16. Naughton GK et al (1995) Emerging development in tissue engineering and cell technology. Tissue Engineering 1(2):211–9
17. Naughton GK, Tolbert WR, Grillot TM (1995) Emerging developments in tissue engineering and cell technology. Tissue Engineering 1(2):211–219
18. Cooley PW, Wallace DB, Antohe BV (2001) Application of ink jet printing technology BioMEMS and MicroFluidic systems[C]. Proc SPIE microfluidics and BioMEMS Conference, San Francisco, CA, 2001
19. Ang BY, Chua CK, Du Z (2000) Study of trapped material in rapid prototyping parts. International Journal of Advanced Manufacturing Technology 16(2): 120–130
20. Ang BY, Chua CK, Du Z (2000) Development of an advisory system for trapped material in rapid prototyping parts. International Journal of Advanced Manufacturing Technology 16 (10):733–738

Chapter 4

Rapid Prototyping Composite and Complex Scaffolds with PAM^2

Giovanni Vozzi, Annalisa Tirella, and Arti Ahluwalia

Abstract

To create composite synthetic scaffolds with the same degree of complexity and multilevel organization as biological tissue, we need to integrate multilevel biomaterial processing in rapid prototyping systems. The scaffolds then encompass the entire range of properties, which characterize biological tissue. A multilevel microfabrication system, PAM^2, has been developed to address this gap in material processing. It is equipped with different modules, each covering a range of material properties and spatial resolutions. Together, the modules in PAM^2 can be used to realize complex and composite scaffolds for tissue engineering, bringing us a step closer to real clinical applications. This chapter describes the PAM^2 system and discusses some of the practical issues associated with scaffold microfabrication and biomaterial processing.

Key words: Rapid prototyping, Microfabrication, Tissue engineering, Biomaterials, Composite scaffolds

1. Introduction

From an engineering point of view, the human body is a composite biphasic material with a hierarchical organization in which the range of elastic moduli is truly astonishing. Bone, for example, has a stiffness of tens of gigapascal while soft tissues, such as the liver and brain, approach a few kilopascal. Moreover, its hierarchical architecture is exquisitely organized at all levels, from the nanometer to several centimeters. Replicating this level of structural variation over several orders of magnitude is one of the main goals of tissue engineering in this century. Traditionally, engineers use a handful of well-characterized and stable materials for construction, and in vitro tissue regeneration therefore poses a huge technical and scientific challenge.

Michael A.K. Liebschner (ed.), *Computer-Aided Tissue Engineering*, Methods in Molecular Biology, vol. 868,
DOI 10.1007/978-1-61779-764-4_4,

Up until the 1990s, tissue engineering usually involved the use of polymer scaffolds made either from FDA-approved synthetic polyesters and their derivatives or from a small range of natural proteins and polysaccharides, all formed as foams, nonwoven fibers, or sponges. Often, the pore distribution of these scaffolds was random, uncontrolled, and unpredictable. Even more critical but less recognized was the mechanical mismatch between biological tissue and these structures, leading to stress shielding-like effects and an effective loss of cell function on scaffolds.

Within the noise, however, a few groups driven by the form-follows-function hypothesis pioneered the concept of custom scaffolds focusing on architectural design. Here, extensive use was and is still being made use of CAD/CAM principles and RP approaches. However, most of these methods have not led to the development of well-organized and multicomponent hierarchical biomimetic scaffolds. We suggest two main reasons for this: the first is that very few of the RP systems available were designed for handling more than one class of material. For example, FDM is limited to polymers which fuse in a narrow range of temperatures and melt to a narrow range of viscosities (1). The Pressure-Activated Microsyringe (PAM) system is only suitable for polymers and suspensions which are soluble or dispersed in volatile solvents with viscosities ranging from 100 to 1,000 cP, and will clog up if suspensions are not well dispersed (2, 3). Secondly, the concept of biomimetic scaffolds which truly mimic surface, mechanical, and architectural properties of biological systems is quite new, and the need for a production process which can handle and assemble fluids, gels, powders, cells, water, nanoparticles, salts, and ceramics together has not yet fully emerged. Pioneering in this sense are the studies by Boland et al. (4) and Mironov et al. (5), which use inkjet printing methods to deposit hydrogels with encapsulated cells. Although the hydrogels are not strictly scaffolds because they do not possess an organized porous three-dimensional architecture, inkjet methods, particularly those using multiple nozzles, come a few steps closer to the truly biomimetic. However, their very specialized jet printing technology is difficult to adapt to new materials and integrate with other fabrication methods.

To overcome the limitations of the RP methods currently available and inspired by biomimetic engineering, we have developed a new modular microfabrication system PAM^2 (Pam-modular microfabrication, pronounced as "*pamquadro*"), which was purposely designed to handle and dispense a multitude of materials with different processing technologies. PAM^2 is an evolving and adaptable system, and can fabricate millimeter- or centimeter-scale 3D scaffolds with a wide variety of biomaterials. Thus, what we define as its processing window, that is, the usable array of materials and material properties with respect to the fidelity (how closely the scaffold geometry approximates the CAD design) and resolution (the smallest line width which can be reproducibly fabricated) of the scaffolds obtained, covers a wide range to suit several tissue engineering applications. PAM^2 is quite simply the integration of several material processing systems, or modules, in one machine, with a single control framework and pull-down menu GUI. Its innovation lies in the system's flexibility; that is, its ability to house more than one processing system and to control the position and actuation of each one independently but in parallel through the GUI. In this chapter, we outline the PAM^2 design philosophy, describe its system architecture, and discuss experimental protocols and tips for scaffold microfabrication.

2. Materials

Table 1 summarizes the different materials processed using the three modules. From the table, the limits of the actuating systems are quite clear. However, the combination of more than one method increases the spectrum of useable materials, viscosities, as well as the range of elastic moduli and spatial dimensions of the scaffolds, as schematically depicted in Fig. 1 (3, 6, 7).

3. Methods

3.1. PAM^2

The principal structure of the PAM^2 system is a three-axis micro-positioner composed of three brushless motors controlled by a microcontroller unit (Ensemble MP10 Controller, Aerotech Inc, USA). Each motor has a spatial resolution of 0.1 μm and a linear velocity up to 50 mm/s (see Note 1). The *Z* axis of PAM^2 can house any number of actuator dispenser, ablation or deposition modules, and further stepper motor units. The basic actuator modules of the PAM^2 system as implemented at present are the PAM, the PAM2, and a laser head. Each of these modules has a different processing window and different working parameters which can be individually modulated through the PAM^2 CAD user interface to obtain scaffolds with a desired fidelity and resolution. A planar substrate, generally a glass slide, is fixed on the horizontal *X* and *Y* axes of the PAM^2 system and moves along a predetermined trajectory during deposition. Figure 2 shows the PAM^2 system with the three actuating modules and the *XY* plane used for deposition. Using the purpose-designed CAD user interface, it is also possible to import or design and draw the trajectories necessary to move the system. Through the software, the working parameters can be modified in real time in order to fabricate well-defined structures. In particular, simple or complex trajectories can be traced for a two-dimensional

Table 1
Materials processed by three modules and limits of each method

Material	Method used	Resolution (micron)	Comments and notes
PLLA	PAM	5	Only useable in a small range of concentrations (see Note 13)
PCL	PAM	5	Only useable in a small range of concentrations (see Note 13)
PLGA	PAM	5	Only useable in a small range of concentrations (see Note 13)
Corethane	PAM	5	Only useable in a small range of concentrations (see Note 13)
PCL with CNT	PAM	100	Needle very easily clogs up due to flow-induced phase separation. Only useable in a small range of concentrations (see Note 14)
Corethane with CNT	PAM	80	(See Note 14)
Alginate	PAM	200	Only certain concentrations below about 6% can be extruded (see Note 15)
Alginate	PAM2	500	PAM2 can exert higher pressures, so high concentrations can be extruded (see Note 16)
PLGA with CNT	Laser system	100	The experiments were performed on spin-coated films of 100-μm thickness (see Note 14)
PLGA with carbon black	Laser system	10	The experiments were performed on spin-coated films of 100-μm thickness (see Note 17)
Agarose	Laser system	80	Several (at least five) consecutive ablations are required as the hydrogel tends to collapse back into the ablated groove (see Note 18)
PLLA with CNT	Laser system	30	The experiments were performed on spin-coated films of 100-μm thickness. See Note 14 for the preparation protocol
PLLA with carbon black	Laser system	10	The experiments were performed on spin-coated films of 100-μm thickness (see Note 17)
PCL	Laser system	20	experiments were performed on spin-coated films of 100-μ thickness. Increasing the polymer concentration with which spin-coated films are obtained the line resolution decreases (see Note 13)

layer, and then 3D scaffolds can be fabricated through a sequential layer-by-layer deposition process (see Note 2).

3.2. PAM Module

The PAM was originally designed to fabricate polyester scaffolds with resolutions close to that of the typical cell diameter (10 μm). Basically, it consists of a stainless steel syringe with a micrometer

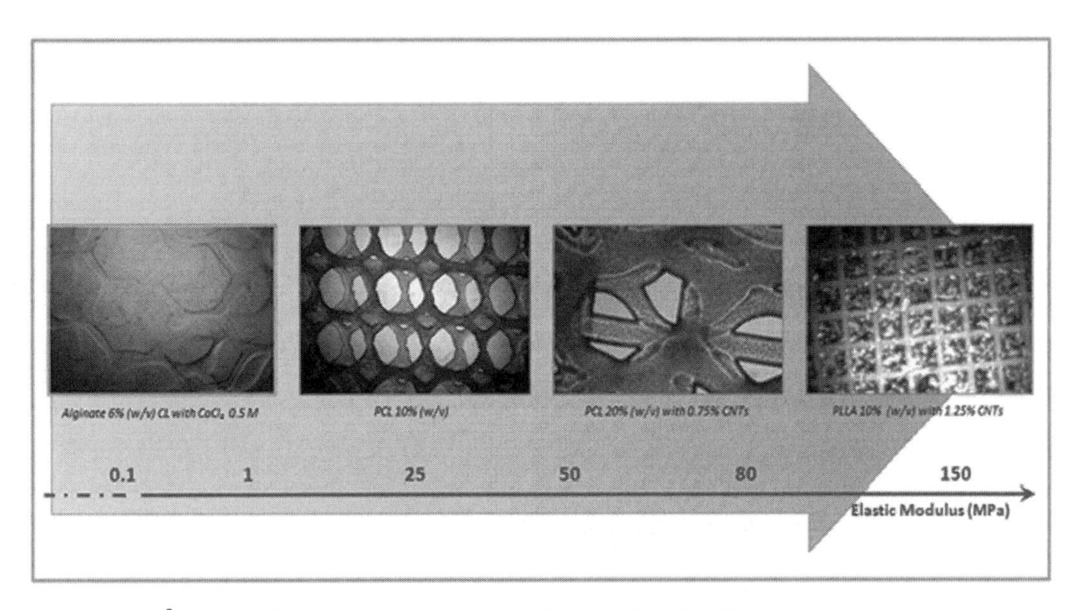

Fig. 1. The PAM^2 system with a laser head, piston-assisted microsyringe (PAM2), and pressure-assisted microsyringe (PAM) mounted on the *Z* axis. The substrate is placed on the *XY* plane and moves as the material is being deposited. The dynamic range of the *XY* deposition plane is 10 $\times$10 cm.

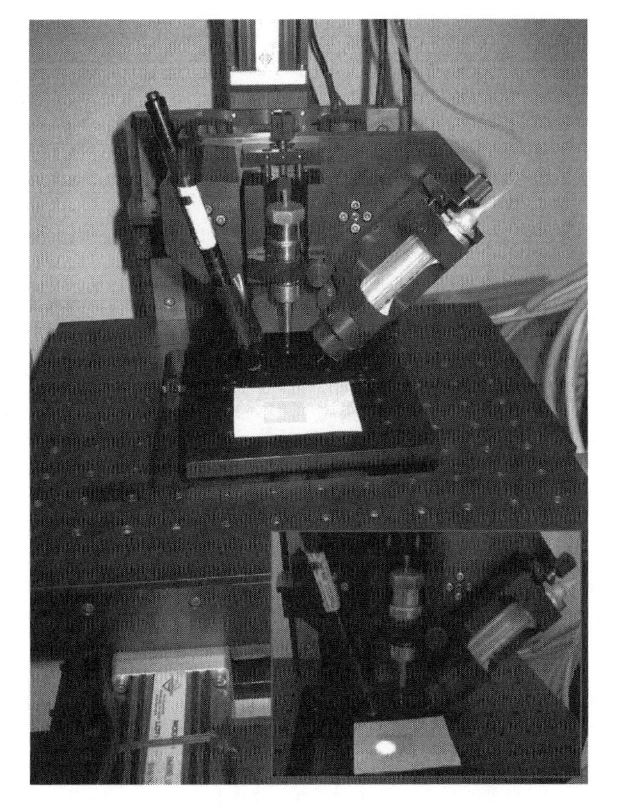

Fig. 2. CAD design of the PAM syringe: reservoir, o-ring, glass needle.

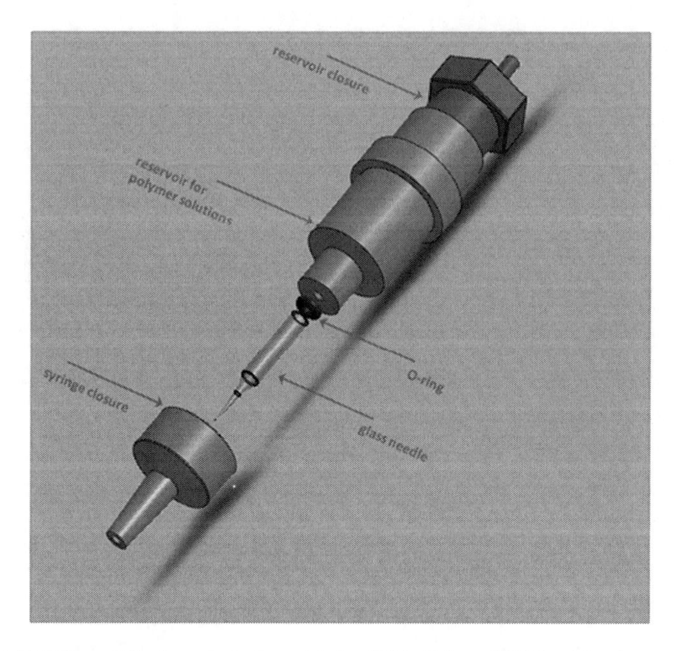

Fig. 3. PAM line width characterization plots for PLLA, PCL, and PLGA (at fixed concentrations and extrusion pressure) as a function of *XY* deposition velocity.

glass capillary needle as the tip as illustrated in Fig. 3. A vertical puller (Patch Pipette Puller, PA-10) is used to pull the tips, which are prepared from soda glass hematocrit capillaries (Globe Scientific, Paramus, NJ) with an outer diameter of 1.5 mm and an inner diameter of 1.15 mm. The tip has a flat end and is gently tapered, a characteristic that lends itself to streamlined flow in the narrowest parts of the capillary. Each capillary is pulled under the same conditions and the internal diameter of the tips is about 20–50 μm (see Note 3). The needle is connected to the syringe barrel and held in place by a small o-ring. A polymer solution is placed inside the syringe, which has a capacity of about 5 ml. The syringe is driven by filtered compressed air, and the pressure is controlled to ±5 mmHg through a software-controlled pressure regulator. Pressures of about 10–500 mmHg can be applied according to the material properties (see Note 4). Through the user interface, it is possible to control the working parameters, which influence the fabrication process: the velocity of the *XY* plane and the applied pressure. In order to control the line width and spatial resolution of the scaffolds, these two working parameters are characterized as a function of the material (and particularly its viscosity) used (see Notes 5 and 6). In general, for a given material viscosity, we fix a minimum pressure, which just enables the material to flow from the needle, and then use the *XY* plane or deposition velocity to modulate the line width of the scaffold. Figure 4 shows typical working plots for three biodegradable polyesters.

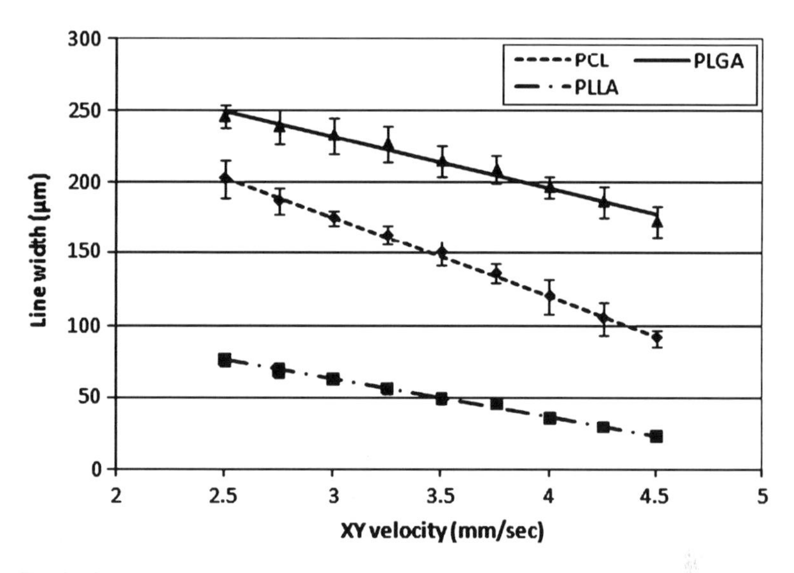

Fig. 4. Line width characterization as a function of *XY* deposition velocity for different alginate solutions using the PAM2 module.

3.3. PAM2 Module

The PAM2 fabrication method was designed to fabricate well-defined 3D scaffolds using materials with high viscosity, typically hydrogels. In terms of system architecture, it can be considered as an evolution of the PAM. The original syringe-based PAM system uses air pressure to extrude materials, but is limited to the use of solutions with viscosities which range from 100 to 1,000 cP. Hydrogels, which have viscosities which may go up to 8,000 cP cannot be safely extruded by the application of air pressure using the PAM system. In order to use highly viscous materials, the application of a much higher mechanical pressure is necessary (see Note 7). Therefore, in PAM2, a controlled stepper motor (Z_2) is added to the 3D micropositioning system to actuate the piston or plunger of a commercial syringe (see Note 8), which is used to control the outflow of material from the needle tip. Since the syringe is sterile and disposable, cells can be incorporated into the working material during the fabrication process (6). In the PAM2 method, the variables which determine the resolution and fidelity of the scaffolds are the material viscosity, the angular velocity of motor Z_2 which drives the piston, and the velocity of the XY plane. Once the material viscosity is established, the latter two variables together determine the line width and fidelity of the scaffolds. Material outflow is determined by the transfer of the Z_2 stepper motor's angular momentum into a linear driving force, which leads to material extrusion. Therefore, the actual quantity of material flowing through the needle for a given angular velocity depends on its viscosity as well as the friction between the needle and the fluid. Reduced outflow of the hydrogel through the syringe indicates the presence of losses due to viscous phenomena or localized tapering. Since material outflow versus angular velocity profiles

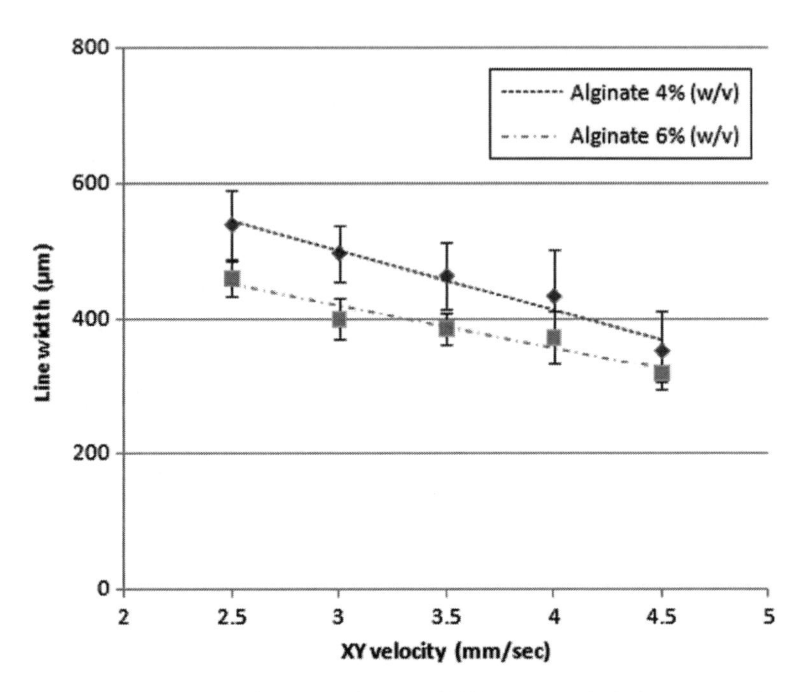

Fig. 5. Line width characterization as a function of *XY* deposition velocity for agarose using the 1.92-μm thulium laser at 2 W.

differ according to material and syringe properties, a preliminary characterization to establish a useful working window has to be performed for each material and needle used. Once the outflow versus angular velocity profile has been determined for a given material, it is extruded at the minimum useful outflow (see Note 9). The second working parameter, the velocity of the *XY* motors, then determines the line width of the extruded material for a given controlled outflow. Typically, line width decreases rapidly as the *XY* deposition velocity increases (Fig. 5).

3.4. Laser Head Module

The laser head module was designed to process scaffolds with an ablative technique using composite materials that the PAM and PAM2 modules are not able to handle because of their high viscosity and the difficulty of obtaining a well-dispersed suspension. During extrusion from the needle, composite materials which contain suspensions or emulsions undergo flow-induced phase separation, resulting in the formation of clots that obstruct the needle. In such cases, classic extrusion-based fabrication methods need to be replaced by alternative processing techniques. The laser system differs fundamentally from PAM and PAM2 because it is an ablative method, so material is not extruded but a substrate is thermally photodegraded. Therefore, the microfabrication head is a Thulium infra red laser (IPG Fibertech, Germany). The working principle of this system is based on the use of polymeric films (see Note 10) of pure or composite material which are ablated by the laser through

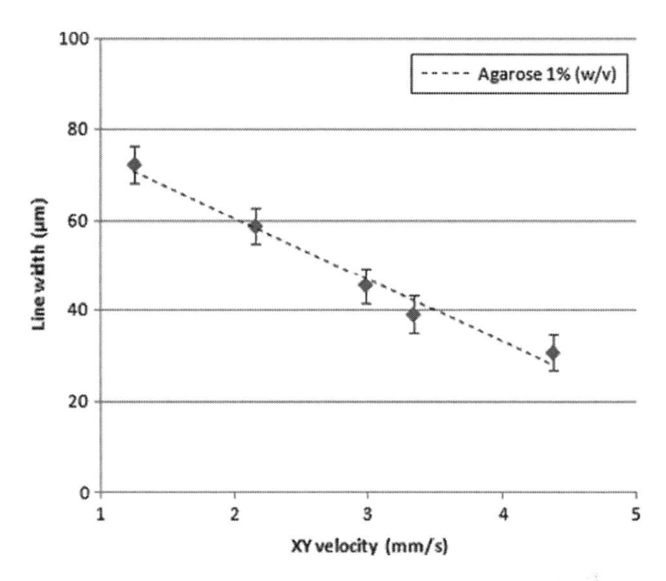

Fig. 6. The spectrum of elastic moduli that can be obtained with PAM^2 using more than one module. The elastic moduli cited have been reported in refs. 3, 6, 7.

sublimation or thermal degradation. The thulium laser has a wavelength of 1.92 µm, and almost all biocompatible polymers have an absorption of over 65% in this region (see Note 11). The laser head has the same dimensions as a PAM syringe and also has an internal focusing system such that the laser spot has a radius of 2 µm. The focal length of the objective is about 18 mm (see Note 12). Laser power can be controlled by an external power supply and the maximum emission is 2 W. Two working parameters are important for modulating the line width of an ablated groove: laser power emission and *XY* motor velocity. However, our experiments show that in order to obtain a well-defined, high-resolution groove, the power should be set at its maximum. In fact, a high power beam quickly sublimates a small area of material while a low laser power requires a longer degradation time and so increases the heat dispersed through the substrate. Therefore, when processing the material, the topology of the structure is modulated only through the deposition or *XY* velocity, as shown in Fig. 6.

3.5. Handling Scaffolds

As the production of RP scaffolds becomes more efficient, they will become more widely available and GMP procedures will need to be applied to standardize their manufacture and use. Scaffold handling and treatment, including procedures such as sterilization, equilibration with media, and cell seeding, are in fact critical to the success of a tissue-engineered construct. As far as synthetic scaffolds are concerned, particularly those derived from the degradable polyesters, one of the most important treatment phases is solvent evaporation. After deposition, if the scaffolds are to be used immediately, they must be kept in a vacuum desiccator for at least 4–6 h. Otherwise, a week's storage in a dark, cool place is sufficient to remove traces of solvent. Before sterilizing, the scaffolds are immersed in deionized water for 2 h, and then rinsed. During this phase, any PVA film applied is dissolved away and the scaffolds float off the support substrate and can be easily handled with tweezers. We use two alternative sterilization procedures: rinsing with 70% ethanol and then exposing to an ultraviolet hood lamp for 10 min each side, or alternatively sterilizing with H_2O_2 gas plasma. The latter method is preferred because it also renders the scaffolds more hydrophilic and amenable to protein adsorption. Before cell seeding, the polyester scaffolds need to be equilibrated with serum-free medium (incubate at 37°C for a few hours, and then rinse in PBS). This is because the polymers tend to be slightly acidic; the medium turns yellow after a few hours. Finally, the scaffolds must be treated with an adhesive protein before cell seeding. We use a variety of methods according to the cell type being employed. One of the simplest and most effective methods is incubation of scaffolds with serum-containing medium (up to 30% serum), followed by rinsing in serum-free medium or PBS. More delicate cells, such as neurons, may need specific adhesive polymers. Hepatocytes usually require a collagen coating (rat tail collagen 1 mg/ml, BD Biosciences): 200 μl of collagen is pipetted over the scaffolds and left to set in the incubator before seeding. Should the scaffolds float or warp during cell culture, a Teflon ring or support (e.g., Scaffdex Cell Crown, Tampere, Finland) may be used to push them down.

Handling and treatment during cell culture is quite different in the case of hydrogel scaffolds. Firstly, the scaffolds are realized in a sterile environment using a glass slide (typically, with dimensions of 12 mm × 12 mm) as a support substrate. The supports are then placed in a 6-multiwell plate and maintained in the incubator. In case of simple hydrogel scaffolds, standard seeding procedures can be used. For cell-incorporated scaffolds, the samples are immediately immersed in the proper cell culture medium and maintained in an incubator.

4. Notes

1. Although the maximum velocity is 50 mm/s, we routinely use lower speeds (10–20 mm/s) in order to maintain mechanical stability of the motors particularly when tracing complicated trajectories. The motors can be programmed to vary the velocity and accelerate or decelerate.
2. 3D scaffolds are built up through sequential layers. The layers are slightly offset with respect to one another to create overhangs between spaces. In order to avoid layers collapse, a small

amount of hydrogel or water-soluble polymer [e.g., 5 g/ml polyvinyl alcohol (PVA, *MW* 30,000–70,000, Sigma Aldrich in water)] can be pipetted over each layer, avoiding the edges. A narrow gel-free edge is necessary to ensure that the layers adhere to one another. When the whole 3D scaffold has been fabricated, the PVA can be dissolved away by washing in water.

3. The capillary must be made of soft soda glass; we use a two-current pulling protocol, in which the first stage is pulled at 19 A and the second stage at 16.7 A. If the end diameter is too small, then the polymer solution will immediately clog up the capillary. The smallest useful diameter is 5 μm, but capillary size does not dictate the resolution to a great extent. In our opinion, the optimum diameter is 20 μm.
4. Pressures over about 500 mmHg are dangerous, and safety glasses are required at all times. If the needle clogs up, a small piece of lint-free paper soaked in chloroform is gently dabbed onto the needle tip to dissolve the polymer.
5. The solution should have about the same viscosity as runny honey. Viscosities are adjusted by changing the solute-to-solvent ratio.
6. The capillary needle should be just above the working surface but not touching. Usually, a trail run is conducted at various *XY* deposition velocities and pressures to characterize the fidelity and resolution of the scaffolds. All our CAD files contain a short preamble deposition trajectory, which enables pressure and needle substrate distance to be established prior to the fabrication of the scaffold.
7. Viscosities of hydrogel solutions obtained with different weight-to-volume ratio were measured with a dynamic rheometer; viscosities' values (8,000 cP) are obtained for 8% (w/v) alginate solutions in PBS (6).
8. Commercial insulin syringes are excellent. The outflow of material also depends on the surface characteristics of the needle; we find that Teflon needles allow higher outflows than steel needles, and short needles are more efficient (smaller angular velocity necessary to drive a given outflow) than long ones.
9. As far as alginate or other pre-cross-linked gels are concerned, it is essential to cross-link the gel immediately; otherwise, the hydrogel will retract forming individual blobs. Moreover, the deposition substrate should be fairly hydrophilic to minimize the formation of these spherical blobs rather than continuous lines. To cross-link alginate, a few drops of 0.5 M $CaCl_2$ (Sigma-Aldrich, St. Louis, MO, USA) are pipetted over the scaffold and after 1 to 5 min (depending on the desired mechanical properties) the solution is aspirated off and rinsed with deionized water (6). Multilayers can be constructed by cross-linking each layer before depositing successive ones.

10. The starting substrate is a spin-coated film of the material to be ablated, placed on the *XY* axis of the PAM^2 system. Usually, the material is a composite of CNT and polymer which is almost paste like. Other materials, such as hydroxyapatite, can also be employed.
11. The ablation process is far more effective if the polymer or material, or at the very least the support, is black because this increases radiative absorption. It is useful, therefore, to incorporate carbon nanotubes, which also increase elastic modulus (3).
12. The laser cannot be aligned visually (safety glasses are in fact essential), so a sacrificial layer (a spin-coated film of the material in question) is used to focus the beam. The focus is selected at the halfway point between the highest and lowest *Z* positions which give rise to ablation.
13. Poly-L-lactide acid (PLLA) (Lactel, NPPharm-F, Bazainville, France) is prepared in a glass flask with a glass stopper as a 50–200 mg/ml solution in chloroform. The solution takes about 24 h to dissolve completely, and we prepare about 5 ml at a time. The flask is sealed with parafilm and can be stored at 4°C for a few weeks. Similarly, both poly-lactide-coglycolide (75:25 PLGA, Lactel) and polycaprolactone (PCL 6,5000 MW, Sigma) are prepared as 50–200 mg/ml solution in chloroform. The polyurethane Corethane (gift from Prfo Tanzi, Polytechnic of Milan) is prepared as a 100 mg/ml solution in dimethylacetamide.

    In general, the more viscous the solution, the better the resolution, but there is a trade-off between the resolution and the high pressure required for deposition. Viscous solutions are also more difficult to handle and pipette.
14. One of the major practical problems of CNTs is their tendency to aggregate, thus rendering processing and fabrication of CNT-based composites and devices rather difficult. We use a dual solvent approach which can be used to obtain CNT–polymer composites with a uniform dispersion of CNT within a polymer matrix (7). First, PCL (MW 65,000, Sigma, Milan, Italy) is dissolved in 1 ml of chloroform to give a 200 mg/ml solution. Then, 25 mg multiwalled CNTs (kind gift from Bayer) are dispersed in 1 ml benzene (Sigma, Milan, Italy) using a pulsed sonicator (Vibracell, Sonics, Milan, Italy) for 1 min. The two solutions are then mixed and sonicated to give a final suspension containing 100 mg/ml PCL and 12.5 mg/ml CNT (8). The suspension should be used immediately after sonicating as the CNT will tend to reaggregate, clogging up the needle. Similarly, the PLLA–CNT composite is prepared by sonicating equal volumes of 200 mg/ml PLLA in chloroform and 25 mg/ml of CNT in benzene, and the PLGA–CNT composite is prepared by sonicating equal volumes of

400 mg/ml PLGA in chloroform and 25 mg/ml of CNT in benzene. Finally, corethane is prepared by sonicating equal volumes of 200 mg/ml corethane in dimethylacetamide and 25 mg/ml of CNT in benzene.

15. Alginic Acid Sodium Salt from Brown Algae in powder form (Sigma-Aldrich, St. Louis, MO, USA) is dissolved in deionized water with weight-to-volume ratio varying from 2 to 6%. Using a capillary needle with a lager diameter (up to 150 μm), this viscous material can be employed with the PAM system. In order to guarantee a well-defined or high-fidelity line, the gel must be cross-linked immediately after the fabrication. Note 9 gives details on the cross-linking procedure.
16. Alginate solutions with concentrations up to 8% (w/v) are obtained by dissolving powdered alginate in Phosphate Buffered Solution (PBS, Sigma-Aldrich, St. Louis, MO, USA). The resulting solution is then autoclaved before loading into a sterile commercial syringe. If necessary, a cell suspension may be added to the gel. The syringe is mounted on the *Z* axis of PAM^2. To cross-link the gel, a 0.5 M calcium chloride in MilliQ water is first sterilized by filtering through a 0.22-μm pore syringe filter, and then applied to each layer of the scaffold as in Note 9. The concentration of this solution is important as it modulates the mechanical properties of the cross-linked hydrogel scaffold (6).
17. The protocol used for carbon black is the same as that used in Note 14. Carbon black is obtained from Sigma-Aldrich, St. Louis, MO, USA.
18. 0.15 ml of 1% w/v agarose (Sigma, Cell Culture Tested) solution in hot deionized water is prepared and then cast in a petri dish of 3 cm of diameter in order to obtain a uniform film of 200 μm.

References

1. Zein I, Hutmacher DW, Tan KC, Teoh SH (2002) Fused deposition modeling of novel scaffold architectures for tissue engineering applications. Biomaterials 23(4):1169–1185
2. Vozzi G, Previti A, De Rossi D, Ahluwalia A (2002) Microsyringe-based deposition of two-dimensional and three-dimensional polymer scaffolds with a well-defined geometry for application to tissue engineering. Tissue Eng 8(6):1089–1098
3. Mattioli-Belmonte M, Vozzi G, Whulanza Y, Seggiani M, Fantauzzi V, Orsini G, Ahluwalia A (2012) Tuning polycaprolactone–carbon nanotube composites for bone tissue engineering scaffold. Materials Science and Engineering: C 32 (2):152–159
4. Boland T, Xu T, Damon B, Cui X (2006) Application of inkjet printing to tissue engineering. Biotechnol J 1:910–917
5. Mironov V, Kasyanov V, Drake C, Markwald RR (2008) Organ printing: promises and challenges. Regen Med 3:93–103
6. Tirella A, Orsini A, Vozzi G, Ahluwalia A (2009) A phase diagram for microfabrication of geometrically controlled hydrogel scaffolds. Bioprinting 1:045002
7. Mariani M, Rosatini F, Vozzi G, Previti A, Ahluwalia A (2006) Characterisation of tissue engineering scaffolds microfabricated with PAM. Tissue Eng 12:547–558
8. Pioggia G, Di Francesco F, Marchetti A, Ferro M, Ahluwalia A (2007) A composite sensor array impedentiometric electronic tongue Part I. Characterization. Biosens Bioelectron 22:2618–2623

Chapter 5

Scaffold Pore Space Modulation Through Intelligent Design of Dissolvable Microparticles

Michael A.K. Liebschner and Matthew Wettergreen

Abstract

The goal of this area of research is to manipulate the pore space of scaffolds through the application of an intelligent design concept on dissolvable microparticles. To accomplish this goal, we developed an efficient and repeatable process for fabrication of microparticles from multiple materials using a combination of rapid prototyping (RP) and soft lithography. Phase changed 3D printing was used to create masters for PDMS molds. A photocrosslinkable polymer was then delivered into these molds to make geometrically complex 3D microparticles. This repeatable process has demonstrated to generate the objects with greater than 95% repeatability with complete pattern transfer. This process was illustrated for three different shapes of various complexities. The shapes were based on the extrusion of 2D shapes. This may allow simplification of the fabrication process in the future combined with a direct transfer of the findings.

Altering the shapes of particles used for porous scaffold fabrication will allow for tailoring of the pore shapes, and therefore their biological function within a porous tissue engineering scaffold. Through permeation experiments, we have shown that the pore geometry may alter the permeability coefficient of scaffolds while influencing mechanical properties to a lesser extent. By selecting different porogen shapes, the nutrition transport and scaffold degradation can be significantly influenced with minimal effect on the mechanical integrity of the construct. In addition, the different shapes may allow a control of drug release by modifying their surface-to-volume ratio, which could modulate drug delivery over time. While soft lithography is currently used with photolithography, its high precision is offset by high cost of production. The employment of RP to a specific resolution offers a much less expensive alternative with increased throughput due to the speed of current RP systems.

Key words: Bio-additive fabrication, Scaffold micro-architecture, Microparticles, Soft lithography, Pore space modulation

1. Facts on Musculoskeletal Disorders

1.1. Severity of the Problem

Musculoskeletal disorders are among the most common disorders encountered in the medical field. They are considered the most common cause of long-term pain and physical disability (1). These disorders affect all age groups, with the prevalence of medical problems

Michael A.K. Liebschner (ed.), *Computer-Aided Tissue Engineering*, Methods in Molecular Biology, vol. 868,
DOI 10.1007/978-1-61779-764-4_5,

increasing with age. As society ages, there is a natural increase in chronic diseases, in general, and musculoskeletal disorders, in particular, that require further treatment and care.

The rate of musculoskeletal disorders found in the adult population is twice that of heart problems and greater than that of all chronic respiratory problems. By the year 2030, the number of people in the United States older than 65 years is projected to double; people over the age of 85 are the most rapidly expanding part of the society (2). Approximately, one-third of the population aged over 75 has significant musculoskeletal problem.

The economic impact of the musculoskeletal disorders is staggering. The annual direct and indirect costs for these disorders in the US in 2004 were around 7.7% of the gross domestic product and are expected to rise up to 20% by 2015.

Musculoskeletal disorders consist of a variety of different diseases that cause pain or discomfort and affecting bones, joints, muscles, and the surrounding structures. Sixty-five percent of musculoskeletal disorders affect mainly the spine in terms of sprains, dislocations, and fractures (2). The burden of musculoskeletal disorders presents a compelling argument for greater understanding and expanding research.

1.2. Challenges for the Medical Field

Nowadays, many people at an advanced age wish to be more active and functional than the previous generation. This increase of the patient's expectations from the health care providers increases the challenges to try to meet the patient's expectations. Furthermore, the current population is the most medically educated population on treatment options related to their ailments.

Musculoskeletal disorders can be subdivided into skin, bone, cartilage, joints, muscle, and ligament diseases. The treatment of such disorders is still a challenging clinical problem. A significant percentage of patients develop arthritic joint problems in the hip or knee with increasing severity over time. A surgical treatment option is to replace the hip or knee joint with an artificial joint in the hopes of receiving a lifelong replacement joint that returns function back to normal. Unfortunately, this cannot be delivered at present as the bone–implant interface destabilizes after about 15–20 years requiring subsequent revision surgery. More than 500,000 joint-replacement surgeries are performed in the US alone per year.

Another challenging field in musculoskeletal disorders is the treatment of vertebral fractures. More than 700,000 vertebral fractures are encountered each year in the United States. With such a high number and taking into consideration the limited longevity of spinal fixation devices, alternatives to metal implants capable of restoring joint function are desperately needed (3–6).

All previously mentioned and many more similar challenging situations drive us to an approach where tissue function and structure need to be restored through the incorporation of biological materials such as cells, growth factors, and biopolymers. The goal is to make regenerative medicine a reality through engineering of functional artificial tissue substitutes.

1.3. Structural Organization of Bone Tissue

Bone is a composite material that consists of five hierarchical levels (Fig. 1) (7); (1) whole bone; (2) architectural level; (3) tissue level (7); (4) lamellar level; and (5) ultrastructural level (8).

1.3.1. Whole Bone Level

The highest level in the structural organization of bone tissue is the organ level, or whole bone level, and it represents the overall shape of the bone (Fig. 2). At this level of structural organization the bone functions on the order of magnitude of the organism, providing structural support and aiding with locomotion. The bone interacts with the body either by tendons, ligaments, muscles, or by other bones.

Optimization at this level is seen mainly in terms of change in the mass rather than the shape of the bone due to several internal and/or external factors. Bone mass is determined mainly by the remodeling that occurs at lower structural organizational levels and adaptation to changes in lifestyle and physical activity.

1.3.2. Architectural Level

The architectural level is primarily related to the micro-architecture of bone tissue, specifically cortical or trabecular bone. These two bone structures play a major role in providing mechanical stability of the bone distributed throughout the osteons and/or trabeculae. At this particular level any remodeling of the organism will be visualized as a change in architecture, specifically the quality of the architecture (9).

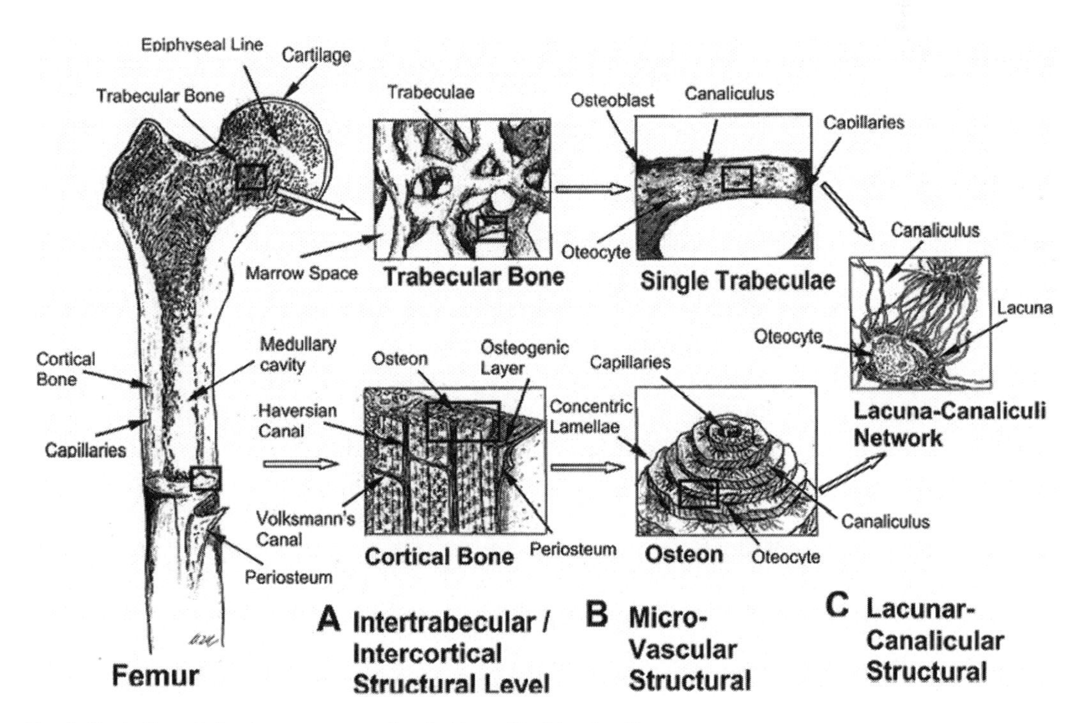

Fig. 1. Illustration of structures corresponding to hierarchical levels of bone.

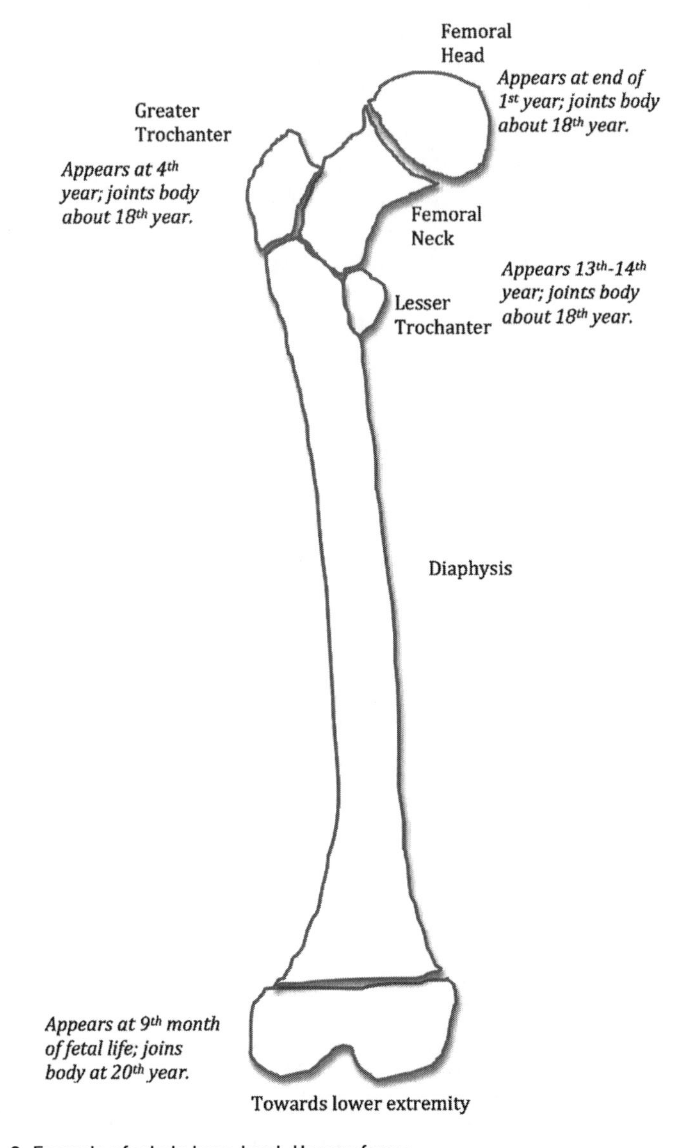

Fig. 2. Example of whole bone level. Human femur.

Trabecular and cortical bones arise depending on the site and loading conditions (Fig. 3). Trabecular bone, or spongy bone, is characterized by high surface area and low density, and it typically occupies the interior part of the bone. It is highly vascular and contains bone marrow where hematopoiesis takes place. On the other hand, cortical bone, or compact bone, is much denser, harder, and stronger than the trabecular bone due to the circular nature of the osteons that make up its structure. It forms the outer shell of the bone. The micro-architecture of the cortical bone consists of microscopic channels in the center of the osteons whereas trabecular bone is highly porous.

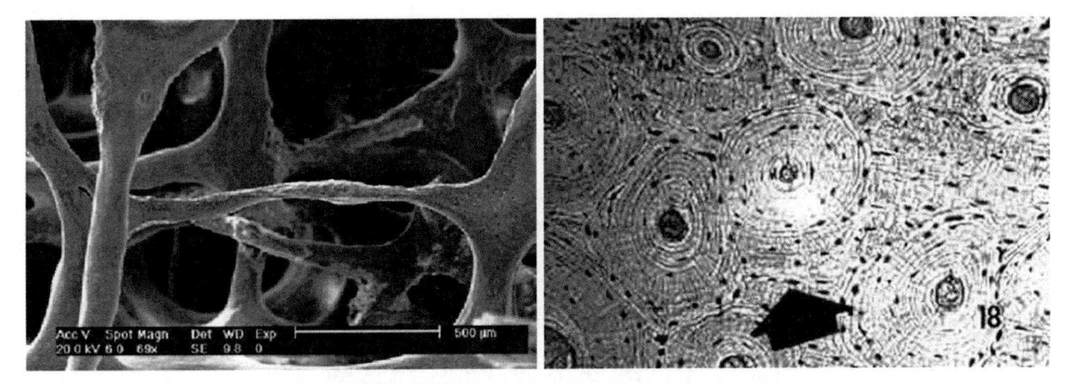

Fig. 3. On left, SEM of trabecular bone illustrating thinning of the trabecular in the center forefront. On right, photomicrograph of cortical bone illustrating the individual osteons and osteocytes.

The architectural level provides support for the whole bone structure and facilitates the bone main function in supporting the body. The strength of this level can be assessed by multiple geometric factors such as trabecular thickness, bone mineral density, and bone surface-to-bone volume ratio. These factors can be evaluated by different imaging techniques.

1.3.3. Tissue Level

The material properties at this level provide support for the geometry of the architectural level above it. Remodeling of the bone at this level affects the material properties of the bone tissue. Tissue proprieties are those that relate directly to the mechanical characteristics of the bone independent of the micro-architecture. Properties such as stiffness, Young's modulus, yield point, and energy to fracture can be dealt with on a fundamental material level. The optimization that occurs at this level is as a result of the modification of the material properties, and is responsible for the apparent properties at the architectural level. Mechanical properties of bone at the tissue level are generally considered "material properties" and mechanical properties of bone at the architectural level are generally considered "structural properties."

1.3.4. Lamellar Level

The lamellar bone consists of multiple parallel layers of collagen laid on top of each other in directions that vary by up to 90° (Fig. 4). These laminations are deposited by the basic multicellular unit (BMU). The sheets of the lamella are on the order of 3–20 µm in thickness. It assists in the bone's ability to resist torsion forces.

Woven bone is produced by osteoid produced by osteoblasts which occurs initially in all fetal bones that will be replaced in adults by lamellar bone. Adults produce woven bone as the first mineralized tissue after fractures or Paget's disease. The woven bone is later remodeled into lamellar bone, which typically takes several years.

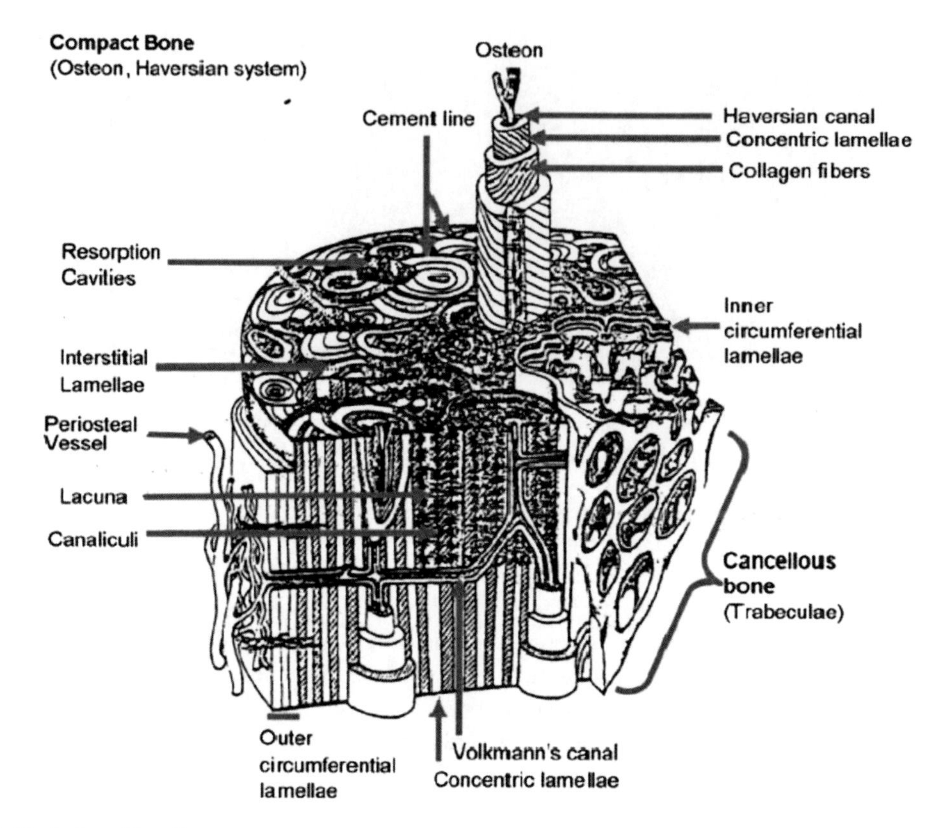

Fig. 4. Illustration of lamellae.

1.3.5. Ultrastructural Level

The lowest structural level of the bone is the ultrastructural level. Both chemical and physical properties can be addressed at this level. The ultrastructural level consists of the minerals that are a part of bone (i.e., calcium, magnesium, and phosphate). The advantage of viewing bone at this level is that it allows the study of an additional function of bone that cannot be addressed until this size. This function is the use of bone as mineral storage for the organism. Mineral storage and the effects of chemistry are the main functional points at the ultrastructural level as is the orientation of collagen in the lamellae.

2. Modulation of the Pore (–) Space

2.1. Scaffold Fluid Environment

Bone is a highly mechanically active 3D tissue consisting of cellular and matrix components. Therefore, the true biological environment of a bone tissue is mainly regulated by the dynamic interaction between cells experiencing mechanical forces and the continuously changing matrix.

As would be expected from a tissue whose main role is to provide structural support, bone cells cultured in vitro respond to a variety of different mechanical signals like fluid flow as one example. Attempts to culture such cells in static conditions typically lead to the formation of a construct with viable cells from the outside and necrotic cells from the inside because of a limited diffusion delivery of nutrients. Direct perfusion of 3D tissue engineered constructs is known to enhance osteogenesis through mechanical stimulation of the seeded bone precursor cells, which can be partly attributed to enhance nutrient and waste transport.

The effects of flow-mediated shear on bone cells have been studied in 2D monolayer cultures. Continuous flow applied to osteoblasts in vitro has been associated with increasing the expression of osteogenic markers (10). Shear stresses between 0.5 and 1.5 Pa affect the production of nitric oxide and prostaglandin E2 as well as osteoblast proliferation. Different fluid flow rates and conditions have been associated with different effects on the bone constructs. Pulsatile and oscillatory flow conditions have been associated with increased gene expression. Although many of the previous results were based on the studies that were performed using 2D cultures, it was suggested that variable flow conditions may also have effect in 3D cultures as well.

Incorporating these findings and our knowledge of bone tissue into a 3D scaffold construct, we realize that permeability and porosity are two main factors that play a key role in shaping the scaffold fluid environment. These terms are sometimes incorrectly used interchangeably in the tissue engineering literature (Table 1). These terms have inherently different meanings where porosity indicates the amount of void space within a structure while permeability is a measure of the ease with which a fluid can flow through the structure from one end to the other. Porosity describes the volumetric ratio of void volume over material volume. It has

Table 1
Terminology comparison of porosity and permeability

Porosity	Permeability
• Volumetric ratio of void volume over material volume	• Ability of fluid to travel through a porous network
• Takes into account trapped pores but does not describe their position	• Does not take into account trapped pores
• Value does not describe architectural arrangement of material	• Value takes into account architectural arrangement of material and the tortuosity of channels in scaffold
• Has effect on apparent scaffold properties	• Does not affect apparent scaffold properties

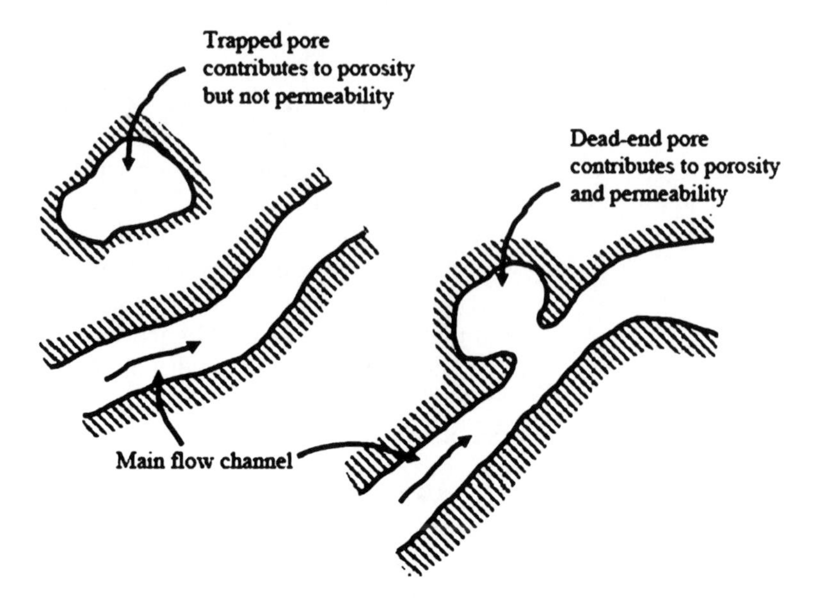

Fig. 5. Permeability and porosity comparison.

an important function whether being used in vitro as well as in vivo. In vitro, porous structures are used to seed cells in construct (scaffold). On the other hand, it is used as a conductive structure in vivo. Permeability describes the ability of fluid to travel through a porous network maintaining nutrient transfer as well as waste products removal from the scaffold. In contrast to porosity, permeability does not take into account the trapped pores. For permeability, only the interconnected pores of an open-cell architecture contribute to its quantitative value. Therefore, high permeability always constitutes a high porosity; however, a high porosity now always results in high permeability (Fig. 5).

2.2. Microparticle Design Approach

Microparticle design can be approached using variant principles including soft lithography as well as rapid prototyping (RP). Soft lithography techniques have a wide range of applications in tissue engineering for microfluidics (11, 12), stamping (13–15), pattering techniques (16), and other methodologies that require high precision and spatial control. Silicon molds generated from masters are inert and sterilizable enabling them to be used multiple times. One of the advantages of such molds is their optical clarity which facilitates the use of photocrosslinkable polymers, increasing the speed of part production and reclamation (17). On the other hand, the cost and time required to prepare these molds are one of their main disadvantages (12).

Conventional techniques utilize a micro-fabricated photolithographic master for which the equipment and cost are prohibitive. The micro-fabricated plates are made from silicon, which is very brittle and fragile. The alternative method uses silicone for soft lithography. Silicone is very flexible and inexpensive. Several

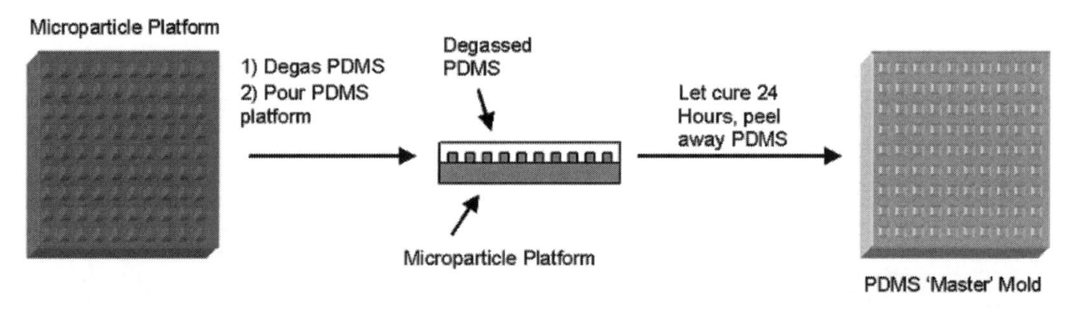

Fig. 6. Schematic procedure for creating PDMS mold using an additive manufactured initial positive plate (not to scale).

processes have been explored in regards to cost reduction. One accepted method involves the selective photocrosslinking of a 2D pattern onto a plate with the use of a mask generated with dark and light spots, this method has been successful in the crosslinking of hydrogels and other photocrosslinkable biomaterials (18–21).

Alternatively, a mold of microparticles can be created from polydimethylsiloxane (PDMS) as illustrated in the Fig. 6. PDMS is the most widely used silicon-based organic polymer. An initial positive plate is for example fabricated through inexpensive 3D phase change printing at high resolution. The viscous PDMS is pored on top of the initial plate to create a negative mold. Because of the flexibility of the PDMS it can be easily removed from the initial positive plate without a risk of destroying either one. The positive plate can then be reused multiple times to generate several negative molds.

The mold is then placed on a glass plate of dimensions 12 cm × 15 cm. In our study, 500 μl of PPF was dissolved in DEF, with 0.5% of the photoinitiator BAPO. This solution was then pipetted onto the PDMS mold. The PPF-DEF was spread into a thin layer over the interior part of the mold. The mold was then covered by another glass plate of equal dimensions, sandwiching the mold between plates and forcing excess material from the mold. Metal clamps were applied to all edges of the glass, holding it in position.

The mold with the metal clamps was subsequently placed in a UV chamber. After 1 h, the clamps were removed and the PDMS was taken away from the glass leaving the microparticles on the glass plate. Then, the individual particles were separated from the glass plate by a razor blade and collected in a container. During each process, there can be several hundred complex shapes fabricated. Nevertheless, this process seems suitable only for research, as it is impractical to mass fabricate particles that way. However, it provides all the flexibility needed to investigate the effect of particle shape on the functionality of the scaffolds built from them (Fig. 7).

Rapid prototyping is a new technique for master mold fabrication. Our microparticles were designed using CAD, resulting in three architectures sharing the same volume but with different

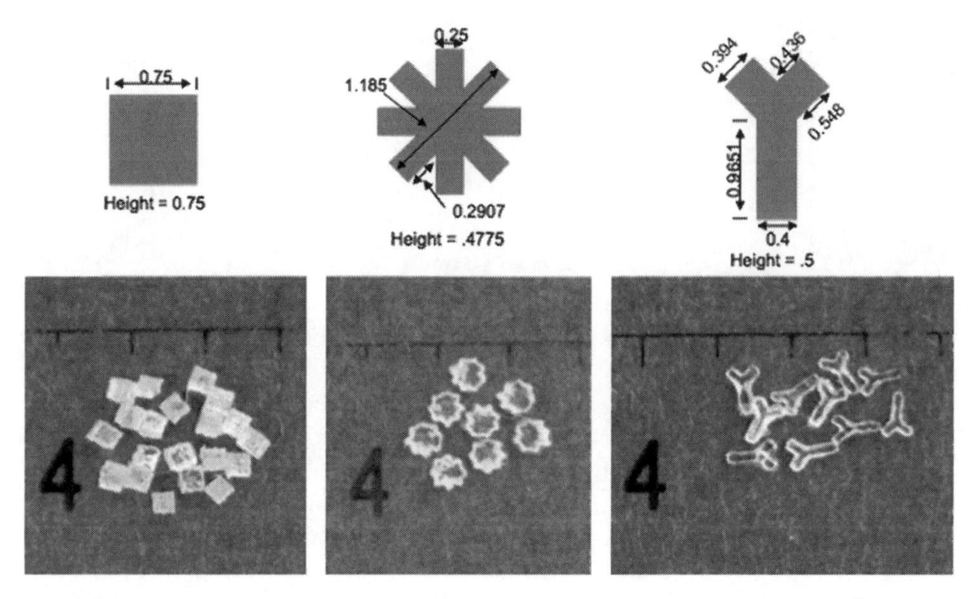

Fig. 7. CAD drawing of designed microparticles. The volume of all microparticles is the same (0.422 mm^3). Surface area of the microparticles varied as a function of geometric complexity. Photographs of the actual fabricated parts below.

geometric complexity. Namely, cube, Y-shape, and an asterisk shape. These shapes were generated at twice the volume of a standard sodium chloride (NaCl) crystal using the PatternMaster (SolidScape, Merrimac, NH), a 3D phase changing printer. The PatternMaster uses two thermoplastic waxes to generate objects in a layer-by-layer fashion. One type of wax is used to build the object while the second one is used as a support material. The wax material is solidified during the fabrication process and becomes solid after ejection in a droplet fashion.

The Y-shape and the asterisk shape have 1.25 and 1.5 times greater surface area than the cube, respectively at the same exact volume. To determine the best method to build these architectures, single versions of each were built to test for geometric inconsistencies. Slice files were prepared using ModelWorks 6.1 with a corresponding build layer height of 12.7 μm in the *z*-direction. Architectures were processed with the three settings available for the PatternMaster at that resolution, 407, 408, and 501; each process corresponds to a specific plotter path and filling method with the inkjet heads. Option 407 builds parts using two rows of build material and four rows of support material and is recommended for parts with delicate features and rounded edges. Option 408 builds with one row of build material and two rows of support and is recommended for intricate parts with small, thin cross-sectional areas. Finally, Setting 501 uses three rows of support and two rows of build and includes a double closed-off row, which aids in the replication of intricate geometry. Process specific patterns are illustrated in the Fig. 8.

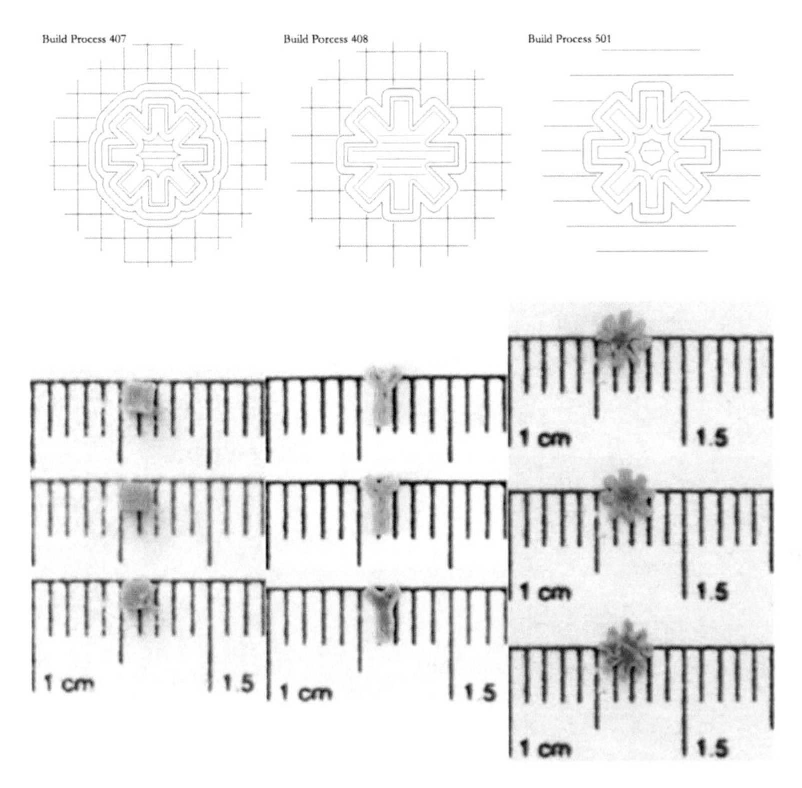

Fig. 8. (*Top*) Build process preparation with ModelWorks (SolidScape, Merrimac, NH). The *lines* correspond to the intended build path of the inkjet heads. *Red* represents the support material and *green* represents the build material. The total width of the star pattern is approximately 0.75 mm. (*Bottom*) Pictures of fabricated particles following the three processes.

Build process 408 resulted in holes at the outer extremities of each arm of the Y-shape. Build process 501 generated an adequate outer approximation except for an inconsistency in the center of the part. The build process 407 generated the exact shape without errors and the Y-shape was built using this process. Each architecture was built using all three processes. The configuration for each, which adequately represented the true shape without gaps or distortion, was accepted. Dimensional modifications were made to each architecture to prevent build inconsistencies and volume loss due to the mold process.

Each architecture was arranged in a 15 × 15 array on a flat platform of global *x* and *y* dimensions of 5 cm × 5 cm with spacing between particles no less than 1.0 mm in each direction. The model was prepared for fabrication in layers determined according to the build process selected for each architecture (Fig. 9). Building plates in a flat, horizontal manner was not easy due to material problems such as curling and delamination between different layers. To solve this problem an extruded rim was added to the bottom of the platform.

Microparticles were successfully generated from the molding process with the photocrosslinkable polymer PPF-DEF. The build

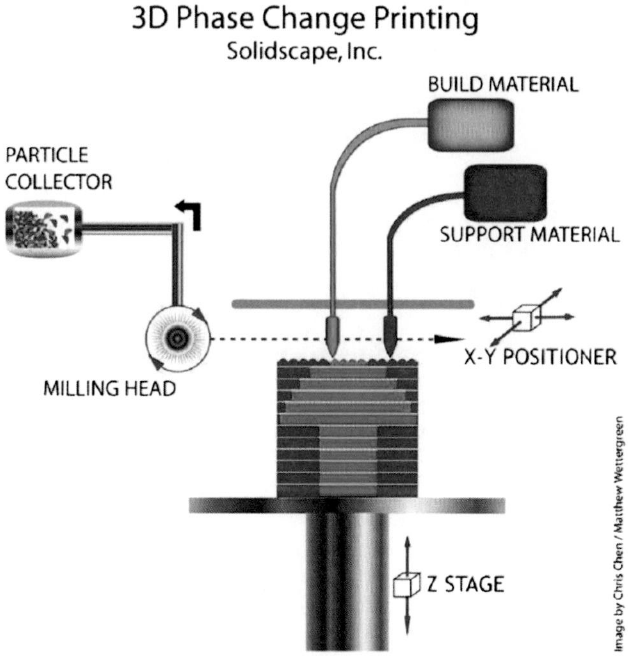

Fig. 9. Wax-based fused deposition modeling.

plates containing the porogens included a surface roughness visible on the vertical surface of the porogens, which exactly matched that of the surface roughness from the RP platforms. An average of three of the 225 architectures per plate were lost during the post-build preparation process, representing 98.7% reclamation rate.

There was a thin film surrounding the particles of residual crosslinked PPF that was apparent while reclaiming those particles. The height of this film was determined by the amount of pressure found between the glass plates and the silicon mold.

Microparticles exhibited geometry from the silicon mold with an average deviation of 56.3% in any 1D. The greatest deviation between the designed dimension and the reproduced dimension was in the length of the trunk of the Y-shape, with shrinkage of 58.4%. This shrinkage is mainly due to the material properties of the polymer used for crosslinking. So, dimensional modification would be a crucial factor in order to obtain particles of a specific size (22).

2.3. Effect of Void Shape on Permeability

The terms porosity and permeability are sometimes used interchangeably. These two terms have two inherently different meanings. While porosity means the void shape in a structure, permeability means the ability of a viscous materials to flow easily through a solid structure.

Many attempts have been made to find a correlation between permeability and other parameters of porous material like porosity, pore size, or tortousity. Agrawal (23) has shown that scaffolds can have different levels of permeability while maintaining the same porosity. O'Brien et al. (24) studied the effect of pore size on the permeability of collagen and glycosaminoglycan (CG copolymers) scaffolds. This study showed that the scaffold pores need to be large enough to allow cells to migrate into the scaffold (>20 μm), but at the same time to be smaller than 120 μm to have sufficiently high specific surface area. Increase in porosity can lead to an increase in permeability only when there is high interconnectivity between pores.

Both scaffold porosity and permeability have an effect on the mechanical properties of the scaffold. For example, a scaffold that has lesser porosity and more solid material has much higher mechanical properties when compared to a scaffold with less solid material and higher porosity.

Permeability is an intrinsic parameter describing the ease of cell movement into the scaffold as well as the diffusion of nutrients in and out of the scaffolds. One way to quantify permeability is to use a constitutive law known as Darcy's law. Darcy's law describes the permeability of mostly materials with a random pore spatial distribution such as soil and sand. The rate of fluid volume discharge is proportional to the hydraulic conductivity (K in units of m^2) of a porous material. Ideally, this measure is describing the scaffold structure independent of sample size and the fluid used. Therefore, the hydraulic conductivity is normalized by the dynamic viscosity of the permeating fluid. It can be expressed mathematically as the following:

$$k = \frac{Q \times L \times m}{P \times A}.$$

The fluid mobility or Darcy's hydraulic permeability coefficient (k, units of m^4/Ns) is also important in defining fluid flow through porous materials and is calculated by the volumetric flow rate (Q), the length of the specimen (L), the sample cross-sectional area, the pressure difference across the scaffold (P), and the dynamic viscosity of the fluid (μ).

Scaffold permeability is dependent on five microstructural parameters (7); (1) porosity; (2) pore size and distribution; (3) tortuosity or interconnectivity (7); (4) fenestration; and (5) pore orientation.

It has been clearly shown that there is no direct correlation between porosity and permeability. However, there is an unexplored relationship between permeability and surface area of void volume. With the ability to *engineer* the porogen's architectures, other models to quantify permeability and their ability to account for higher percentages of variance between samples were investigated.

2.4. Permeability Models for Predictability

Numerous modeling approaches exist to predict fluid transfer properties. Capillary, hydraulic radius, drag, and phenomenological models were applied to porous solids based on the given information in the studies for porosity, pore distribution, pore architecture, and pore volume.

2.4.1. Capillary Models

Scheidegger's "capillary model" is simply an expansion of the Poiseuille's equation for laminar flow inside a straight pipe, which accounts for an n number of channels of uniform length and diameter. Instead of calculating the volumetric fluid flow, the Scheidegger model expresses the fluid conductivity in terms of permeability. The "straight capillary model" is also considered a variation of the Carman–Kozeny (25) equation. The "parallel capillary" model makes a further assumption to account for orthogonal flow in the scaffold. The assumption (which builds upon the previous model of orthogonal flow only) is that 1/3 of the flow occurs orthogonally; thus the model takes the form:

$$k = \frac{\phi \bar{\delta}^2}{96}.$$

where ϕ is equal to the porosity and δ is equal to the average pore diameter. The final capillary model is the "serial model."

$$k = \frac{\phi \bar{\delta}^2}{96 T^2},$$

where

$$\text{Tortuosity } (T) = \frac{\mathrm{d}s}{\mathrm{d}x} = \frac{s}{x}.$$

2.4.2. Hydraulic Radius

Hydraulic radius models of fluid flow properties build upon the capillary models to assume that hydraulic conductivity has dimensions in m^2; therefore, a length characteristic (termed the hydraulic radius) should exist which is linked to the channels that the porous medium is thought to be analogous to. One simple measure is of the ratio of the volume to the surface of the void space. This model has been shown to work well for packed solids, which, depending on the porosity of the studied solid, can be an appropriate model to use (25). In the Kozeny–Carman relation the hydraulic radius model takes the form of:

$$k = \frac{\phi^3}{5 S_0^2 (1 - \phi)^2},$$

where:

ϕ = porosity.

S_0 = specific surface area (m^2/m^3).

2.4.3. Drag Models

The drag model of permeability assumes that the volume fraction conducive to flow is higher than the structure containing it. Therefore, the structure is treated as an obstacle to the flow. The drag on the walls is estimated from Navier-Stokes and the summation of the drag is assumed to be the resistance to flow (26). The advantage of the drag model is the ability to model the porous nature as fibers or spheres. Unlike other models, permeability is a function of flow rate. Utilizing the Darcy substitution for the fiber model, the equation takes the form:

$$k = \frac{3}{16}\left(\frac{\phi\delta^2}{1-\phi}\right)\left(\frac{2-\ln(\delta Q_{A}\rho/\mu\phi)}{4-\ln(\delta Q_{A}\rho/\mu\phi)}\right),$$

where:

$$\mu = \text{viscosity}\ (\text{N}\ \times \text{s/m}^2).$$
$$Q_A = \frac{\text{flow rate}}{\text{unit area}}(\text{m}^3/\text{s}).$$
$$\rho = \text{apparent density}\ (1-\phi).$$
$$\delta = \text{average fiber diameter}\ (\text{m}).$$

2.4.4. Phenomenological Models

The final model briefly discussed here is a phenomenological model where a permeation factor, K, is related to actual geometric features of the perfused object expressed as parameters. This particular equation includes empirically derived parameters and assumes an inverse natural log relationship between the parameters and the permeation factor, K (25). The form of the phenomenological model is:

$$K = e^{(3.84V_1 + 0.2V_2 + (0.56x10^{-6})TD - 8.09\phi - 2.53)},$$

where:

$$K = \text{water permeation factor}\ (\text{m/s}).$$
$$V_1, V_2 = \text{volume of two classes of pores}\ (\text{m}^3).$$
$$TD = \text{threshold diameter}\ (\text{m}).$$
$$\phi = \text{modified total porosity}.$$

In order to evaluate the abilities of each permeability model to explain or predict the hydraulic conductivity or tissue engineering scaffolds, we conducted a permeation experiment. These models were applied to porous solids generated using rapid prototyped porogens (see previous Subheading 2.2 on porogen fabrication) with the goal of modeling permeability as a function of porogen surface area. RP is an attractive choice in this regard because the process lends itself to repeatable architectures with standardized geometric properties.

The straight capillary model most closely predicted the permeability of the NaCl scaffold for all three porosities but poorly predicted the permeability of the Y-shaped samples. The NaCl scaffold represented cubic-shaped porogens, which is typical of salt crystals that are used in a porogen leaching method for solid scaffolds. The serial capillary model grossly underpredicted permeability for both architectures and, similar to the straight capillary model, reports permeability values for the Y-shape, which are lower than the permeability of the NaCl at similar porosities. The hydraulic radius model predicted similar but low permeability values for all architectures although the permeability trend of the Y-shape versus the NaCl was correctly predicted.

Drag theory underpredicted the NaCl greater than any other model but was the second closest prediction of permeability for the Y-shape porous solid. The phenomenological model was the only model that overpredicted permeability and did so for all samples, even the 76% porous NaCl samples. Incidentally, the phenomenological model was the closest approximation for both porosities for the Y-shape porogen architecture and the 55 and 76% NaCl porogen.

Overall, the measurement of the permeability of these architectures demonstrated that there is an increase in the permeability of the porous solids for porogens with higher surface area and more complex geometry (Table 2), the first which is supported by research by Scheidegger (26) and the latter which is counter intuitive. Predictive models for permeability based on the above equations showed that for randomly porous scaffolds, such as those evaluated in this study, the phenomenological models most closely predicted the resulting permeability. These results may be applied to the optimization of the scaffold permeability to explore what effect, if any, permeability of the scaffold plays on a resulting tissue engineered scaffold in vivo. This study also highlights the need for more complex imperical models for predicting the permeability of scaffolds based on porogen shapes, orientation, and other geometric features.

Table 2
Hydraulic permeability measured and approximated utilizing the various discussed permeability models

	Comparison of models to experimental results				
	55% Y-shape	**64% Y-shape**	**55% NaCl**	**64% NaCl**	**76% NaCl**
Measured permeability	1.71E-04	5.47E-04	2.38E-05	9.25E-05	3.33E-04
Straight capillary	1.04E-08	1.21E-08	2.08E-05	2.42E-05	2.87E-05
Serial capillary	9.47E-09	1.10E-08	5.57E-08	6.48E-08	7.70E-08
Hydraulic radius	3.88E-08	7.05E-08	5.52E-09	1.00E-08	2.68E-08
Drag theory	2.20E-07	3.22E-07	1.27E-15	1.86E-15	3.31E-15
Phenomenological	7.64E-03	1.58E-02	9.31E-04	4.49E-04	1.70E-04

3. Outlook: Heterogeneous Scaffold Design

Tissue engineering aims to restore tissue structure and function through incorporation of different biological materials such as cells, growth factors, and biopolymers into a 3D extracellular material "scaffold." The goal for this scaffold is to dissolve regenerate into a functional biological tissue over time, thereby overcoming known problems associated with permanent implants.

The ability to design scaffolds that mimic the natural tissue at the micro-structural level as well as at the macro-structural level and be able to serve as temporary surrogates for the native ECM is a crucial factor in scaffold design. If the scaffolds cannot fulfill the biological function from the time it is placed inside a patient to when it is completely degenerated if failed and may lead to secondary complications and revision surgery.

Scaffolds can be designed by either homogeneous or heterogeneous materials. Homogeneous scaffolds are composed of only one material. On the other hand, heterogeneous scaffolds are composed of two or more different materials distributed continuously or discontinuously (27). These materials have varying degrees of composition and microstructure and have the ability to be assembled together into one part heterogeneous scaffolds, also known as functional gradient materials (FGM).

With the advancement of design and manufacturing technology it has become possible to design scaffolds composed of heterogeneous materials. These scaffolds have the ability to exhibit different compositions as well as microstructures in a continuous manner which leads to multiple functionalities and design requirements satisfaction (28). Heterogeneous scaffolds often have better overall performances when compared with the homogeneous scaffolds (29–31). Moreover, the differing material composition in the heterogeneous scaffolds produces a gradient in its properties allowing them to fulfill critical functional requirements that cannot be met by homogeneous scaffolds (28).

Different models of heterogeneous scaffold design have been reported in the literature. Kumar and Dutta reported a model to build multi-material objects based on R_m sets and R_m classes as representation of material objects (32). Local composition control (LCC) approach has been described by Jackson et al. (33, 34). Yang and Qian (28) have reported a B-spline-based approach for heterogeneous objects designing and analysis.

Several conventional scaffold fabrication techniques have been developed. These techniques are defined as processes that create scaffolds with a continuous pore structure lacking any long-term range channeling micro-architecture. Examples of these techniques are solvent-casting particulate leaching, gas foaming, fiber bonding, phase separation, melt molding, freeze drying, and solution casting. Certain drawbacks have been reported in conventional scaffold design fabrication including the inability of precisely controlling pore size, geometry, and spatial distribution. A major drawback is the required high volume fraction of particles (>70%) in order to achieve an interconnected porous network.

Solid freeform fabrication (SFF), also known as RP, or layered manufacturing (LM) can offer a possibility to fabricate scaffold with customized external shape and reproducible internal morphology. Solid freeform fabrication is a computerized fabrication technique, in which 3D objects can be generated by CAD. In SFF, parts are being fabricated by depositing material layer by layer in contrast to the previously mentioned conventional fabrication techniques.

All SFF processes require that the CAD model of the manufactured part to be expressed as a series of cross-sectional layers that implemented in SFF machine. Different methods of SFF are reported including three dimensional printing (3DP), stereolithography (35), fused deposition modeling (FDM), 3D plotter, and jet printing.

In context for particle fabrication, we presented a novel approach of using LM in addition to mold making with PDMS for the fabrication of biodegradable particles with complex geometries. While most additive fabrication techniques are not capable of making parts from biopolymers, this mold-making process embodies a convenient alternative with enough flexibility for research purposes.

References

1. Bos GW, Poot AA, Beugeling T, van Aken WG, Feijen J (1998) Small-diameter vascular graft prostheses: current status. Arch Physiol Biochem 106(2):100–115
2. BMUS (2008) The burden of musculoskeletal diseases in the United States. 1:1–19. Online report, http://:www.boneandjointburden.org/
3. Wettergreen MA, Bucklen B, Sun W, Liebschner MAK (2005) Computer-aided tissue engineering in whole bone replacement treatment. In: Proceedings of the ASME International mechanical engineering congress and exposition. Orlando, Florida
4. Kaech DL, Jinkins JR (2002) Concluding remarks: the present and future of spinal restabilization procedures. Elsevier, Amsterdam, The Netherlands
5. Kandziora F, Pflugmacher R, Kleemann R, Duda G, Wise DL, Trantolo DJ, Lewandrowski KU, (2002) Biomechanical analysis of biodegradable interbody fusion cages augmented with poly (propylene glycol-co-fumaric acid). Spine 27 (15):1644–1651
6. Riggs BL, Melton LJ 3rd, (1995) The worldwide problem of osteoporosis: insights afforded by epidemiology. Bone 17(5 Suppl):505S–511S
7. Laschke MW (2006) Angiogenesis in tissue engineering: breathing life into constructed tissue substitutes. Tissue Eng 12:2093–2104
8. Liebschner MAK, Wettergreen M (2003) Optimization of bone scaffold engineering for load bearing applications. Topics Tissue Eng 22:1–39
9. Liebschner MA, Muller R, Wimalawansa SJ, Rajapakse CS, Gunaratne GH (2005) Testing two predictions for fracture load using computer models of trabecular bone. Biophys J 89 (2):759–767
10. Mullender M, El AJ, El Haj AJ, Yang Y, van Duin MA, Burger EH, Klein-Nulend J (2004) Mechanotransduction of bone cells in vitro: mechanobiology of bone tissue. Med Biol Eng Comput 42(1):14–21

11. Beebe DJ, Moore JS, Bauer JM, Yu Q, Liu RH, Devadoss C, Jo BH (2000) Functional hydrogel structures for autonomous flow control inside microfluidic channels. Nature 404 (6778):588–590
12. Tan W, Desai TA (2003) Microfluidic patterning of cells in extracellular matrix biopolymers: effects of channel size, cell type, and matrix composition on pattern integrity. Tissue Eng 9(2):255–267
13. Xia Y, Qin D, Whitesides GM (1996) Microcontact printing with a cylindrical rolling stamp: a practical step toward automatic manufacturing of patterns with submicrometer-sized features. Adv Mater 8(12): 1015–1017
14. Tien J, Nelson CM, Chen CS (2002) Fabrication of aligned microstructures with a single elastomeric stamp. Proc Natl Acad Sci USA 99(4):1758–1762
15. Folch A, Toner M (1998) Cellular micropatterns on biocompatible materials. Biotechnol Prog 14(3):388–392
16. Rogers-Foy JM, Powers DL, Brosnan DA, Barefoot SF, Friedman RJ, LaBerge M (1999) Hydroxyapatite composites designed for antibiotic drug delivery and bone reconstruction: a caprine model. J Invest Surg 12(5):263–275
17. Timmer MD, Carter C, Ambrose CG, Mikos AG (2003) Fabrication of poly(propylene fumarate)-based orthopaedic implants by photo-crosslinking through transparent silicone molds. Biomaterials 24(25):4707–4714
18. Vozzi G, Flaim C, Ahluwalia A, Bhatia S (2003) Fabrication of PLGA scaffolds using soft lithography and microsyringe deposition. Biomaterials 24(14):2533–2540
19. Chiu DT, Jeon NL, Huang S, Kane RS, Wargo CJ, Choi IS, Ingber DE, Whitesides GM (2000) Patterned deposition of cells and proteins onto surfaces by using three-dimensional microfluidic systems. Proc Natl Acad Sci USA 97(6):2408–2413
20. Kane RS, Takayama S, Ostuni E, Ingber DE, Whitesides GM (1999) Patterning proteins and cells using soft lithography. Biomaterials 20(23–24):2363–2376
21. Xia Y, Whitesides G (1998) Soft lithography. Angew Chem Int Ed 37:550–575
22. Wettergreen MA, Timmer MD, Mikos AG, Liebschner MAK (2003) Modification of porogen surface to volume ratio can be used to optimize the permeability of scaffolds used in tissue engineering. In 5th international bone fluid flow workshop. Cleveland, OH, 2003
23. Agrawal CM, McKinney JS, Lanctot D, Athanasiou KA (2000) Effects of fluid flow on the in vitro degradation kinetics of biodegradable scaffolds for tissue engineering. Biomaterials 21(23):2443–2452
24. O'Brien FJ, Harley BA, Yannas IV, Gibson LJ (2005) The effect of pore size on cell adhesion in collagen-GAG scaffolds. Biomaterials 26(4): 433–441
25. Dullien FAL (1979) Porous media: fluid transport and pore structure. Academic Press, New York, NY
26. Scheidegger A (1974) The physics of flow through porous media. University of Toronto Press, Toronto, p 353
27. Wu XJ, Liu WJ, Wang MY (2007) Modeling heterogeneous objects in CAD. Comput Aided Design Appl 4:731–740
28. Yang P, Qian XA (2006) B-spline-based approach to heterogeneous objects design and analysis. Comput Aided Design 39:95–111
29. Park SM, Deckard CR, Beaman JJ (2001) Volumetric multi-texturing for functionally graded material representation. In: Sixth ACM Solid Modeling Symposium 2001. Ann Arbor, MI, USA
30. Pratt MJ (2003) 3D modeling of material property variation for computer aided design and manufacture. In: Proceedings of the SIAM conference on mathematics of industry: challenges and frontiers, 2003, Toronto, Ontario, Canada
31. Wang MY, Lee CS, Wang X, Mei Y (2005) Design of multi-material compliant mechanisms using level set methods. J Mech Design 127(5):941–956
32. Kumar V, Dutta D (1998) An approach to modeling and representation of heterogeneous objects. ASME J Mech Design 120(4): 659–667
33. Jackson T (2000) Analysis of functionality graded material representation methods. Massachusetts Institute of Technology, Cambridge, MA
34. Jackson T, Liu H, Patrikalakis NM, Sachs EM, Cima MJ (1999) Modeling and designing functionality graded material components for fabrication with local composition control. Mater Design 20(2–3):63–75
35. Krams R, Wentzel JJ, Cespedes I, Vinke R, Carlier S, van der Steen AF, Lancee CT, Slager CJ (1999) Effect of catheter placement on 3-D velocity profiles in curved tubes resembling the human coronary system. Ultrasound Med Biol 25(5):803–810

Chapter 6

Scaffold Informatics and Biomimetic Design: Three-Dimensional Medical Reconstruction

Jorge Vicente Lopes da Silva, Tatiana Al-Chueyr Pereira Martins, and Pedro Yoshito Noritomi

Abstract

This chapter briefly describes the concepts underlying medical imaging reconstruction and the requirements for its integration with subsequent applications as BioCAD, rapid prototyping (RP), and rapid manufacturing (RM) of implants, scaffolds, or organs. As an introduction to the problem, principles related to data acquisition, enhancement, segmentation, and interpolation are discussed. After this, some available three-dimensional medical reconstruction software tools are presented. Finally, applications of these technologies are illustrated.

Key words: Biomimetic design, Scaffold, Informatics, Rapid prototyping, Medical reconstruction, BioCAD

1. Introduction

The reconstruction of missing bones, as implants used in cranioplasty, has a long history dated from 3,000 years ago in the ancient civilizations (1). Since then, the humanity tries to solve problems related to the effective substitution or replacement of unhealthy or missing tissues in human body.

The search for solutions to these problems gave rise to a new exciting field of research called tissue engineering (TE). The TE is an investigative area that intends to understand, mimics, and augment the regenerative power of human tissue in order to heal an injured tissue or substitute missing or unhealthy organs.

Tissue engineering is a convergence of diverse science and technology fields, some of them reasonably new and others not so much. Among these fields are biology, material science (mainly biomaterials from many sources and types), medicine, and engineering. Vacanti and Vacanti (2) refer to the biblical and historical

Michael A.K. Liebschner (ed.), *Computer-Aided Tissue Engineering*, Methods in Molecular Biology, vol. 868,
DOI 10.1007/978-1-61779-764-4_6, © Springer Science+Business Media, LLC 2012

roots to invoke the concepts of TE and, probably, provide the first description of what could be termed TE of structural tissue referring to a report dated from 1908. According to the WTEC Panel Report on: Tissue Engineering Research (3), the term TE was coined at 1987 in the National Science Foundation Meeting and lately defined in an NSF workshop as:

The application of principles and methods of engineering and life sciences toward fundamental understanding of structure–function relationships in normal and pathological mammalian tissues, and the development of biological substitutes to restore, maintain, or improve tissue function.

This chapter comprises five sections. This first section presents some considerations to develop medical image reconstruction. Subheading 2 introduces some of the medical image reconstruction tools and techniques and some commercial and free software for this area. The third section presents the needs and the new concepts of BioCAD in a qualitative manner without further details. The fourth section explains the integration of medical image reconstruction as the foundation for subsequent applications with an example of autoadaptive prosthesis. Subheading 5 highlights some conclusions.

2. Medical Image Reconstruction

2.1. Introduction to Medical Image Processing

The medical mission differs from the other forms of image processing in several aspects, such as the importance of the user who interacts with the software. Whereas a three-dimensional (3D) reconstruction provides the clinician with powerful tools for image analysis and measurement, to detect and screen for the primary findings relies on the skills of the human visual system and on the experience of the human operator.

A common approach used in the development of image processing software is to organize data flow into pipelines (4). Frequently called visualization network, the pipeline consists of three basic parts: data objects, objects to operate data and direction of data flow. The first class data objects represent data information as 2D image pixels or 3D polygonal surface. Objects to handle data, such as filters and mappers, on the other hand, allow data edition and adjustments. Finally, the direction of data flow shows how data objects, filters, and mappers are related. Figure 1 illustrates a generic pipeline.

Most common medical reconstruction filters can be divided into three categories, according to their functionalities:

– *Enhancement*: basic tasks involved in filtering and preprocessing the data before detection and analysis are performed either

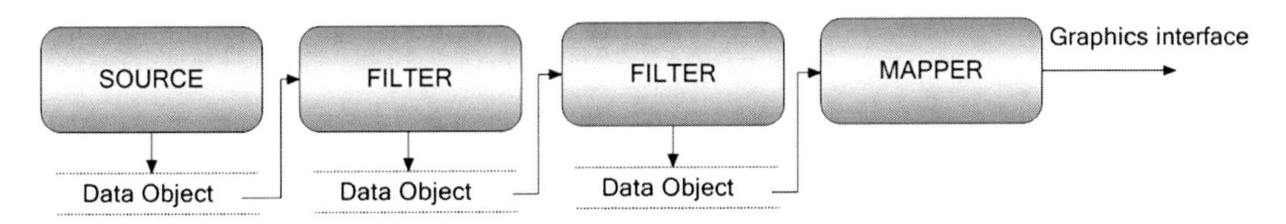

Fig. 1. Data objects are connected with algorithms (filters) to create a generic pipeline. The *arrows* point out the direction of data flow.

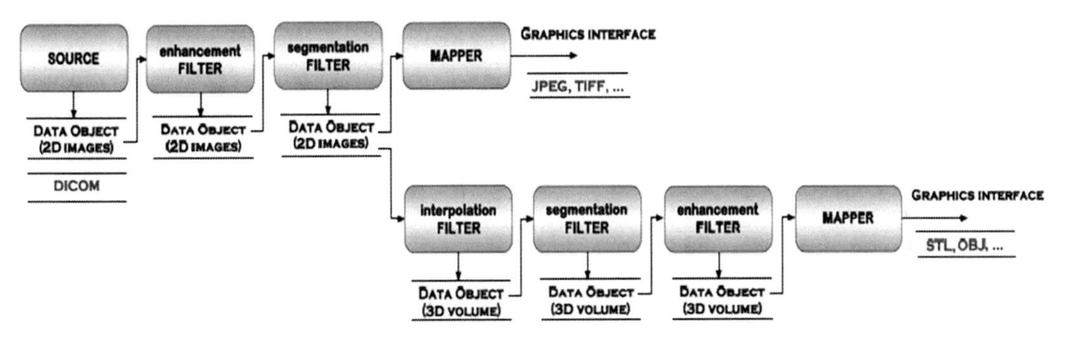

Fig. 2. Data objects are connected with algorithms (filters) to create the medical reconstruction pipeline.

by the machine or the human operator. Enhancement filters are usually applied for noise removal, contrast adjustment, and feature improvement.

- *Segmentation*: this is the task of partitioning an image (2D array or volume) into adjoining regions with cohesive properties. Various approaches exist, such as statistical, structural, and hybrid methods. Special filters are usually applied for detecting medical conditions or events.
- *Interpolation*: task of aligning multiple images, allowing joining different data, and creating a more powerful diagnostic tool than any single image alone.

A traditional medical reconstruction software, therefore, allows data interpolation—generating based on a sequence of 2D images a 3D model—and offers filters for enhancement, segmentation, and mapping of the managed data. Figure 2 illustrates a three-dimensional imaging reconstruction pipeline. In spite of the pipeline flow, adequate applications allow the human operator to manipulate, measure, and export the data to other file formats. Usually, 2D data can be exported to JPEG and PNG, while 3D data can be exported to stereolithography (STL), file format used in rapid prototyping (RP) equipments or standards like initial graphics exchange specification (IGES) or standard for the exchange of product model data (STEP).

In general, source data in medical image reconstruction is a sequence of 2D images, acquired by means of proper equipment.

Therefore, before explaining how enhancement, segmentation, and interpolation filters usually work, it is important to understand how source data is acquired and the limits of the existing technique of reconstruction.

2.2. Acquisition

Since Godfrey Hounsfield invented the computerized tomography (CT) introducing it into medical practice in 1971, the medical imaging modalities, resolution, and definition are growing respectively in quantity and quality. There are several noninvasive techniques for acquiring morphological characteristics of human anatomical region in a detailed and accurate manner. Some of the most often used methods are: Computerized tomography—presented in Fig. 3, magnetic resonance imaging (MRI), positron emission tomography (PET), single positron emission-computerized tomography (SPECT), micro-computerized tomography (MCT), and ultrasound (US). Each of them is specialized to highlight different structures or functions of the human body.

A large amount of data can be obtained from such scan equipments, normally being provided in the form of two-dimensional (2D) projections, frequently mentioned as slices of the desired anatomical region. These images are commonly saved digitally in the international standard Digital Imaging and Communications in Medicine (DICOM) file format simplifying the interoperability among different medical systems with no extra costs for data conversion. This file format contains the necessary data for three-dimensional reconstruction and information of the patient. Figure 3 shows acquisition and 3D reconstruction processes.

Appropriate reconstruction software, based on 2D DICOM files, can generate the correspondent 3D anatomical structures. Using this type of software, internal and external anatomy can be shown simultaneously or separately, and the data can be assessed qualitatively and quantitatively. In this process of reconstruction,

a

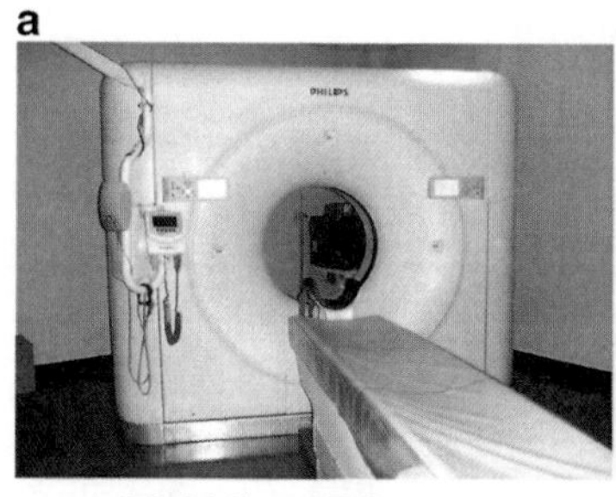

Multislice CT Scanner is used for scanning patients

b

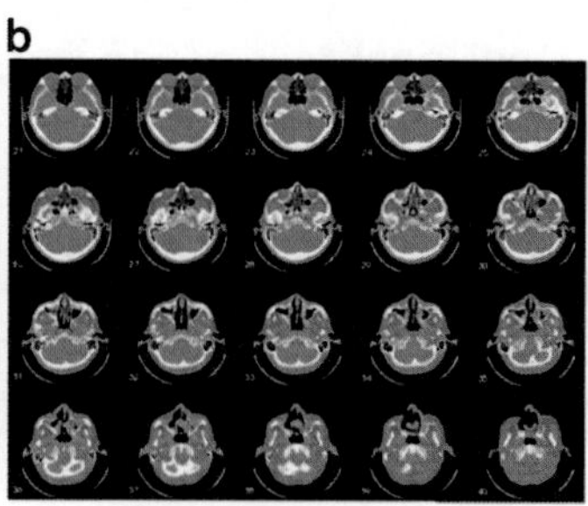

Sequence of 2D axial patient's head slices generated by CT, in DICOM format

c

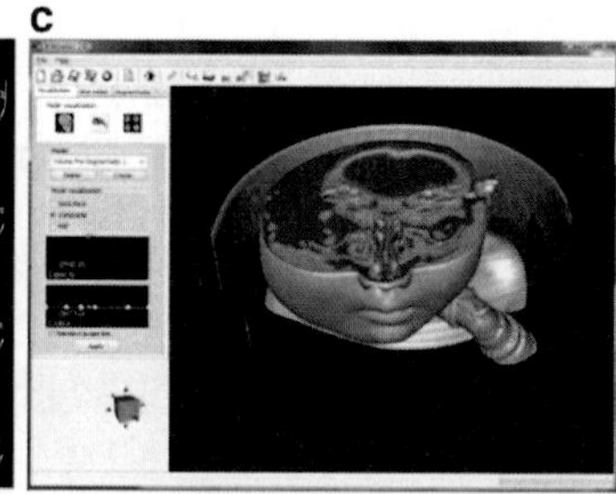

InVesalius software 3D reconstruction, based on 2D DICCOM images

Fig. 3. Medical image acquisition and reconstruction process overview.

the accuracy and resolution of the original images are fundamental. In addition, as only sample values of the projections are available in practical medical imaging equipments, the reconstruction can only yield approximate images even under the most favorable circumstances.

Commonly, the acquisition of medical data is not optimized for 3D reconstruction and presentation. Indeed, on CT acquisition, to improve the signal to noise relationship, thicker slices are usually requested by clinicians. The result of this process is better contrast at the cost of higher partial-volume artifact and poor spatial resolution in the effective *z*-direction. Another practice that difficult 3D reconstruction is to obtain a sequence of slices that is not adjoining. This is often done to reduce the radiation exposition (X-ray) to the patient and the time required to perform a scan, but causes irreparable problems to 3D virtual reconstruction: small gaps are introduced between slices. Since most presentation is 2D, these practices seldom affect the radiologist. However, they can be critical to 3D reconstruction.

The creation of effective medical reconstruction tools needs fundamental understanding of the source of the image data, the technology involved, and the physical principles from which the image values are obtained. This chapter provides a quick reference to the topic. One of the most familiar forms of 3D medical imaging modality is X-ray computerized tomography, and therefore we present filters and segmentation tools that are commonly used for this type of images. The reader interested in the development and specialized use of medical image reconstruction is encouraged to continue the exploration in specialized publications.

2.3. Medical Imaging Reconstruction Principles

2.3.1. Enhancement Filters

For image processing, not only in the medical field, it is usually necessary to perform several operations (such as noise reduction and edge preserving) before performing higher-level processing steps. Rather than describe the theory behind each of the filters presented, this section focuses on its effects and applications considering a users' point-of-view. Figure 4 illustrates application of five different filters on a CT slice image. It is important to notice that if enhancement filters are not applied carefully, they can prejudice original image. Also, occasionally, it might be interesting to apply filters only on parts of an image.

Medical images, acquired from digital imaging modalities, are typically recorded in 4,096 levels of intensity according to Hounsfield scale. Each dot (pixel) of the image, in this case, is represented by a 12-bit value. Unfortunately, the acquired image may not always be satisfactory for clinicians' analysis and 3D reconstruction. Some images may be underexposed while others may be overexposed. To improve readability, one frequent adjustment is to set different brightness and contrast values, called in medical imaging of window and level. The effect of the level feature adjustment can

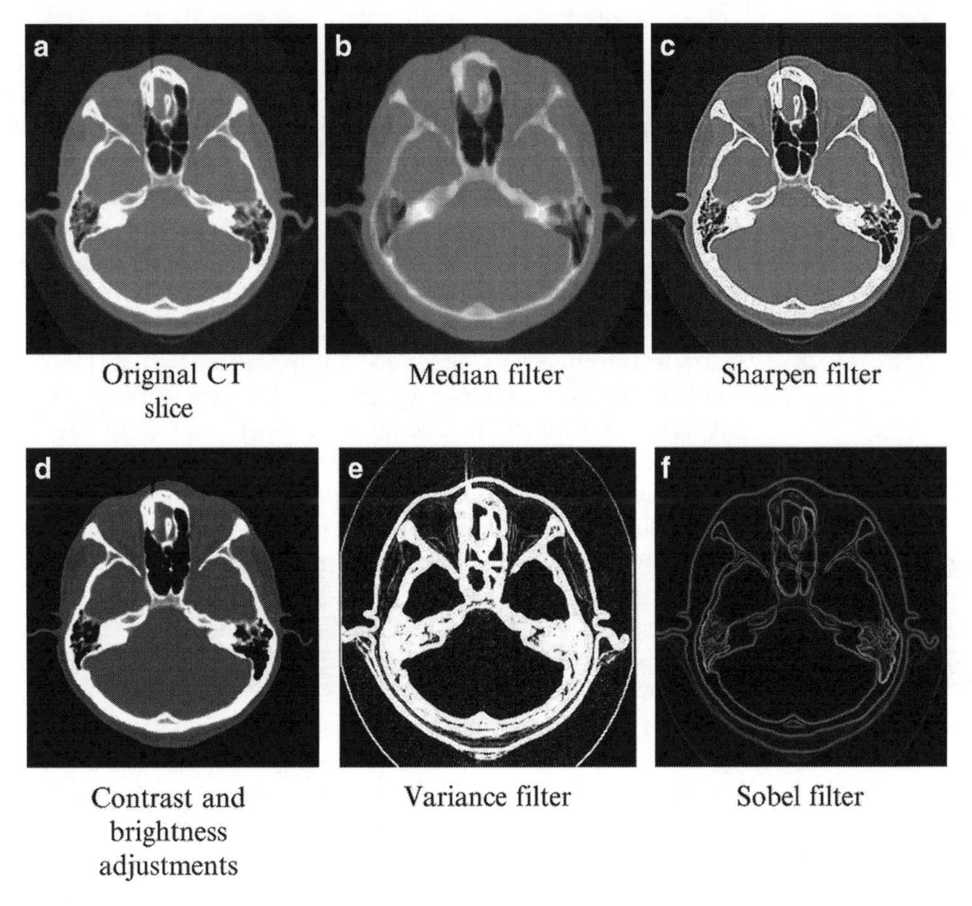

Fig. 4. Results of the application of five different enhancement filters in an original slice. (**a**) Original CT slice; (**b**) median filter applied to original slice; (**c**) sharpen filter, generated by convolution of original image with appropriate kernel; (**d**) brightness (level and window) and contrast of original image changed; (**e**) statistical variance filter applied to original image; (**f**) Sobel kernel convoluted to original image.

be seen on Figure 4. It allows enhancing image characteristics that originally are not seen.

Apart from adjustments, such as window and level, there are linear and nonlinear filtering techniques that might be applied in several situations to enhance the original image. The median filter, for instance, is a nonlinear digital filtering technique, often used to remove noise from images. The idea is to examine a sample of the input and decide if it is representative of the signal. This is performed using a window consisting of an odd number of samples. Median filtering is a common step useful to reduce data drop-out noise (commonly referred to as intensity spikes, speckle, or salt and pepper noise) caused by errors in data transmission. The corrupted pixels are either set to the maximum value or have single bits flipped over. Median filter manages to blur the image, removing "salt and pepper" like appearance. Despite of this, its edge-preserving nature is useful, in cases where edge blurring is undesirable.

In contrast to median filter operation, it is possible to apply a linear sharpen filter to the image, simply by convolving it with a suitable kernel. The basic idea of convolution is that a window of some finite size and shape (kernel) is scanned across the image, producing an output pixel value that is the weighted sum of the input pixels within the window. This process is fairly inexpensive in computations but can produce artificially over-sharp or over-noisy images if not used carefully.

The Sobel operator is one of the existing edge detection algorithms. Technically, it is a discrete differentiation operator, computing an approximation of the gradient of the image intensity function. The Sobel operator is convolved with the original data, being inexpensive to apply it. However, the gradient estimation which it produces is relatively crude, in particular for high frequency variations in the image.

Finally, another common filter is variance filter. It is related to calculate the variations in a moving window over the image. It is an indicator of texture and can be calculated efficiently. The variance filter is widely used and its effect is presented on Fig. 4.

2.3.2. Segmentation Filters

Segmentation can be defined as the partitioning of a dataset into adjoining regions (or subvolumes) whose member elements (pixels or voxels) have common, cohesive properties. This clustering of voxels, or conversely, the partitioning of volumes, is a precursor to identifying these regions as objects and subsequently classifying or labeling them as anatomical regions with similar physiological properties. Segmentation is the means to isolate structures of interest on raw medical image data—affecting 2D or 3D datasets and later foresee the anatomy or pathology in question. The use of tools for the region of interest selection makes it possible to plan interventions and treatments, addressing the medical condition.

The simplest form of segmentation involves thresholding the image intensity. Thresholding is inherently a binary decision made on each pixel independently of its neighbors. Intensity above the threshold yields one classification, below the threshold another. This simple operation can be surprisingly effective, especially in data, such as computerized tomography, where the pixel value has real-world significance. The pixel value (gray level) in a CT image is represented in Hounsfield units matching to the attenuation of X-rays measured in a particular sample of tissue. Since bones are much more radiopaque than other human tissues, specially its cortical part, they can be easily separated in a CT image by setting the threshold between the Hounsfield scale for bones. An example is shown in Fig. 5 where a surface was reconstructed at two different threshold intervals in a CT scan data set of a patient's head.

In most applications of thresholding, the ideal thresholding is not known beforehand, but depends instead upon the particular imaging modality, tissue-type, and parameters of image acquisition.

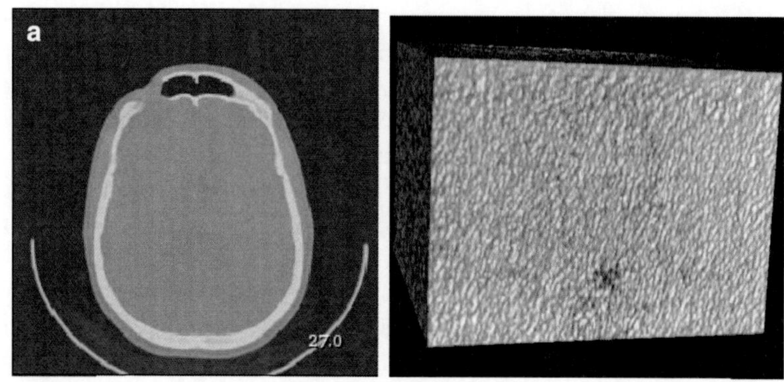

Original acquired data set, threshold filter is set between 0 and 4095 doesn't affect it.

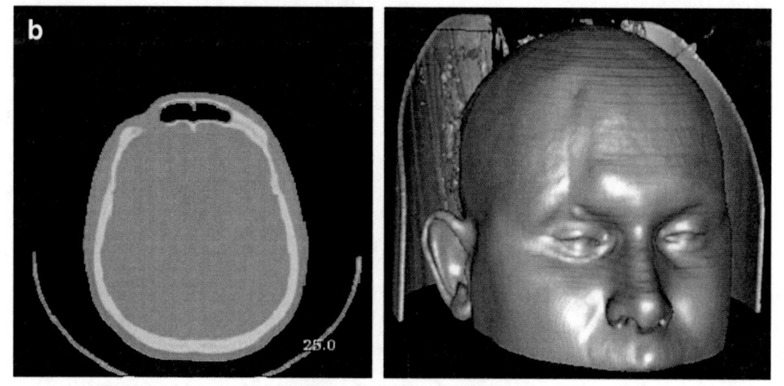

Threshold between 762 and 4095 – dark level pixels ((black) are removed both from the 2D representation and the 3D reconstruction.

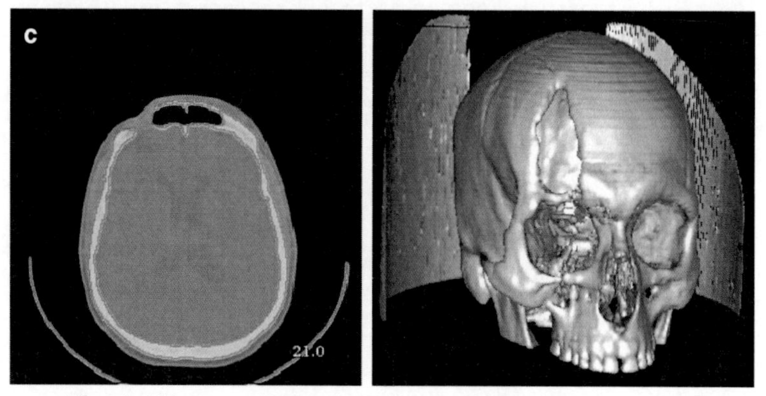

Threshold between 1295 and 4095 –only bright pixels are shown.

Fig. 5. Effects of thresholding filter both on the forty-fifth CT slice (a *red mask* shows the effect, based on the background original image) and on the correspondent 3D reconstruction. Images generated with InVesalius 2 software, having a 4,096 gray scale level CT exam.

In such cases, an optimal threshold might be chosen manually by looking at the image itself or at a histogram of intensity values from the tissues to be segmented. Figure 5 shows threshold filter applied to a patient's head CT data.

Although many useful operations can be done to an image using point operators, on individual pixels, most image processing tasks require the simultaneous consideration of multiple pixels. Techniques that combine multiple pixels in a linear and space-invariant manner are known as linear filtering. A basic mean of taking into account spatial proximity and connectedness as well as alikeness of intensity values is the use of a region-growing flood-fill technique.

The basic premise of a region growing algorithm is to set a start point in the image, called seed. The seed represents the region of interest, and based on this start point the algorithm expands that region into neighboring pixels according to some criteria. There are a number of different region growing algorithms that use different criteria to select which neighboring pixel has to be included in the segmented region. The confidence connected approach to region growing, as described in the ITK Software Guide (5), first starts at the seed region and computes the mean m and standard deviation σ of the region intensities. Neighboring pixels are included in the segmented region if their intensities fall between the interval $(m - f\sigma)$ and $(m + f\sigma)$, where f is a user-defined multiplier that essentially determines the "sensitivity" of the algorithm. Figure 6 shows the results of the application of a seed point (225, 210, 70) and a multiplier value $f = 2.2$ for the segmentation of a specific region of the brain for two different slices.

By applying statistical properties and including known or estimated values for intensity mean, variance and scale, one can extend the concept of threshold-connected components to include confidence-connected components. These statistical methods of linking pixels of voxels in a dataset can be used to delineate regions or boundaries, depending on the application. This is a loose description

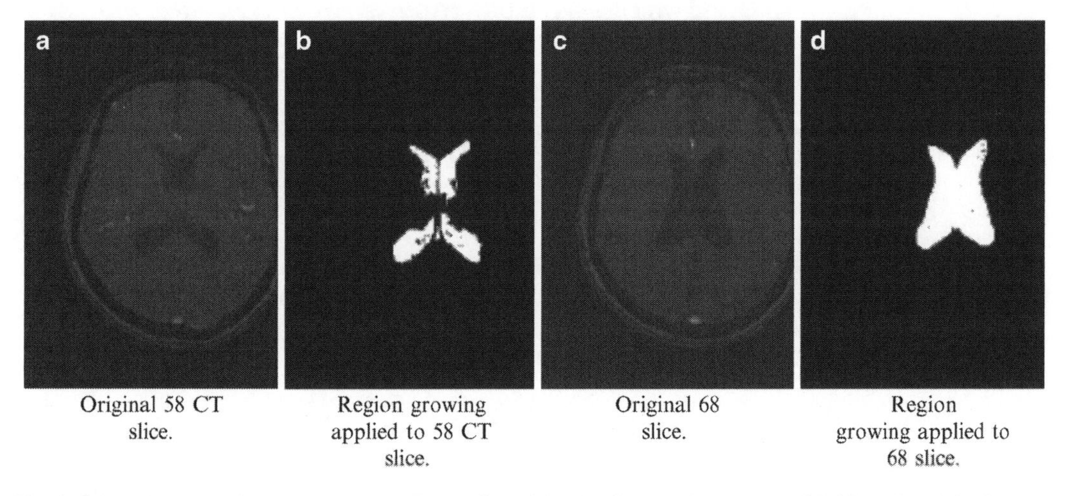

Fig. 6. Selected results of the segmentation. From left to right, the first two images are of Z-Slice 58 and the last two images are of Z-Slice 68.

of the idea of "fuzzy-connectedness." More sophisticated approaches to region growing can be found on specific books (5).

There is no perfect segmentation of any dataset; the correctness of the granularity of a partitioning of any picture or volume depends on the application and the insight wanted. Just as there are no perfect segmentation results, there is no preferred partition technique. Each type of dataset presents unique challenges that must be addressed. A creative and resourceful software developer will carefully select from the many available tools, often combining them in a constructive process to achieve the desired goal. Therefore, segmentation is a challenging task for software implementers and often being an interactive task among users and the medical image reconstruction system.

2.3.3. Interpolation

As described previously, the slices acquired through medical imaging equipments are only a part of the patient's region of interest. Biomedical images are sampled anisotropically, with the distance between consecutive slices significantly greater than the in-plane pixel size.

For visualizing the medical images in 3D space or perform other operations on them, however, isotropic discretization is often needed. To achieve a desired discretization, interpolation is always performed to decrease the interslice distance in medical image processing. In volume estimation, interpolation is also performed after segmentation to achieve higher accuracy. Figure 7 shows the superposition of slices, illustrating the need to rebuild data between.

There are several interpolation methods, such as linear, cubic spline, modified cub spline, and sinc-based functions. Whereas linear interpolation might be accurate to represent a femur, for instance, it usually is not satisfactory to reconstruct abdominal images, which have more organs in when compared with other biomedical images.

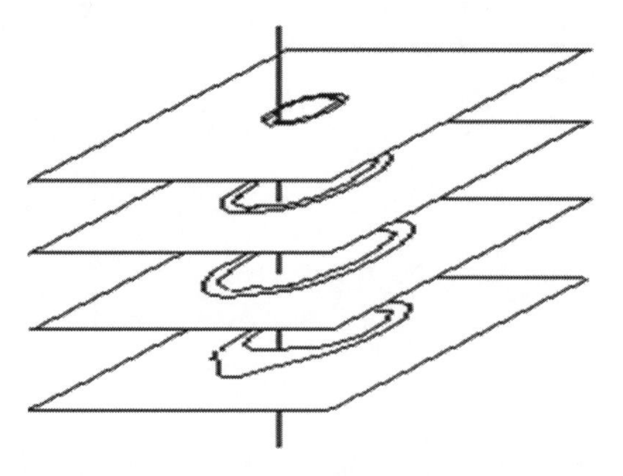

Fig. 7. Interpolation of several slices to generate 3D matching anatomical model.

Usually, when there is a large distance between consecutive images, linear interpolation acts as well as, even slightly better than the other methods, in the sense of signal-to-noise ratio. Also, the computational load of the linear algorithm is half of the other methods.

To define the most adequate interpolation method, it is important to select the region of interest. Specific literature offers more information on this topic (4, 5).

2.4. Medical Image Reconstruction Softwares

There are several medical image reconstruction softwares available both on Internet and on market. Among the most commercially known softwares are Mimics (Materialise NV), Analyze (Mayo Clinic), and Vitrea (Vital Images). These softwares provide user support and several facilities. There are also free alternatives. Osirix is one of the most famous free software in the area, but runs only on MacOS platform. A very promising free software in the area of 3D medical image reconstruction, developed since 2001 by Renato Archer Information Technology Center—CTI, is InVesalius (6). Currently, it runs on both Windows and Linux and is available at no cost in the Brazilian Public Software Portal (7). A new beta version is already available, including MacOS platform.

Some screenshots of InVesalius can be seen on Fig. 8. These pictures illustrate how vast is the application of 3D medical imaging software. For instance, they can be applied from diagnosis to treatment in the majority of medical specialties, veterinary, archeology (mummy's reconstruction for example), and in paleontology (fossil separation from the rock in which it is embedded) to name a few. All these current applications, integrated with technology as RP, bring possibility to manipulate safely an anomaly of a patient or rare objects like fossils. It is expected with the ongoing research that reconstruction will be of great value for computer-aided tissue engineering (CATE) as the basis for tissue modeling and specification.

Medical reconstruction software usually provides visualization, segmentation among various specific tools. This class of software is commonly used for generating 2D images for papers and medical reports, or for generating 3D models for CAD and RP.

In spite of not having license cost, free medical software allows engineers to customize it to specific needs. As most of these software use toolkits in its developments, such as Visualization Toolkit (4) and Insight Toolkit (5), customization might not be difficult.

3. BioCAD Basic Concepts and Modeling

Probably, the first references to the term BioCAD are related to noun of companies or specific products for the commercial domain in the 1990s. In spite of the term BioCAD has been used for more

a

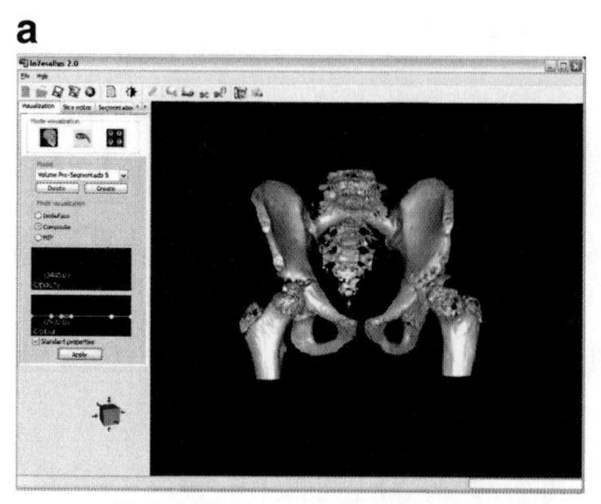

Pelvis bone segmentation.

b

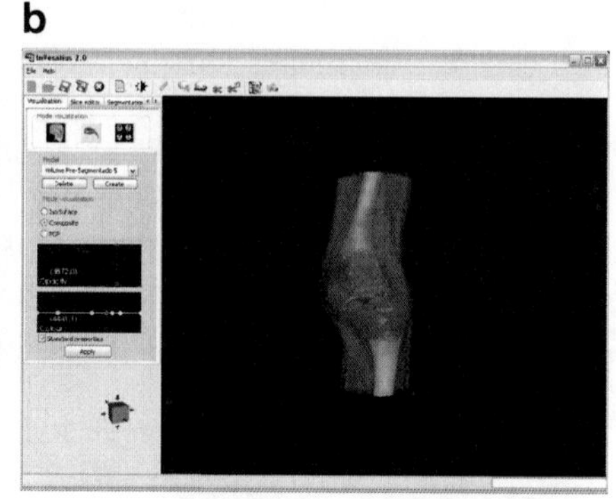

Knee analysis considering both muscles and bone structures.

c

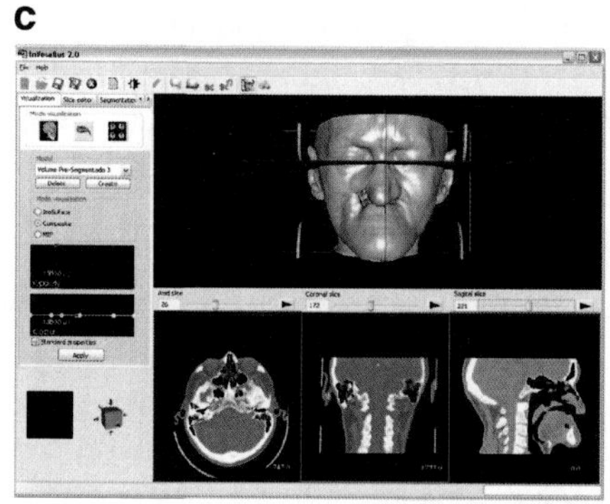

Simultaneous 2D and 3D analysis of facial reconstruction.

d

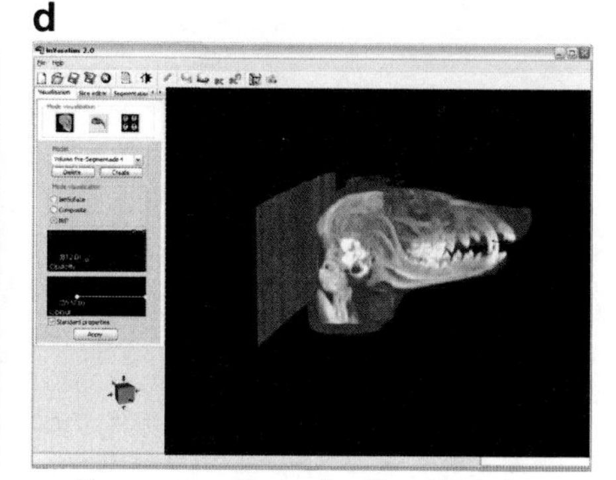

Reconstruction of a Brazilian fox head

Fig. 8. Reconstruction of different structures with InVesalius medical image reconstruction software.

than one decade, only in the last 3 years it has been used to currently describe the research in BioCAD modeling. After 2005, Sun et al. (8) have published a wide range definition of BioCAD and its area of application in the bioengineering and tissue engineering. Before it, the closest use of the term was used in 2003 by Alves et al. (9) referring to a hybrid system for 3D modeling for use in reverse architectural design but with different meaning.

According to Sun et al. (8), BioCAD—spelt as Bio-CAD by this author—is a concept that describes a set of tools, techniques, and procedures adapted from engineering to be used for bioengineering purposes. The concept involved is to provide solutions to the biological and anatomical modeling in the same way that engineering tools provide solutions for the product design discipline and other areas where computational modeling is important. In short, BioCAD is "the effort to model human body parts in a CAD based virtual environment" (8). This section discusses briefly the bioengineering and BioCAD applications used for solving this kind of demand.

The availability of new technology is improving the techniques applied for solving medical and general health problems. This new frontier of technology applications and science research is becoming much more complex due to the multidisciplinary approach demanded and the intrinsic complexity of dealing with individual variations and naturally developed characteristics like anatomical geometry.

In the front edge of research applied to the digital representation of human anatomy, there is a challenge when the demand is for acquisition and computer virtual representation of anatomical geometry of body parts either for hard or soft tissues.

The bioengineering and BioCAD concepts are relatively new when compared with other traditional research topics. Besides, both of them belong to a multidisciplinary role, what makes the consolidation of this body of knowledge even more difficult to gather.

In an ideal point of view, BioCAD provides the computational and technological tools to develop bioengineering and tissue engineering as applied sciences. This means that, in the future, using BioCAD it will be possible to construct living tissue solutions for living organism problems, replacing missing or malfunctioning parts in the same way that modern engineering replaces mechanical parts. Besides, TE, together with BioCAD advances, will make possible to substitute lost parts of body, like limbs or internal organs, completely from the scratch.

Nowadays, applications of BioCAD are mainly based on medical image reconstruction of the real anatomical geometry acquired from medical 3D digital imaging examinations like CT scanning. These techniques are based on reverse engineering of internal and external anatomical structures used as reference for generating high complex mathematical surface geometries that represents the original anatomy more adequate for CAD/CAE manipulation.

However, the use of conventional reverse engineering techniques to reconstruct the complexity of anatomical geometries may be efficient; it is not effective, because the resulting solid geometry and surface organization do not represent essential anatomic landmarks, which are crucial for future usage of this geometry in bioengineering applications. So, it is mandatory to develop new approach for bioengineering representation of anatomical geometries, which is the first step of BioCAD conception.

According to this point of view, BioCAD technique must take into account anatomical references and landmarks of the body structure in representation in order to keep the most important information of this body part, at the same time that it provides simplifications demanded for the usage of computer applications in future bioengineering solutions.

This BioCAD approach results in 3D geometries, described in terms of surfaces that are strategically generated, complex enough to represent fundamental anatomic landmarks and simple enough to be used as the basis of any computer project or bioengineering analysis.

A sort of this technique has been developed at the Renato Archer Information Technology Center, a Brazilian federal institute where a multidisciplinary bioengineering group is dealing with 3D digital references of anatomical structures to convert them into effective BioCAD models to streamline future applications. The expertise of this group is leading to some BioCAD definitions and protocols that have been proposed in many supporting applications provided by the group by means of a broad cooperation with professionals of the healthy sector.

One of the BioCAD definitions generally applied is the identification of anatomic landmarks of any anatomical geometry before starting to recover it from its reference to the final 3D representation. The BioCAD model generated is highly adaptable and sustains high fidelity to the original structure, while it has surfaces simple enough to make it easy to use as a base for prosthetic design or other device development and even for analysis in bioengineering with finite element methods (FEM).

Another characteristic of BioCAD technique is the care with the size and complexity relationship. Specially for biological structures, the more the size reduces, the more complex and interesting structures become, so it must be set a convenient trade-off solution that ensures that the biological phenomena can be efficiently represented and the amount of details are kept simpler in terms of model to be analyzed in a computer system.

These are some of the main techniques in BioCAD applications, there are much more, but most of them are exclusive, which means that the analyzed problem must be seen individually and the solution, in terms of BioCAD techniques, can be standardized until it reaches some steps, but for other steps it demands particular solutions.

The effort to specify and implement a BioCAD, differently from a regular CAD system, has to consider the many peculiarities of TE as microstructure—geometry and size of pores, its continuity and distribution as a 3D mesh, biomaterial and cell distribution—together with the macro geometry—anatomy of the tissue or organ.

Today is still missing a complete BioCAD specification and, mainly, its implementation in terms of a software tool commercially available. As a result, anatomical structures still have to be modeled as a combination of many different tools and commercial systems.

4. Medical Image Reconstruction and Integration with Other Technologies

The fantastic advances and availability of the computational systems in terms of hardware and software developments in the last three decades permitted the evolution of a myriad of technology areas that are helping medical science to grow in an unthinkable pace. With the evolution of technologies, more profound researches are being enabled in the area of TE promising to change the paradigm from the necessity to have a donor to the production of scaffolds, living tissues, or organs directly from a machine called CATE. The main technologies that are enabling the progress of CATE are quickly described:

1. Acquisition of internal patient and medical image reconstruction presented in the last section.
2. Computer-aided design, manufacturing, and engineering (CAD/CAM/CAE) is a well-established set of integrated technologies intensively used in industry in general that allows the design process, simulation and production of goods. These technologies are helping human health applications not only with the production of precise and specialized medical equipments, and high quality implants, but also paving the way for the more challenging applications as in medicine, demanding new concepts and tools.
3. Rapid prototyping and rapid manufacturing technologies (RP/RM) comprise a set of more than 30 different processes commercially available and an equivalent number in research. In short, these processes are based on the construction of physical models in a layer-by-layer way.

The first approach to use RP/RM in medical applications was the possibility to create any complex structure. Nowadays, the great appeal is the possibility to control the deposition of distinct materials freeing the bio-designer to think about macro and microcomplex and hybrid structures typical of human tissue. It is a relatively new technology with the first commercial machine presented in 1986 at Autofact Fair in Chicago. Currently, the most common applications of RP/RM in health are commercially available as direct production of metallic implants (EBM—electronic beam manufacturing of Arcam, DMLS—direct metal laser sintering of EOS GmbH, SLM—selective laser melting of MCP-HEK), hearing-aid systems (SLA—stereolithography and SLS—selective laser sintering of 3D Systems), braces for orthodontics (SLA-3D Systems), and guides for dental implants (SLA and SLS). The most promising applications are the production of specialized scaffold by commercial companies (Therics, Envisiontech, etc.) and organ printing, object of research today.

Figure 9 presents a diagram with some possibilities of integration of the medical acquisition and reconstruction of medical images in the production of biomodels by means of RP for surgical planning and diagnosis; rapid manufacturing of implants, scaffolds, and constructs (defined as scaffold plus cell seeding); and the new concept of bioprinting for TE solutions. The conventional

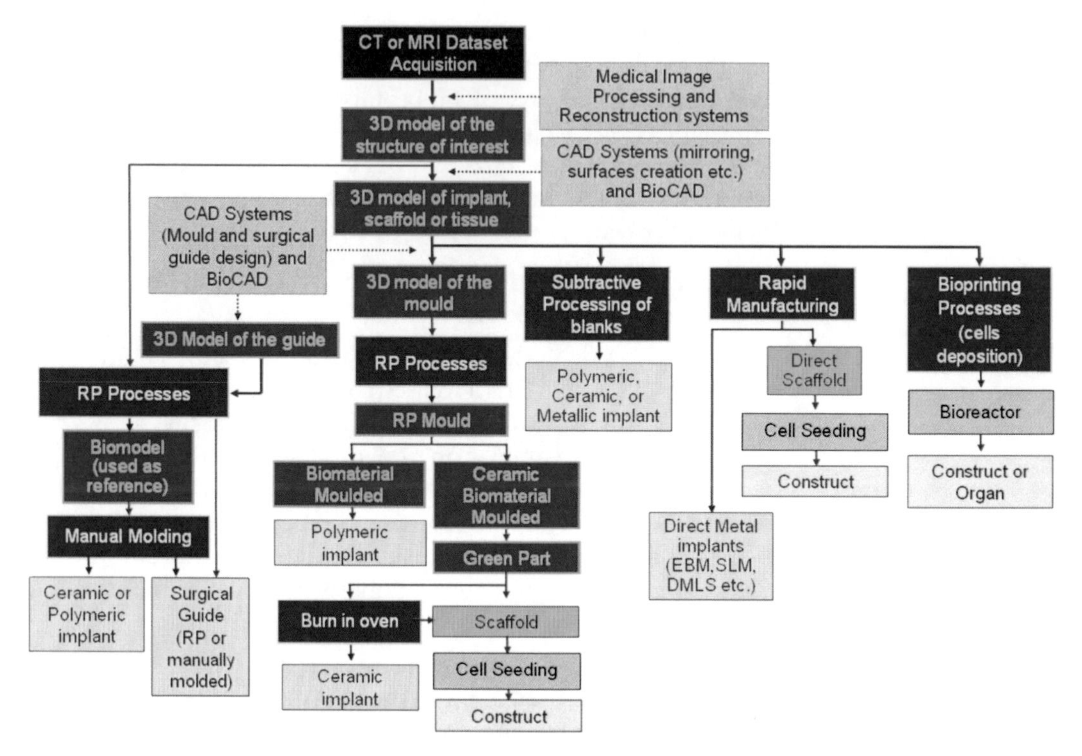

Fig. 9. Medical image acquisition and reconstruction systems as enabler for a broad class of applications in medical tissue reconstruction and replacement. Adapted from Silva (10).

production of implants by means of conventional subtractive methods is one of the branches represented in this diagram.

4.1. Example of Integration of Medical Imaging Reconstruction, BioCAD Concepts, and RP

The example of application is a case study developed to support surgical simulation and planning of a patient infant, 11 years old, victim of traffic accident with fracture on cranium bone and exposition of the brain through an opening of approximately 12 cm. This case was developed with the cooperation of Roland Clinic (SP, Brazil). The patient did not present any cephalic trauma or mobility sequel. The prescribed surgery was a cranioplasty, but with an additional complication for the prosthetic device to adapt to the bone growth of the patient avoiding future deformations in the skull.

For this case, BioCAD procedures have demanded the TC scanning of the patient and the 3D reconstruction, providing the 3D references. Using this initial 3D model, a BioCAD tool for mirroring the original skull was used to generate reference for the design of the prosthetic device as a copy of the symmetric skull curvature. Growth pattern of the skull bones was investigated and an innovative solution for the construction of the prosthetic device was proposed considering this information. The final solution has used a three parts sliding device, in order to provide mechanical resistance and auto-adaptation to the size changing due to natural growing process.

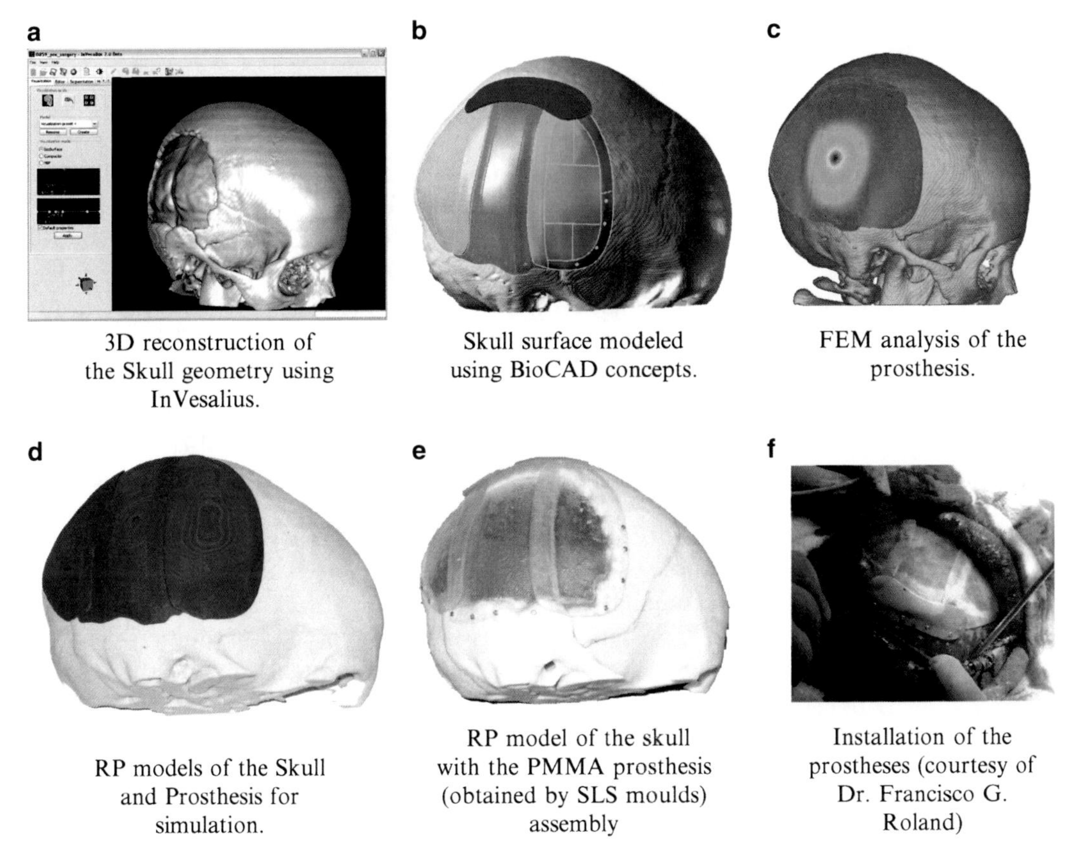

Fig. 10. BioCAD process for design of a personal auto-adaptable prosthesis for cranioplasty applied to an ingrowth patient.

The complete cycle can be seen in the Fig. 10. This figure shows that the BioCAD solution for prosthesis design has been followed by a bioengineering analysis using 3D FEM, to verify the required stress and strain resistance for some critical mechanical conditions. After this, the virtual design has been materialized using RP machines, providing the first models for handling tests and surgery planning. Finally, molds were generated to give form to the prosthetic parts in polymetilmetacrilate (PMMA) polymeric biomaterial. This prosthetic device, made of biomaterial, was implanted in the head of the patient in a successful surgery. The patient will be checked regularly until adult age.

Cases like this are becoming even more usual in the high technology society and the use of bioengineering and BioCAD to improve the solutions for them are becoming a common place. Hence, the development of new BioCAD computational tools that are best fitted to the bioengineering demands and more efficient to provide individual solutions, achieving the time and quality for a convenient response for the most of the medical applications is the new frontier for the researchers.

5. Conclusions

The medical image reconstruction of human tissues and organs based on data acquired from one of the imaging modalities is the basis for many pipelined applications today like CAD/CAM/CAE, RP/RM. It is responsible to reproduce the internal anatomy with high fidelity by means of 3D virtual models. These models can reduce subjectivity on examinations, facilitates interpretations, and avoid mental reconstruction of a huge set of 2D images.

The technology for medical image reconstruction is expanding in two directions: firstly, the macro direction with high quality images, and, secondly in the micro direction bringing precise information of the microstructure of tissues and organs.

On the other hand, a new class of machines arises innovative ways to think about rapid manufacturing of "spare parts" for the humankind in the form of scaffolds and even organs. These applications are much beyond the current use of reconstruction and RP to produce tailored prosthesis. It is typically a multidisciplinary approach. In between the medical reconstruction and these rapid manufacturing machines is a new concept called BioCAD. It is growing day-by-day as a real necessity and much development is still necessary to have available specifications for the production of living tissues and scaffolds for its manufacture.

Another branch of reconstruction in medical applications is the reverse engineering technology. This theme is presented in other chapters of this book. It is of wide use in the industry to copy existing physical objects geometry in a computational model that can be edited or for quality inspection processes by acquiring a point cloud of the surfaces. The most common technologies are based in acquiring points with a contact probe, a laser scan, or structured light. The last is the most appropriate to acquire external human structures for medical applications because of the speed of acquisition, quantity of points, and offers no risk in contact with eyes. It has been used especially for the acquisition of the geometry of terminal limbs in amputees for orthesis adaptation. Another interesting application is the possibility to copy healthy structures like ear, nose, orbit, and many others to create molds in RP processes and reproduce the missing region in soft material like silicone for cosmetic reconstructions.

This chapter covers briefly some of the issues and approaches used in common methods for medical image reconstruction. Additionally, the perfect integration of medical image reconstruction as an enabler for many applications as BioCAD, CAD/CAM, FEM, and CATE are considered of great importance for future developments in the TE and bioengineering domains.

References

1. Sanan A, Haines SJ (1997) Repairing holes in the head: a history of cranioplasty. Neurosurgery 40:588–603
2. Vacanti CA, Vacanti JP (2000) The science of tissue engineering. Orthop Clin N Am 31 (3):351–356
3. WTEC Panel on Tissue engineering Research, Final Report, Larry V. McIntire (panel chair), January 2002. Available at http://www.wtec.org/loyola/te/final/te_final.pdf
4. Schroeder W, Martin K, Lorensen B (2006) The Visualization Toolkit An Object-Oriented Approach To 3D Graphics, 4th Edition. Kitware, Inc. publishers
5. Ibañez L, Schroeder W, Ng L, Cates J (2003) The ITK Software Guide, 1st edn. Kitware, Inc.
6. Martins TAP, Barbara AS, Silva GBC, Faria TV, Dalava BC, Silva JVL (2007) InVesalius: three-dimensional medical reconstruction software. In: 3rd international conference on advanced research in virtual and rapid prototyping, vol 1. Leiria, Portugal, September 2007, pp 135–142
7. Brazilian Public Software Portal (2007) February, 07, 2012 http://www.softwarepublico.gov.br
8. Sun W, Starly J, Nam J, Darling A (2005) Bio-CAD modeling and its applications in computer-aided tissue engineering. Comput Aided Design 37:1097–1114
9. Alves NMF, Bártolo PJS, Ferreira JMGC (2003) The use of Biocad and rapid prototyping for building restoration. In: VR@P – international conference on advanced research in virtual and rapid prototyping. Leiria, Portugal, 1–4 October 2003, pp 193–199
10. Silva JVL (2005) Rapid prototyping applications in the treatment of craniomaxillofacila deformities: utilization of biomaterials. In: Invited conference presentation in the IV annual meeting of the brazilian materials research society (SBPMat). Recife, Brazil, 16–19 October 2005

Chapter 7

The Development of Computer-Aided System for Tissue Scaffolds (CASTS) System for Functionally Graded Tissue-Engineering Scaffolds

Novella Sudarmadji, Chee Kai Chua, and Kah Fai Leong

Abstract

Computer-aided system for tissue scaffolds (CASTS) is an in-house parametric library of polyhedral units that can be assembled into customized tissue scaffolds. Thirteen polyhedral configurations are available to select, depending on the biological and mechanical requirements of the target tissue/organ. Input parameters include the individual polyhedral units and overall scaffold block as well as the scaffold strut diameter. Taking advantage of its repeatability and reproducibility, the scaffold file is then converted into .STL file and fabricated using selective laser sintering, a rapid prototyping system. CASTS seeks to fulfill anatomical, biological, and mechanical requirements of the target tissue/organ. Customized anatomical scaffold shape is achieved through a Boolean operation between the scaffold block and the tissue defect image. Biological requirements, such as scaffold pore size and porosity, are unique for different type of cells. Matching mechanical properties, such as stiffness and strength, between the scaffold and target organ is very important, particularly in the regeneration of load-bearing organ, i.e., bone. This includes mimicking the compressive stiffness variation across the bone to prevent stress shielding and ensuring that the scaffold can withstand the load normally borne by the bone. The stiffness variation is tailored by adjusting the scaffold porosity based on the porosity–stiffness relationship of the CASTS scaffolds. Two types of functional gradients based on the gradient direction include radial and axial/linear gradient. Radial gradient is useful in the case of regenerating a section of long bones while the gradient in linear direction can be used in short or irregular bones. Stiffness gradient in the radial direction is achieved by using cylindrical unit cells arranged in a concentric manner, in which the porosity decreases from the center of the structure toward the outside radius, making the scaffold stiffer at the outer radius and more porous at the center of the scaffold. On the other hand, the linear gradient is accomplished by varying the strut diameter along the gradient direction. The parameters to vary in both gradient types are the strut diameter, the unit cell dimension, and the boundaries between two scaffold regions with different stiffness.

Key words: Computer-aided design, Tissue scaffolds, Functionally graded, Selective laser sintering, Rapid prototyping

Michael A.K. Liebschner (ed.), *Computer-Aided Tissue Engineering*, Methods in Molecular Biology, vol. 868,
DOI 10.1007/978-1-61779-764-4_7,

1. Introduction

The loss or failure of organs or tissues due to trauma or ageing is one major concern in human health care as it is a costly and devastating problem (1, 2). This problem has led to the development of tissue engineering (TE), which aims to create biological substitute to repair or replace the failing organs and tissues (2–4).

One of the approaches in TE is to seed cells on porous scaffold to grow them and form new tissue that can replace the damaged tissue (1). Biodegradable scaffolds are highly engineered structures that act as a temporary support for cells to facilitate the regeneration of the target tissue. The challenge in such scaffolding is to design and fabricate customizable biodegradable scaffolds with desirable properties that can promote cell adhesion and support cell growth, proliferation, and differentiation, and eventually the formation of the extracellular matrix (ECM) for the tissues (5).

It has been found that different cell types prefer different scaffold characteristics, namely, pore size, porosity, and also surface properties of the scaffold (6). Thus, a scaffold should be designed with pore size and porosity values according to what is needed by the specific cell to better accommodate the growth of the cell.

Besides the design of internal architecture, scaffold properties are highly influenced by the processing technique and the biomaterial the scaffold is made of. There are a variety of processing methods to fabricate TE scaffolds, such as fiber bonding, phase separation, and particulate leaching (7). Unfortunately, most of these processes have limitations in controlling the pore size, porosity, and pore interconnectivity. Recently, rapid prototyping (RP) techniques have been found to be promising because of their flexibility and outstanding manufacturing capabilities. In TE application, RP has distinct advantages of being able to fabricate scaffolds with controllable micro- and macroarchitecture (8).

It is important to have a control over the scaffold microstructures, such as pore size, porosity, and also pore geometry, as these properties directly affect cells' ingrowth. For example, in the case of growing bone tissue, a scaffold with pore size 2–6 μm and porosity value lower than 35% results in no tissue ingrowth. A scaffold with pore size 15–40 μm and porosity of 46% results in fibrous tissue ingrowth while the one having pore size of 60–100 μm and 48% porosity successfully grows bone tissue (7).

Human organ or tissue consists of more than one cell type or phenotype and different types of cells have shown to prefer different scaffold environment for the cells to grow, proliferate, and function well. The different cell types or phenotypes have their own specific functions and are important for the tissue or organ to function properly. Take the example of skin, which consists of two layers (epidermis and dermis). The epidermis is composed primarily of keratinocytes, which is tough and impermeable to toxic substances and harmful organisms and also regulates the loss of water from the body. On the other hand, the dermis is relatively acellular and mainly composed of collagen fibrils as ECM to provide strength, flexibility, and elasticity to the skin (9). Besides, hard tissue like bone possesses gradient in terms of mechanical properties, such as stiffness and strength, across the tissue to accommodate load transfer and strain energy absorption (10).

In order to construct tissues with different cell types or mechanical properties gradient, for example osteochondral region consisting of bone and cartilage part, a successful scaffold is likely to have two different regions with different microstructures, including pore size and porosity. There should also be an intermediate region between the two regions to allow for smooth transition to avoid scaffold delamination and also to better facilitate the stress transfer (11). This structural difference in the scaffold is called structural gradient. The scaffold with such structural gradient is called a functionally graded scaffold (FGS) since each region is unique and related to the typical functions the different cells are performing. The structural gradient is meant to facilitate the cells in performing their functions.

Catering to different requirements for different cell types, i.e., designing scaffold with pore size and porosity gradient, could make the design stage complicated and tedious. To date, there are a few available computer-aided design (CAD) systems to ease the design stage. One example of such system is the computer-aided system for tissue scaffolds (CASTS). By using CASTS, users can design a customizable scaffold with ease. The essence of the system is a collection of cellular units that can be assembled together into three-dimensional scaffolds. Users can acquire the defect image from a patient's body as an input to the system. After the cellular units' assembly is ready, a Boolean operation is done between the defect image and the assembled cellular units. The result will be a scaffold consisting of a stacked cellular units with the external shape of the patient's organ/tissue defect (12). However, the currently available CASTS system only allows the construction of scaffold with uniform pore size, which may not be suitable for regenerating a heterogeneous tissue, which is composed of different types of cells and possesses mechanical properties gradient. Therefore, further improvement to CASTS system is considered necessary to allow designers to create FGSs with ease.

As has been discussed earlier in this section, there are two types of requirements, namely, biological and mechanical requirements. Biological requirements arise from the structures of the native tissues themselves, having different layers with unique functions of each layer for the tissue or organ to function normally. This can be seen from both hard and soft tissues. Bone as a hard tissue possesses cortical and cancellous parts (13), skin as a soft tissue possesses dermis and epidermis (9), while cartilage, also soft tissue, owns zonal gradients of ECM and chondrocyte phenotypes (14). Thus, FGSs in terms of pore size and porosity are needed to facilitate the cells to proliferate and differentiate in the appropriate layer to achieve healthy tissue regeneration.

From a mechanical standpoint, human tissues have different mechanical properties distributed spatially with respect to anatomical site. Cortical and cancellous bones have different mechanical properties for smooth stress transfer and strain energy absorption (15). To guide the new tissue to behave like the original tissue, some kinds of mechanical cues are needed (16). Therefore, to tailor the appropriate mechanical cues, the scaffold should have gradient in terms of mechanical properties, too. This can be achieved by varying the structure, such as pore size and porosity, of the scaffold.

Accounting for both biological and mechanical concerns, continuous, instead of discrete, FGSs are needed to help regenerate defected tissues into healthy tissues. Currently, the challenges are the design and fabrication processes of such scaffolds. RP processes are preferred to conventional processes as the fabrication means due to the high degree of control and consistency over the scaffold macro- and microarchitecture (8). Such accuracy and consistency are due to the use of integrated CAD system into RP. However, fabricating FGS presents new challenges in automating the design process, as manual CAD designing of FGS is a very complicated task. Thus, a system that can automate the FGS design process is needed, and this is where this research contributes.

2. Materials

Polycaprolactone or PCL is synthetic polymer that has been approved by Food and Drug Administration (FDA) for use in the human body (17). It is biocompatible and its degradation products are nontoxic and completely metabolized in vivo. Although it has a long degradation time (in the order of 2–3 years (18)), it can be compatibly blended with other polymers, such as polylactic-acid (PLA), polylactic-glycolic-acid (PLGA), or polyhydroxybutyrate (PHB), to tailor the degradation time (19, 20). Furthermore, it has also been shown that PCL scaffold supports cell growth in bone applications (17). As such, PCL is chosen to be the biomaterial used in this work.

PCL is a semicrystalline polymer that falls into polyester family (17). It degrades by hydrolytic mechanism in vivo by surface erosion. The degradation products are then taken up by macrophages and degraded intracellularly (21). The degradation products include caprolactone and 6-hydroxyhexanoic acid (22).

PCL powder is used throughout this research and was purchased from Solvay Interox Pte Ltd., UK, under the brand name CAPA® 6501. The powder is irregular in shape as shown in Fig. 1 with molecular weight of 50,000. The material specifications of PCL are summarized in Table 1. The PCL powder is used as received, to be sintered using selective laser sintering (SLS), a form of RP technology. The suitable sintering parameters for processing PCL powder using SLS are listed in Table 2.

Fig. 1. SEM micrograph of PCL powder.

Table 1
PCL material specifications

Particle shape	Irregular
Particle of size <100 μm	99%
Powder density (g/cm^3)	1.15
Glass transition temperature (°C)	−60
Melting temperature (°C)	60

Table 2
SLS parameters for sintering PCL powder

Laser power	3 W
Laser scan speed	3,810 mm/s (150 in./s)
Part bed temperature	40°C

3. Methods

Even with the incorporation of RP systems, designing customizable scaffolds with desired internal and external architecture could be tedious and time consuming. Therefore, a system, which has a wide range of designs to choose from and can generate and customize the scaffold according to patients' specific needs, is required. This is where the CASTS, which was built in previous works by Cheah (23), contributes to satisfy these requirements.

The platform of CASTS is Pro/ENGINEER, a three-dimensional CAD/computer-aided manufacturing (CAM) system. CASTS possesses three modules, namely, the input module, the designer's toolbox, and the output module.

3.1. Input Module

Besides the standard hardware and software for computer systems, CASTS requires imaging software, such as Materialise's interactive medical image control system (MIMICS), for its input module. This imaging software is for converting raw patient data from magnetic resonance imaging (MRI) or computer tomography (CT) scan to surface files of initial graphics exchange specification (IGES) format, which is a neutral data format for file transfer to dissimilar system.

3.2. Designer's Toolbox

The designer's toolbox is to provide the CASTS user with a variety of useful tools to customize the scaffold according to patient's specific needs. The tools include the following.

1. Parametric library, where the user can select a basic cellular unit/packing configuration, which will be the basis of the scaffold block. Besides that, the user can also decide whether the presence of triangulation or node is required.
2. Sizing routines, where the user can specify the dimension of each cellular unit and also the overall dimension of the scaffold block in the x, y, and z directions.
3. Automated algorithm to link the choices/values inputted by the user (from tools 1 and 2) and the CAD file of the scaffold. Besides generating the scaffold according to user's specified size, necessary outputs, such as porosity, pore size, and surface-to-volume ratio (scaffold internal architecture data), are also generated.
4. Scaffold internal architecture data to allow the user to obtain critical information about the scaffold generated, such as porosity, pore size, and surface-to-volume ratio. These values are calculated based on the dimensions inputted by the user in tool 2.

3.3. Output Module

The output of the designer's tool module is a scaffold block in the PRT (24) file format. To customize the scaffold to patient's specific shape, the surface files of IGES format from the input module are then used. After the customization is done, the part file must be converted into STL file format for scaffold fabrication using SLS. The STL file format is basically a series of triangular facets, each of which is represented by three vertices and a unit normal vector.

CASTS system consists of polyhedral unit cells that can be assembled together to form a scaffold block. The polyhedra used in CASTS are convex polyhedra, whose shapes are not too complex in order to be easily modeled in CAD environment. The assembly process was done by repeating the unit cells and mating their faces with the faces of their adjacent unit cells. Some of the polyhedral unit cells with the resulted scaffold blocks are shown in Fig. 2.

The existing algorithm of CASTS can only allow scaffold designer to create scaffolds with uniform pore size or discrete pore size gradients. Discrete gradients have distinct weaknesses, especially at the interface between two regions with different pore sizes, which do not mimic the native tissues with their compatible and strong bond at tissue interfaces. Therefore, to overcome the weak interface region, a new and novel method to obtain scaffolds with continuous pore size gradient is proposed.

The continuous gradient can be either in a radial or linear (axial) direction or both. Radial pore size gradient is useful in the case of regenerating a section of long bones, such as femur, which consist of cortical part in the outer radius and cancellous part in the inner radius. On the other hand, the gradient in linear direction can be used in generating scaffolds for tissue interfaces, such as the bone–cartilage interface.

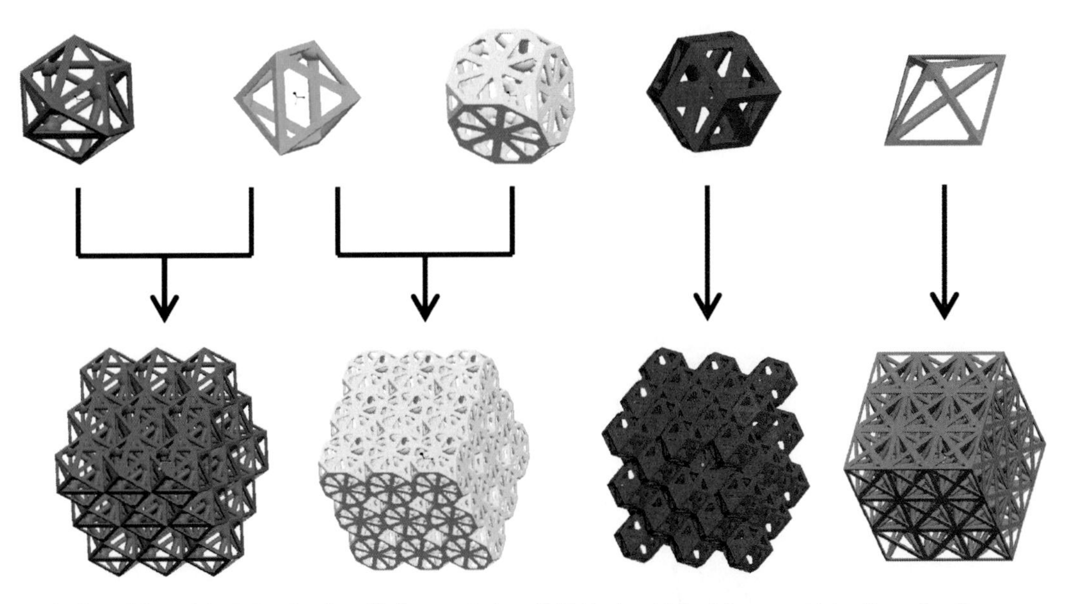

Fig. 2. Five of the polyhedral unit cells with the resulted scaffold blocks obtained from combining the unit cells.

The structural gradients of a long bone, which are the cortical and the cancellous, occur radially, where the cortical is located at the outer ring and the cancellous at the inner ring.

The methodology to achieve the porosity in radial direction is to assemble the pores in the arrangement of concentric rings in a cylinder (refer to Fig. 3a). When pores with equal sizes are arranged as a ring and their centers are connected together, they form a regular polygon (N-sided polygon), as seen in Fig. 3b (six-sided polygon). The regular polygon has N vertices, where N is the number of pores at that particular ring. Four main parameters for the generation of such structure are the pore size, radial distance, radian increment, and the number of pores at a particular ring.

The formulae for the parameters of the ring structure are explained as follows.

A. *Radial distance of the pores.*

The tabulation of the radial distance is important to ensure that the pores at a ring do not overlap with the pores at the neighboring rings. The radial distance, R_n, is measured from the center of the pore of the first ring to the center of the pore at the current (nth) ring (as shown in Fig. 4).

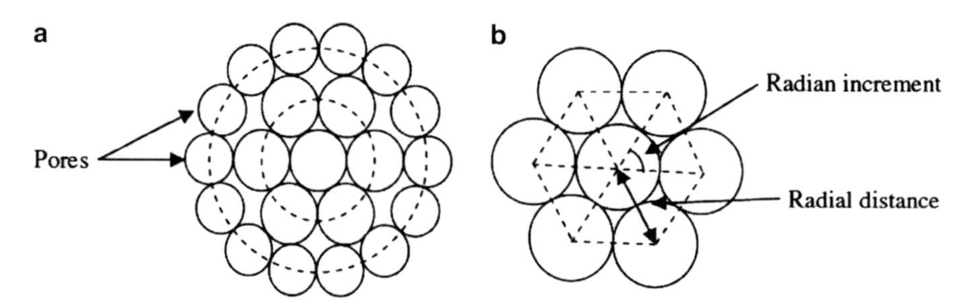

Fig. 3. Pores assembled in concentric rings of a cylinder.

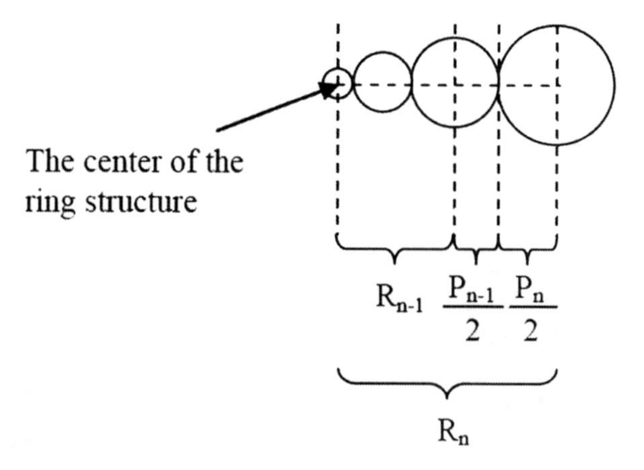

Fig. 4. Radial distance of a pore.

$$R_n = R_{n-1} + \frac{P_{n-1}}{2} + \frac{P_n}{2}, \tag{1}$$

where R_n = radial distance of pores at *n*th ring, R_{n-1} = radial distance of pores at *(n−1)*th ring, P_{n-1} = diameter of pores at *(n−1)*th ring, and P_n = diameter of pores at *n*th ring.

B. *Radian increment.*

Trigonometry rules can be applied to calculate the radian increment θ_n (refer to Fig. 5). In this case, the position of each pore is specified in polar coordinates.

$$\sin\left(\frac{\theta_n}{2}\right) = \frac{P_n}{2R_n}$$

$$\theta_n = 2\arcsin\left(\frac{P_n}{2R_n}\right), \tag{2}$$

where θ_n = radian increment of pore n.

C. *Number of pores at nth ring.*

When the angle in a circle (2π) is divided by the radian increment (θ_n), the predicted number of pores might have the excess of the radian increment, and this excess creates the presence of voids (refer to Fig. 6). Therefore, to avoid the generation of voids, the

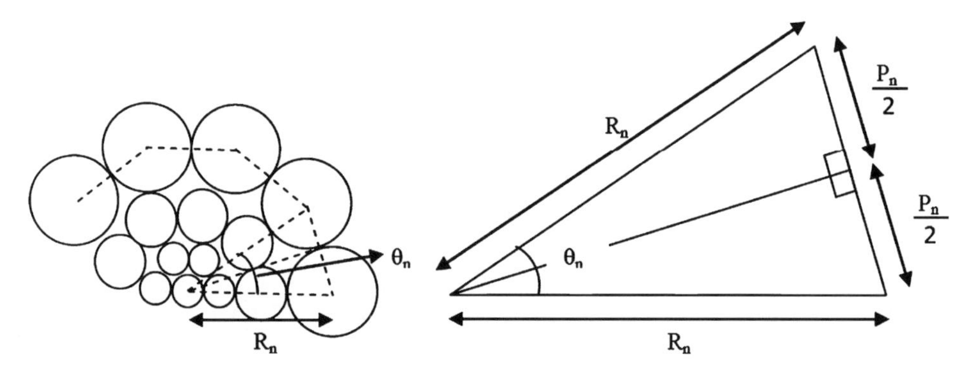

Fig. 5. Radian increment.

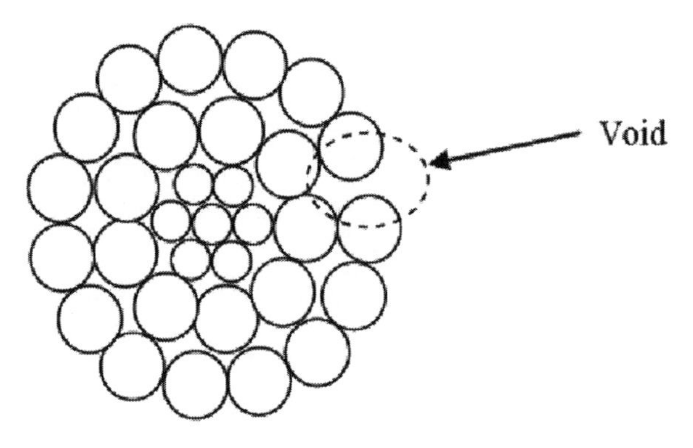

Fig. 6. Presence of void in the ring structure.

calculated number of pores (N) is rounded up or down, depending on the needed application. When N is rounded down, it results in increasing pore size toward the outer radius while rounding N up results in decreasing pore size.

For increasing pore size in the outward radial direction:

$$N_{\mathrm{n}} = \left\lfloor \frac{2\pi}{\theta_{\mathrm{n}}} \right\rfloor, \tag{3}$$

where $N =$ the number of pores at nth ring.

For decreasing pore size in the outward radial direction:

$$N_{\mathrm{n}} = \left\lceil \frac{2\pi}{\theta_{\mathrm{n}}} \right\rceil. \tag{4}$$

D. *Number of layers at nth ring in the cylinder.*

The number of layers at nth ring in the cylinder can be calculated by rounding up the value obtained from dividing the cylinder height with the pore size.

$$L_{\mathrm{n}} = \left\lceil \frac{h}{P_{\mathrm{n}}} \right\rceil, \tag{5}$$

where $L_{\mathrm{n}} =$ the number of layers in the cylinder at nth ring in the cylinder and $h =$ the height of the cylinder.

The isometric and plan views of the assembled pores are presented in Fig. 7, which shows decreasing pore sizes toward the outer radius. Each sphere represents a pore, whose radius is a function of the cylinder radius.

The pore size distribution across the ring structure for radially decreasing pore diameter is shown in Fig. 8. It is shown that the

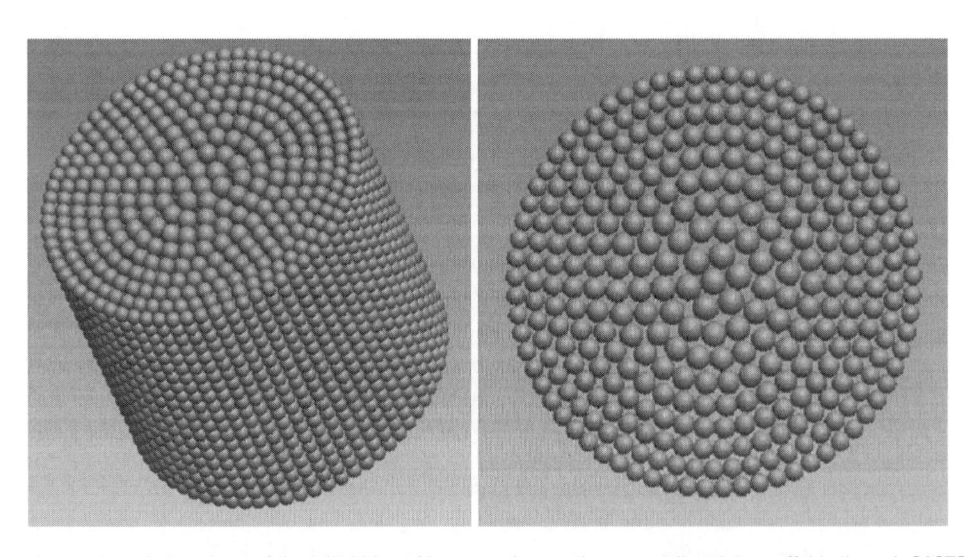

Fig. 7. Isometric and plan views of the initial idea of incorporating continuous gradient into scaffolds through CASTS.

pore size gradually decreases from the centre of the ring structure (pore size = 1 mm) to the perimeter (radial distance = 15.74 mm, pore size = 0.75 mm).

The approximate porosity values are calculated for four types of unit cells as an example and plotted against the radial distance in Fig. 9. In this case, the spheres' diameter is taken to be the polyhedral unit cells' pore size. The porosity calculation is based on the mathematical formulas described in Cheah (2004) (23). The strut size is kept at a constant value of 0.3 mm but with decreasing pore size, according to Fig. 8, for all types of unit cells. Unit cell 1 is octahedron and tetrahedron configuration, 2 is cuboctahedron and octahedron, 3 is triangular prism, and 4 is square prism. The porosity values are shown to decrease with radial distance. This is so, as when the pore size decreases and the strut size is constant, the empty space in the scaffold will be less. Thus, the porosity decreases with decreasing pore size.

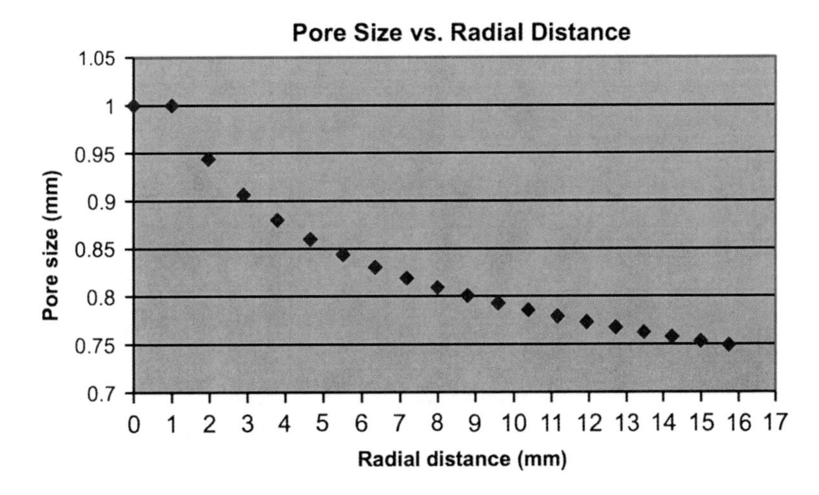

Fig. 8. The pore size distribution with respect to the radial distance in the ring structure.

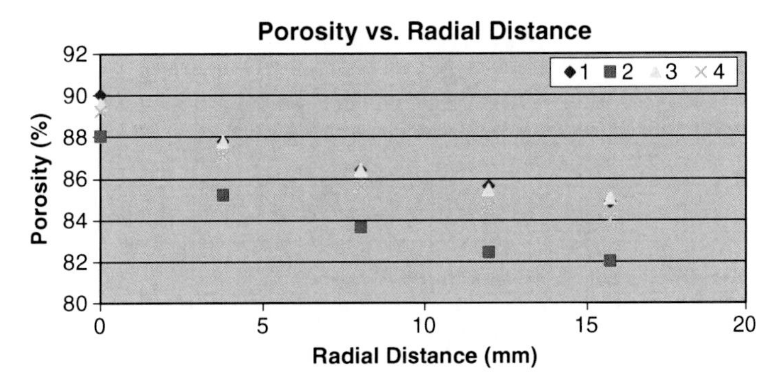

Fig. 9. The porosity distribution with respect to the radial distance in the ring structure.

4. Conclusion

The current ring structure acts as a basis for further development to be implemented with polyhedral-based scaffold structure. An improvement can be made on the algorithm to control the gradient.

In the current ring structure, the gradient is purely controlled by the mathematical relationship between the first pore size (the one at the center of the ring structure), radial distance, and radian increment (refer to Fig. 10a). The term gradient refers to the change in the scaffold structures, such as pore size and porosity, with respect to their position in the scaffold. However, the actual scaffold's gradient is to be adapted according to the mechanical (such as strength and stiffness) and biological (such as porosity, pore size, and pore interconnection) function of the regenerated tissue. For a clearer picture, the possible method is shown in Fig. 10b, which is to define the ring radial distance, where all the pores are of size P_1, and the region, where the pores are of size P_2. Between the two regions, there is a transition region, where the pore size is graded continuously from P_1 to P_2 according to the mathematical relations. The pore size and porosity in P_1 and P_2 regions should be adjusted according to what is required by the cells seeded on the scaffold, mechanically and biologically.

The approach to relate the scaffold structural gradient to the biological and mechanical requirements for the FGS can be broken into two parts. The first is to obtain empirical relationships between the stiffness and strength of the FGS with the strut dimensions, pore size, porosity, and the pore geometry, i.e., the mechanical requirements are translated into geometrical function of strut dimensions, pore size, porosity, and pore geometry. The second part is the biological requirements that are to be correlated to the geometrical definition in terms of strut dimensions, pore size, porosity, and pore geometry. Since both mechanical requirements and biological requirements are accounted for in the same domain, establishing the suitable FGS design that satisfies both requirements can then be achieved.

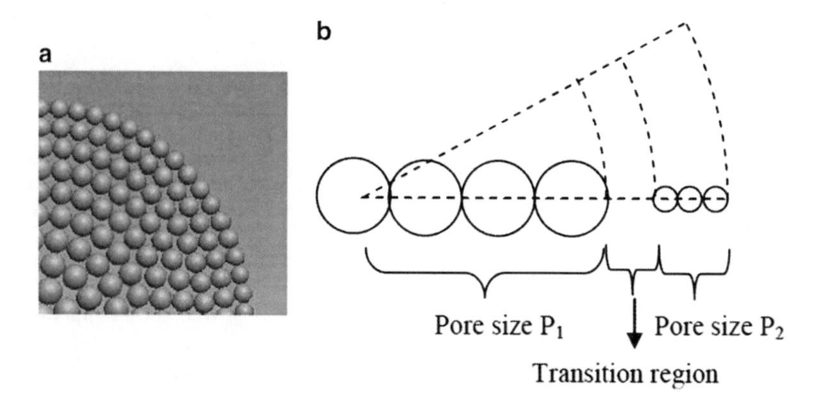

Fig. 10. (**a**) The current ring structure, (**b**) proposed idea for controlling the pore size gradient.

References

1. Langer R, Vacanti JP (1993) Tissue engineering. Science 260(5110):920–926
2. Risbud M (2001) Tissue engineering: Implications in the treatment of organ and tissue defects. Biogerontology 2(2):117–125
3. Tan KH, Chua CK, Leong KF, Cheah CM, Gui WS, Tan WS, Wiria FE (2005) Selective laser sintering of biocompatible polymers for applications in tissue engineering. Bio-Med Mater Eng 15(1–2):113–124
4. Vacanti CA (2006) History of tissue engineering and a glimpse into its future. Tissue Eng 12 (5):1137–1142
5. Gross KA, Rodríguez-Lorenzo LM (2004) Biodegradable composite scaffolds with an interconnected spherical network for bone tissue engineering. Biomaterials 25(20):4955–4962
6. Zeltinger J, Sherwood JK, Graham DA, Müeller R, Griffith LG (2001) Effect of pore size and void fraction on cellular adhesion, proliferation, and matrix deposition. Tissue Eng 7(5):557–572
7. Yang SF, Leong KF, Du ZH, Chua CK (2001) The design of scaffolds for use in tissue engineering. Part 1. Traditional factors. Tissue Eng 7(6):679–689
8. Yang SF, Leong KF, Du ZH, Chua CK (2002) The design of scaffolds for use in tissue engineering. Part II. Rapid prototyping techniques. Tissue Eng 8(1):1–11
9. Parenteau NL, Hardin-Young J, Ross RN (2000) Skin. In: Lanza RP, Langer R, Vacanti J (eds) Principles of tissue engineering. Academic, San Diego, USA, pp 879–890
10. Martin RB, Burr DB, Sharkey NA (1998) Skeletal tissue mechanics, Springer-Verlag, New York, xiv, p 392
11. Sherwood JK, Riley SL, Palazzolo R, Brown SC, Monkhouse DC, Coates M, Griffith LG, Landeen LK, Ratcliffe A (2002) A three-dimensional osteochondral composite scaffold for articular cartilage repair. Biomaterials 23 (24):4739–4751
12. Naing MW, Chua CK, Leong KF, Wang Y (2005) Fabrication of customised scaffolds using computer-aided design and rapid prototyping techniques. Rapid Prototyping J 11 (4):249–259
13. Ford RG, Miyamoto Y, Nogata F (1999) Lessons from nature. In: Miyamoto Y, Kaysser WA, Rabin BH, Kawasaki A, Ford RG (eds) Functionally graded materials: design, processing and applications. Kluwer, Boston, pp 7–28
14. Woodfield TBF, Malda J, de Wijn J, Peters F, Riesle J, van Blitterswijk CA (2004) Design of porous scaffolds for cartilage tissue engineering using a three-dimensional fiber-deposition technique. Biomaterials 25(18):4149–4161
15. Guo XE (2001) Mechanical properties of cortical bone and cancellous bone tissue. In: Cowin SC (ed) Bone mechanics handbook. CRC, Boca Raton, FL, pp 10.1–10.23
16. Kreeger PK, Shea LD (2002) Scaffolds for directing cellular responses and tissue formation. In: Dillow AK, Lowman AM (eds) Biomimetic materials and design. Marcel Dekker, New York, NY, pp 283–309
17. Ciapetti G, Ambrosio L, Savarino L, Granchi D, Cenni E, Baldini N, Pagani S, Guizzardi S, Causa F, Giunti A (2003) Osteoblast growth and function in porous poly [var epsilon]-caprolactone matrices for bone repair: a preliminary study. Biomaterials 24(21):3815–3824
18. Gunatillake PA, Adhikari R (2003) Biodegradable synthetic polymers for tissue engineering. Eur Cell Mater 5:1–16
19. Edlund U, Albertsson AC (2001) Degradable polymer microspheres for controlled drug delivery. In: Albertsson AC (ed) Degradable aliphatic polyesters. Springer, Berlin, pp 67–112
20. Stridsberg KM, Ryner M, Albertsson AC (2001) Controlled ring-opening polymerization: polymers with designed macromolecular architecture. In: Albertsson AC (ed) Degradable aliphatic polyesters. Springer, Berlin, pp 41–65
21. Engelberg I, Kohn J (1991) Physico-mechanical properties of degradable polymers used in medical applications: A comparative study. Biomaterials 12(3):292–304
22. Hakkarainen M, Albertsson AC (2002) Heterogeneous biodegradation of polycaprolactone – Low molecular weight products and surface changes. Macromol Chem Physic 203 (10–11):1357–1363
23. Cheah CM, Chua CK, Leong KF, Cheong CH, Naing MW (2004) An automatic algorithm for generating complex ployhedral scaffolds for tissue engineering. Tissue Eng 10(3–4):595–610
24. Hollister SJ, Lin CY, Saito E, Lin CY, Schek RD, Taboas JM, Williams JM, Partee B, Flanagan CL, Diggs A, Wilke EN, Van Lenthe GH, Muller R, Wirtz T, Das S, Feinberg SE, Krebsbach PH (2005) Engineering craniofacial scaffolds. Orthod Craniofac Res 8(3): 162–173

Chapter 8

Computer-Designed Nano-Fibrous Scaffolds

Laura A. Smith and Peter X. Ma

Abstract

Nano-fibrous scaffolding mimics aspects of the extracellular matrix to improve cell function and tissue formation. Although several methods exist to fabricate nano-fibrous scaffolds, the combination of phase separation with reverse solid freeform fabrication (SFF) allows for scaffolds with features at three different orders of magnitude to be formed, which is not easily achieved with other nano-fiber fabrication methods. This technique allows for the external shape and internal pore structure to be precisely controlled in an easily repeatable manner, while the nano-fibrous wall architecture facilitates cellular attachment, proliferation, and differentiation of the cells. In this chapter, we examine the fabrication of computer-designed nano-fibrous scaffolds utilizing thermally induced phase separation and reverse SFF, and the benefits of such scaffolds over more traditional tissue engineering scaffolds on cellular function and tissue regeneration.

Key words: Nano-fibers, Bone tissue engineering, Scaffolds, Solid freeform fabrication, Phase separation

1. Introduction

Tissue engineering is an interdisciplinary field, which aims to restore, maintain, or replace the function of failing tissue (1, 2). To achieve this three approaches have emerged: implantations of isolated cells or cell substitutes to replace the needed cell function; targeted delivery of tissue inducing substances, such as growth factors; and the use of three-dimensional scaffolding to support cell growth. For large defects, the use of scaffolding to direct cell growth is required.

When designing scaffolds for tissue engineering applications many factors need to be considered. They include the scaffold morphologies (porosity, pore size, and interpore connectivity), mechanical properties, degradation, cellular interaction, and neo tissue genesis (2). As our ability to precisely control scaffolding structures and properties has improved, so have the variety and the quality of the tissues engineered.

Michael A.K. Liebschner (ed.), *Computer-Aided Tissue Engineering*, Methods in Molecular Biology, vol. 868,
DOI 10.1007/978-1-61779-764-4_8,

Recent efforts in scaffold design have focused on emulating the extracellular matrix of the native tissue. In bone tissue, type I collagen is a major component of the organic extracellular matrix. During bone development, type I collagen forms the base of the extracellular matrix on to which many other proteins and molecules are adsorbed or secreted as the tissue matures (3). Type I collagen consists of three collagen polypeptide chains wound around one another to form a rope-like superhelix which then assemble into higher-order collagen fibrils ranging in diameter from 50 to 500 nm. Many biomimic scaffolds attempt to mimic the size scale of these fibrils in order to enhance cellular interactions and tissue development (2).

In this chapter, we examine the use of thermally induced phase separation (TIPS) in combination with computer-designed reverse solid freeform fabrication (SFF) and poragen incorporation to fabricate scaffolds with complex architectures on multiple size scales and the benefits of such design on cell behavior and tissue formation.

2. Thermally Induced Phase Separation

TIPS has been used to create porous synthetic polymer scaffolds for over a decade (4–6). In this process, a temperature reduction results in the separation of the polymer solution into two phases: one having a low polymer concentration (polymer-lean phase); and one having a high polymer concentration (polymer-rich phase). Once phase separation has occurred, the solvent is removed by extraction, evaporation, or sublimation, while the polymer-rich phase solidifies to form a network of pores called a polymer foam. The foams can possess better mechanical properties (moduli 20 times greater) than scaffolds with the same porosity produced by salt leaching (7). However, the pore structure of these foams is typically not well controlled and the size may not be ideal for certain tissue engineering applications.

The structure of the polymer foams can be altered through varying the polymer, solvent, polymer concentration, and the phase separation temperature. This process can be used to fabricate several types of structures, including microtubules (8), ladders (9), and nano-fibers (10). However, bicontinuous nano-fibrous networks can only be formed from biodegradable polymers under certain conditions (10). The selections of solvent and phase separation temperature are vital to nano-fibrous foam formation. The nanofibers formed by this process have been found to range in diameter from 50 to 500 nm (same range as native type I collagen) and can create a material with a porosity as high as 98% (10) (Fig. 1).

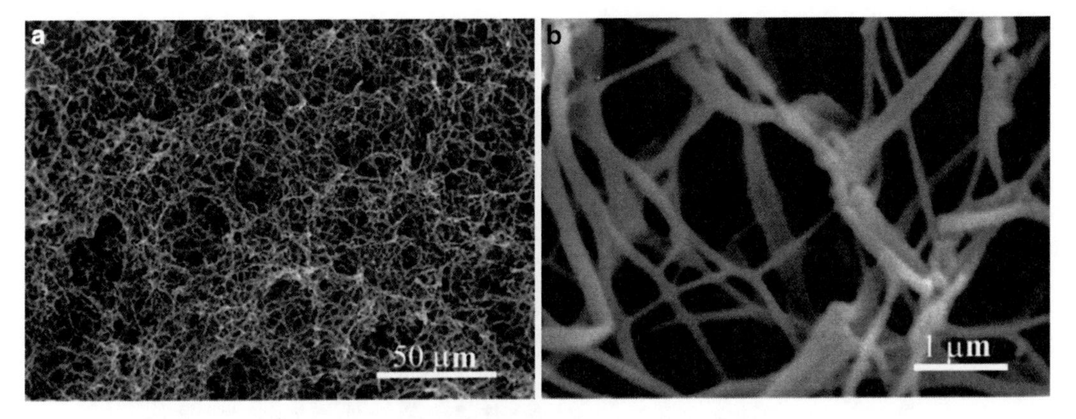

Fig. 1. SEM micrographs of poly(L-latic acid) fibrous matrix prepared from 2.5% (wt/v) PLLA/THF solution at a gelation temperature of 8°C: (**a**) °—500; (**b**) °—20 K. (From Ma and Zhang (10), copyright 1999 and reprinted with permission of Wiley & Sons, Inc.).

Unlike other methods of generating nano-fibers, TIPS does not require the expensive equipment or complicated synthesis schemes. TIPS can easily be adapted from a small laboratory setting to a large-scale industrial production. TIPS can also be used in combination with other processing techniques to fabricate scaffolds with complex three-dimensional structures, well-defined pore morphologies, and biological surface chemistries. For instance, the use of SSF and porogen-leaching techniques have been used to increase scaffold porosity and create intricate internal pore structures (11, 12). The surface chemistry of these scaffolds can be altered to improve cellular interactions and function through the addition of biological molecules, such as gelatin (13–15) or the incorporation of minerals such as hydroxyapatite (7, 16). The nano-fibrous scaffolds generated using TIPS are also compatible with micro-/nanosphere drug delivery strategies to improve tissue formation (17, 18). The flexibility of TIPS to work in combination with many other technologies is what enables it to meet the requirements for a diverse array of tissue engineering applications.

3. Reverse Solid Freeform Fabrication

Solid freeform fabrication is one of the scaffold fabrication techniques compatible with TIPS (12). In this method, a model of the scaffold is designed using computer software. Depending on the software package, the scaffold morphology can be based on geometric shapes or patient scans, such as computer tomography or magnetic resonance imaging (12, 19, 20) (Fig. 2). The model is then converted to stereolithography data and uploaded to the computer-aided manufacturing equipment for fabrication.

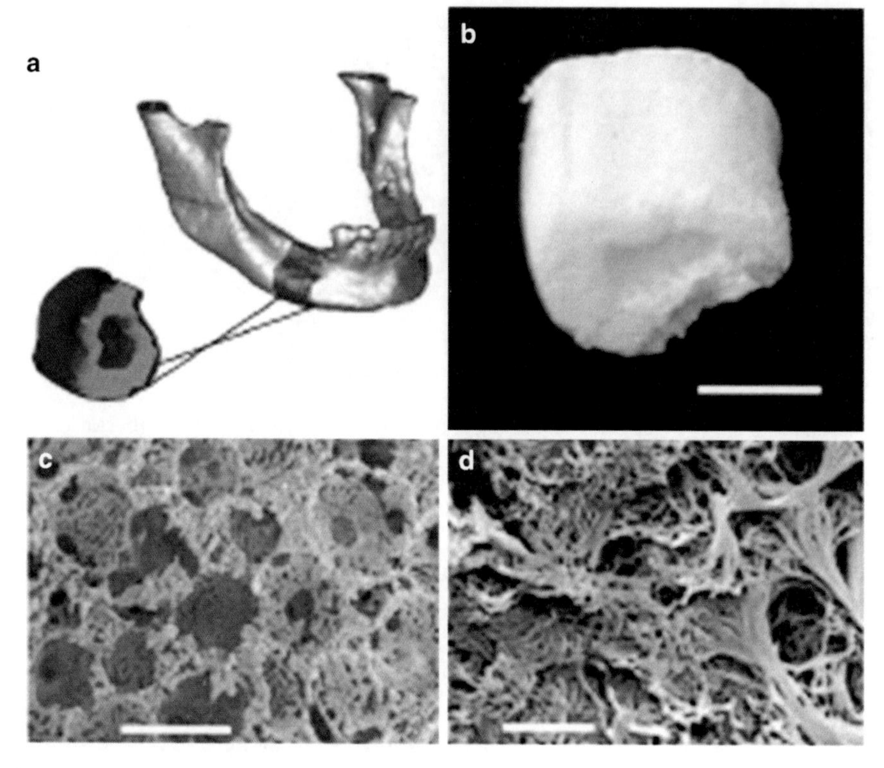

Fig. 2. Scaffolds created from three-dimensional reconstructions of CT-scans or histological sections. (**a**) human mandible reconstruction from CT-scans. Purple segment shows the reversed image of the bone fragment to be engineered, (**b**) resulting NF scaffold of the mandible segment (scale bar; 10 mm), (**c**) SEM micrographs of the interconnected spherical pores within the mandible segment (scale bar; 500 mm), and (**d**) the NF pore morphology of a spherical pore (scale bar; 5 mm). (From Chen et al. (12), copyright 2006 and reprinted with the permission of Wiley & Sons, Inc.).

The fabrication unit creates a negative mold of the scaffold. Once the support material is removed, the polymer solution is cast into the mold and phase separated. After completing the TIPS and solvent removal, the mold is dissolved, freeing the nano-fibrous scaffold. Beyond the incorporation of TIPS, this computer-aided design and manufacturing process (reverse SFF) allows for the use of materials that require solvents, which would dissolve the printer heads on most direct SFF units. Thus, it greatly expands the types of polymers and ceramics that can be utilized in computer-designed scaffold fabrication. Porogen incorporation and leaching techniques, such as for paraffin spheres (Fig. 2c) (12), sugar spheres (16), salt or sugar particles (21–24), and solidified solvent crystals (12), can be employed to provide additional porosity to the scaffold. This leads to better cellular distribution throughout the scaffold, tissue integration, improved mass transport, and reduced amounts of breakdown products as the scaffold degrades. It also allows for reproducible generation of both the external shape and internal pore morphology at the macroscale, which is not easily achievable with other methods. However, the accuracy and complexity of these scaffolds are often limited by the pixel resolution of the fabrication unit and the precision of positioning the nozzle of the fabrication unit.

4. Benefits of Computer-Designed Nano-Fibrous Scaffolds

Through incorporating SFF with TIPS intricate architectures with three different orders of magnitude can be formed: on macro (millimeter or larger-sized external shapes), micro (micrometer sized internal pores), and nano (nanometer-sized fibers) scales. At the macro level, this technique allows for computer-aided design of patient specific scaffolds, which will tightly fit into the patient defects without damaging the internal scaffolding structures or leaving gaps between the scaffold and the host tissue, allowing for the best integration of new tissue and reducing scar tissue formation. At the micro level, the pores allow for cell penetration throughout the scaffold, enabling even tissue formation in the entire defect. The micropores also enhance mass transport and blood vessel formation, improving cell and tissue function due to improved nutrient supply and metabolic waste removal. At the nano level, the fibrillar structure has been shown to influence cell attachment, proliferation, migration and differentiation of numerous cell types (25), improving cell function and tissue formation within the scaffold.

4.1. Biological Effects of Nano-Fibers

Although nano-fibers have an effect on the behavior of numerous cell types, we focus the discussion on bone formation due to space limitation. Cell attachment is an important first step in the tissue formation process since most cells require anchorage to survive, proliferate, and differentiate. Nano-fibrous materials have shown increased cellular attachment over more traditional scaffolding materials (26, 27). These nano-fibrous materials adsorb increased levels of attachment proteins (fibronectin, vitronectin, and laminin) compared to smooth materials (27, 28) and increased expression of integrins that are active in cellular attachment to these proteins have also been observed in cells cultured on nano-fibrous materials compared to smooth materials (26, 28). However, increase $\alpha 2$ and $\beta 1$ integrin expression was maintained on nano-fibrous scaffolds even after cellular collagen formation was blocked with inhibitors but not on the smooth material, (26) indicating that integrin binding could be occurring directly with the nano-fibers and not the adsorbed or secreted proteins on the material.

Another important aspect of cellular attachment that nano-fibers have an effect on is cellular morphology, which is thought to more closely resemble in vivo morphology on nano-fibrous materials than on smooth materials (26, 28–31). After 24 h of culture,

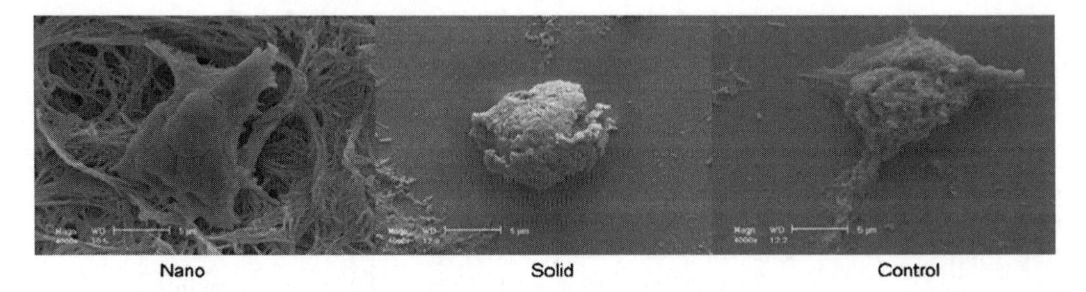

Fig. 3. SEM micrographs of mouse embryonic stem cells after 12 h under differentiation conditions on nano-fibrous matrix (nano), flat films (solid), gelatin-coated tissue culture plastic (control), scale bar = 5 μm. (From Smith et al. (28), copyright 2009 and reprinted with the permission of Mary Ann Liebert, Inc.).

the differences in morphology are striking with osteogenically differentiating mouse embryonic stem cells. The cells on the nano-fibrous material extend processes to the nano-fibers, while the cells on the smooth material do not spread well or extend processes and the cells on the gelatin-coated tissue culture plastic spread well but do not extend processes (Fig. 3). Mouse osteoprogenitors cultured on nano-fibrous materials have also exhibited fewer stress fibers and focal adhesions than cells on smooth materials (29). Alterations to the surface chemistry of the nano-fibrous materials did not alter the stress fiber or focal adhesion formation in these cells indicating the cells are potentially interacting directly with the nano-fibers and not adsorbed proteins. Additionally, these morphology changes between nano-fibrous and smooth materials have been linked to RhoA and Rac expression (29, 32), regulators of cytoskeleton assembly known to affect cellular functions in several cell types.

Once cells attach, they proliferate to fill the available space in the scaffold. Several osteogenic cell types have shown increased proliferation on nano-fibrous material than on smooth material (12, 31). More rapid proliferation of the desired cell type within the scaffold in vivo would lead to a reduction in scar tissue formation and increased formation of the desired tissue.

Beyond their role in attachment, integrins play a vital role in cellular differentiation. Paxillin and focal adhesion kinase phophorylation, components in intergin-activated differentiation pathways, were observed at increased levels in osteoblasts cultured on nano-fibrous materials compared to smooth materials (26). In addition, blocking α2 and α5 integrin activity in osteogenically differentiating embryonic stem cells decreased both the mesodermal and osteogenic differentiation of the cells, while only blocking α2 integrin had a significant effect on the mesodermal and osteogenic differentiation of osteogenically differentiating embryonic stem cells on the smooth materials (28).

As indicated by the integrin data on differentiation, the nano-fibrous and smooth materials create substantially different microenvironments for the cells, which alter their differentiation.

This has been dramatically illustrated with embryonic stem cells, which under osteogenic conditions express five times less nestin, a neural marker, and 50% more Runx2, a mesenchymal marker, when culture on nano-fibrous material compared to smooth material (28).

More committed cell types have also shown increased commitment to the osteogenic lineage on nano-fibrous materials than smooth materials. For instance, increased expression of alkaline phosphatase, bone sialoprotein, and osteocalcin have been observed in cells cultured on nano-fibrous materials compared to cells cultured on smooth materials after three or more days of osetogenic culture (12, 26, 29, 33). The increased expression of bone sialoprotein on nano-fibrous material compared to smooth material has been linked to the effect of nano-fibers on the RhoA Rock signaling pathway (29).

With increasing culture time, cells cultured on nano-fibrous materials appear healthier and form better tissue compared to cells cultured on smooth-walled materials (Fig. 4) (12). With a combination of in vitro and in vivo culture, nano-fibrous scaffolds have resulted in vascularized and mineralized tissue with embedded osteocyte-like cells (34). Cell cultured on the nano-fibrous materials have also been shown to secrete extracellular matrix components like collagen type I evenly throughout the matrix (34, 35). As the secreted extracellular matrix matures, it begins to mineralize. Three-dimensional nano-fibrous scaffolds have been found to contain up to 13 times more calcium than their smooth-walled counterparts (12, 26). The mineral content of the computer-designed (reverse SFF) nano-fibrous scaffolds was found to be more evenly spread throughout the scaffolds than on the smooth-walled scaffolds, where it was located principally near the exterior of the scaffold (Fig. 4d, e).

In a critical size calvarial bone defect model, nano-fibrous scaffolds lead to increased bone formation compared to smooth-walled scaffolds 8 weeks after implantation (36). Similar to the in vitro data, the nano-fibrous scaffolds contained collagen and mineral deposits throughout the scaffold while the smooth-walled scaffold contained collagen and mineral at its periphery (36). The tissue within the nano-fibrous scaffolds appears more developed than the tissue within the smooth scaffolds (Fig. 5) (36).

5. Conclusion

The increasing understanding of regenerative biology and the development of advanced scaffold structures have led to the rapid expansion and progress in tissue engineering over the past 15 years. However, the engineered tissues at the present time are still not of

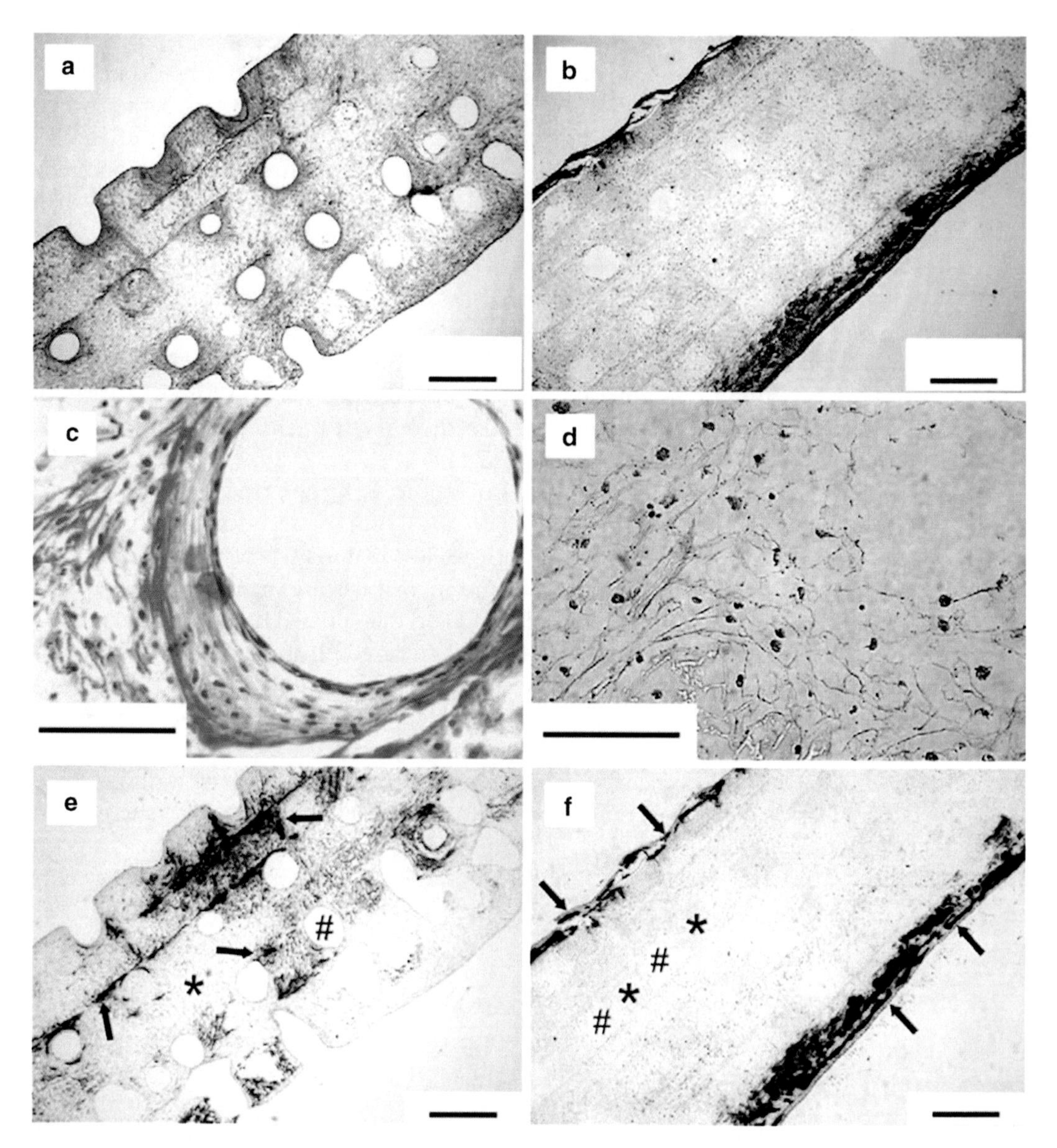

Fig. 4. In vitro response of NF and SW PLLA scaffolds after seeding with MC3T3-E1 osteoblasts and cultured under differentiation conditions for 6 weeks. Shown are histological sections of representative areas within the scaffold. H&E staining showing (**a**) overview of an NF scaffold, (**b**) overview of a SW scaffold, (**c**) center region of an NF scaffold, and (**d**) center region of an SW scaffold. Von Kossa's silver nitrate staining showing (**e**) center region of an NF scaffold, and (**f**) center region of an SW scaffold. Scale bars of (**a**), (**b**), (**e**), and (**f**); 500 mm. Scale bars of (**c**) and (**d**); 100 mm. *Asterisk* denotes the PLLA scaffold, *hash* a scaffold pore. *Arrows* in von Kossa-stained sections denote mineralization. NF scaffolds retained a small amount of the histological dyes and therefore are more visible than the SW scaffolds in the pictures. (From Chen et al. (12), copyright 2006 and reprinted with the permission of Wiley & Sons, Inc.).

the quality for the aimed clinical applications. The combination of advanced scaffolding fabrication technologies, such as TIPS and SFF, allowing for the fabrication of complex scaffolds with controlled structural features at multiple size scales, is an important

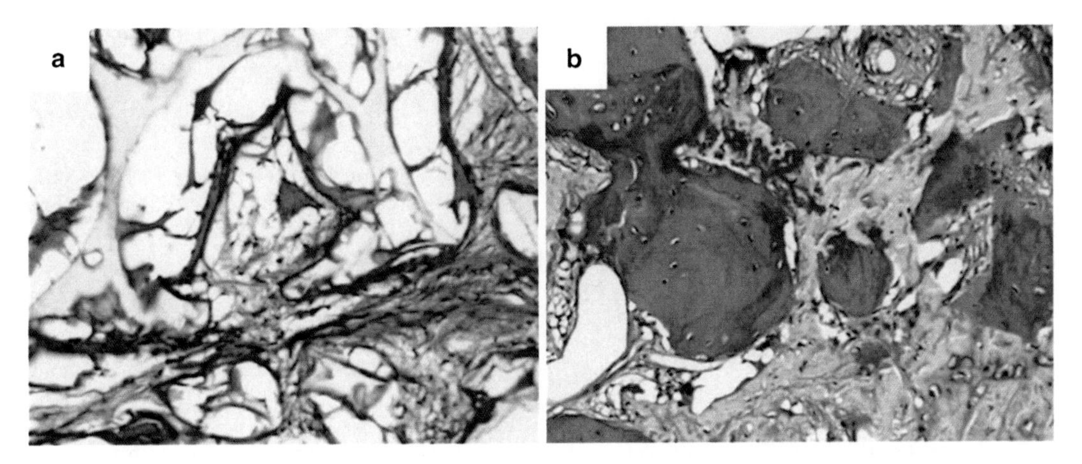

Fig. 5. Histological images of H&E stained solid-walled scaffolds (**a**) and nano-fibrous scaffolds (**b**) implanted in calvarial defects for 8 weeks. Scale bar = 100 μm. (From Woo et al. (36), copyright 2009 and reprinted with the permission of Mary Ann Liebert, Inc.).

advance. This strategy utilizes the complimentary advantages of the two techniques, while attempting to ameliorate their drawbacks. TIPS offers the ability to produce nano-fibrous pore wall architecture which has been shown to improve cellular behavior and function over more traditional smooth-walled scaffolds, while SFF offers easily customizable control of external and internal scaffold shape. Both are compatible with many forms of scaffold modification so that scaffolds made utilizing TIPS and SFF can be further modified to meet the needs of various tissue engineering applications.

References

1. Langer R, Vacanti J (1993) Tissue engineering. Science 260:920–926
2. Ma PX (2008) Biomimetic materials for tissue engineering. Adv Drug Deliv Rev 60:184–189
3. Kadler K (2004) Matrix loading: assembly of extracellular matrix collagen fibrils during embryogenesis. Birth Defects Res C Embryo Today 72:1–11
4. Nam YS, Park TG (1999) Porous biodegradable polymeric scaffolds prepared by thermally induced phase separation. J Biomed Mater Res 47:8–17
5. Lee SH, Kim BS, Kim SH, Kang SW, Kim YH (2004) Thermally produced biodegradable scaffolds for cartilage tissue engineering. Macromol Biosci 4:802–810
6. Zhang RY, Ma PX (1999) Poly(alpha-hydroxyl acids) hydroxyapatite porous composites for bone-tissue engineering. I. Preparation and morphology. J Biomed Mater Res 44:446–455
7. Ma PX, Zhang RY, Xiao G, Franceschi R (2001) Engineering new bone tissue in vitro on highly porous poly(alpha-hydroxyl acids)/hydroxyapatite composite scaffolds. J Biomed Mater Res 54:284–293
8. Ma PX, Zhang RY (2001) Microtubular architecture of biodegradable polymer scaffolds. J Biomed Mater Res 56:469–477
9. Ma PX, Zhang RY (1999) Porous poly(L-lactic acid)/apatite composites created by biomimetic process. J Biomed Mater Res 45:285–293
10. Ma PX, Zhang RY (1999) Synthetic nano-scale fibrous extracellular matrix. J Biomed Mater Res 46:60–72
11. Chen VJ, Ma PX (2004) Nano-fibrous poly(L-lactic acid) scaffolds with interconnected spherical macropores. Biomaterials 25:2065–2073
12. Chen VJ, Smith LA, Ma PX (2006) Bone regeneration on computer-designed nano-fibrous scaffolds. Biomaterials 27:3973–3979

13. Liu XH, Smith LA, Wei G, Won YJ, Ma PX (2005) Surface engineering of nano-fibrous poly(L-lactic acid) scaffolds via self-assembly technique for bone tissue engineering. J Biomed Nanotechnol 1:54–60
14. Liu XH, Won YJ, Ma PX (2005) Surface modification of interconnected porous scaffolds. J Biomed Mater Res A 74A:84–91
15. Liu XH, Won YJ, Ma PX (2006) Porogen-induced surface modification of nano-fibrous poly(L-lactic acid) scaffolds for tissue engineering. Biomaterials 27:3980–3987
16. Wei G, Ma PX (2006) Macroporous and nanofibrous polymer scaffolds and polymer/bone-like apatite composite scaffolds generated by sugar spheres. J Biomed Mater Res A 78:306–315
17. Wei G, Jin Q, Giannobile W, Ma PX (2007) The enhancement of osteogenesis by nano-fibrous scaffolds incorporating rhBMP-7 nanospheres. Biomaterials 28:2087–2096
18. Jin Q, Wei G, Lin Z et al (2008) Nanofibrous scaffolds incorporating PDGF-BB microspheres induce chemokine expression and tissue neogenesis in vivo. PLoS One 3:e1729
19. Lee M, Dunn J, Wu B (2005) Scaffold fabrication by indirect three-dimensional printing. Biomaterials 26:4281–4289
20. Lin C, Kikuchi N, Hollister S (2004) A novel method for biomaterial scaffold internal architecture design to match bone elastic properties with desired porosity. J Biomech 37:623–636
21. Ma PX, Langer R (1999) Fabrication of biodegradable polymer foams for cell transplantation and tissue engineering. In: Morgan J, Yarmush M (eds) Tissue engineering methods and protocols. Humana Press Inc, Totowa, NJ, pp 47–56
22. Zhang RY, Ma PX (2000) Synthetic nano-fibrillar extracellular matrices with predesigned macroporous architectures. J Biomed Mater Res 52:430–438
23. Taboas J, Maddox R, Krebsbach P, Hollister S (2003) Indirect solid free form fabrication of local and global porous, biomimetic and composite 3D polymer-ceramic scaffolds. Biomaterials 24:181–194
24. Mattioli-Belmonte M, Vozzi G, Kyriakidou K et al (2008) Rapid-prototyped and salt-leached PLGA scaffolds condition cell morpho-functional behavior. J Biomed Mater Res A 85: 466–476
25. Smith LA, Beck J, Ma PX (2007) Fabrication and tissue formation with nano-fibrous scaffolds. In: Kumar C (ed) Nanotechnologies for tissue, cell and organ engineering. Wiley-VCH, Weinheim, Germany
26. Woo KM, Jun JH, Chen VJ et al (2007) Nano-fibrous scaffolding promotes osteoblasts differentiation and biomeneralization. Biomaterials 28:335–343
27. Woo KM, Chen VJ, Ma PX (2003) Nano-fibrous scaffolding architecture selectively enhances protein adsorption contributing to cell attachment. J Biomed Mater Res A 67:531–537
28. Smith LA, Liu X, Hu J, Wang P, Ma P (2009) Enhancing the osteogenic differentiation of mouse embryonic stem cells by nanofibers. Tissue Eng 15:1855–1864
29. Hu J, Liu X, Ma PX (2008) Induction of osteoblast differentiation phenotype on poly(L-lactic acid) nanofibrous matrix. Biomaterials 29:3815–3821
30. Schindler M, Ahmed I, Kamal J et al (2005) A synthetic nanofibrillar matrix promotes in vivolike organization and morphogenesis for cells in culture. Biomaterials 26:5624–5631
31. Shih YV, Chen CN, Tsai SW, Wang YJ, Lee OK (2006) Growth of mesenchymal stem cells on electrospun type I collagen nanofibers. Stem Cells 24:2391–2397
32. Nur-E-Kamal A, Ahmed I, Kamal J, Schindler M, Meiners S (2005) Three dimensional nanofibrillar surfaces induces the activation of Rac. Biochem Biophys Res Commun 331:428–434
33. Li MY, Mondrinos MJ, Gandhi MR, Ko FK, Weiss AS, Lelkes PI (2005) Electrospun protein fibers as matrices for tissue engineering. Biomaterials 26:5999–6008
34. Shin M, Yoshimoto H, Vacanti JP (2004) In vivo bone tissue engineering using mesenchymal stem cells on a novel electrospun nanofibrous scaffold. Tissue Eng 10:33–41
35. Yoshimoto H, Shin YM, Terai H, Vacanti JP (2003) A biodegradable nanofiber scaffold by electrospinning and its potential for bone tissue engineering. Biomaterials 24:2077–2082
36. Woo KM, Chen VJ, Jung H et al (2009) Comparative evaluation of nano-fibrous scaffolding for bone regeneration in critical sized calvarial defects. Tissue Eng 15:2155–2162

Chapter 9

A Neural Network Technique for Remeshing of Bone Microstructure

Anath Fischer and Yaron Holdstein

Abstract

Today, there is major interest within the biomedical community in developing accurate noninvasive means for the evaluation of bone microstructure and bone quality. Recent improvements in 3D imaging technology, among them development of micro-CT and micro-MRI scanners, allow in-vivo 3D high-resolution scanning and reconstruction of large specimens or even whole bone models. Thus, the tendency today is to evaluate bone features using 3D assessment techniques rather than traditional 2D methods. For this purpose, high-quality meshing methods are required. However, the 3D meshes produced from current commercial systems usually are of low quality with respect to analysis and rapid prototyping. 3D model reconstruction of bone is difficult due to the complexity of bone microstructure. The small bone features lead to a great deal of neighborhood ambiguity near each vertex. The relatively new neural network method for mesh reconstruction has the potential to create or remesh 3D models accurately and quickly. A neural network (NN), which resembles an artificial intelligence (AI) algorithm, is a set of interconnected neurons, where each neuron is capable of making an autonomous arithmetic calculation. Moreover, each neuron is affected by its surrounding neurons through the structure of the network. This paper proposes an extension of the growing neural gas (GNN) neural network technique for remeshing a triangular manifold mesh that represents bone microstructure. This method has the advantage of reconstructing the surface of a genus-n freeform object without a priori knowledge regarding the original object, its topology, or its shape.

Key words: Self-organizing map neural network, Microstructure, Meshing, 3D model reconstruction, Neural network

1. Introduction

Metabolic bone diseases are at the forefront of scientific and biomedical research worldwide (1, 2). In metabolic bone diseases, such as osteoporosis that are marked by increased bone fragility, micro-architectural deterioration of bone tissue occurs, leading to micro-fractures (3, 4); therefore, early diagnosis is a key to intervention. Traditional techniques based on bone mineral density (BMD) tests provide 2D results at the bone macroscale level

Michael A.K. Liebschner (ed.), *Computer-Aided Tissue Engineering*, Methods in Molecular Biology, vol. 868,
DOI 10.1007/978-1-61779-764-4_9, © Springer Science+Business Media, LLC 2012

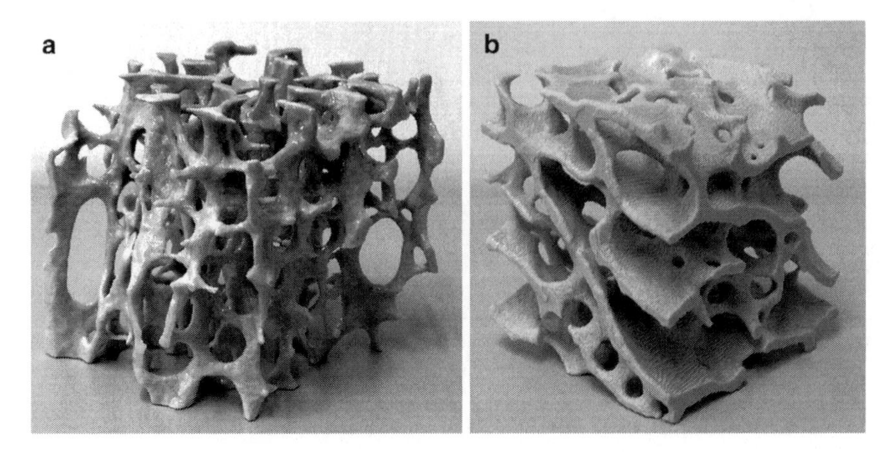

Fig. 1. Rapid prototyping models of bone microstructure: (**a**) specimen from lumbar spine and (**b**) specimen from femoral bone.

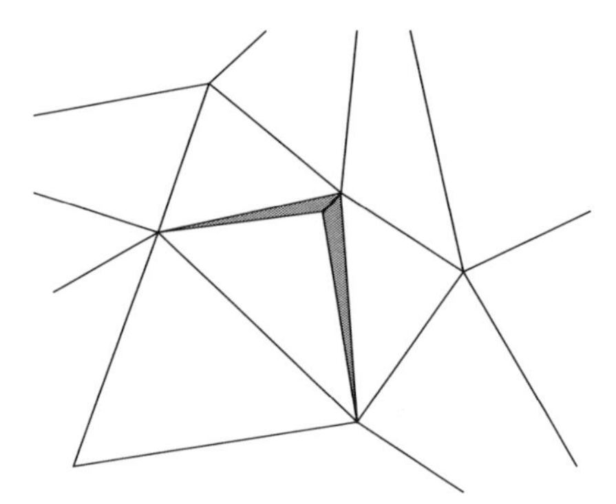

Fig. 2. Low-quality mesh; bad elements are *highlighted*.

and thus cannot accurately assess 3D micro-architecture at the microscale level of the bone. Response to osteoporosis treatment with various medications aimed at microstructural improvement and fracture prevention can be only partly assessed using 2D techniques. At the microstructural level, bone is constructed from thin rods and plates, called trabeculae, arranged in semi-regular 3D patterns and forming highly anisotropic and heterogenic material as shown in Fig. 1. Diagnostic abilities rely on high technology and advanced methods of 3D scanning, modeling, and analyzing the bone microstructure.

Mesh reconstruction from μCT/μMRI images produced by commercial software often results in a highly distorted triangulation, as shown in Fig. 2, not suitable for finite element mechanical analysis, and therefore requires remeshing and mesh optimization. Remeshing processes are time consuming and sometimes demand manual user interventions.

Hence, a new approach based on neural network (NN) was adopted from a different field and adapted to suit remeshing for microstructures. NN artificial techniques attempt to imitate the operation of the human brain (5). NN models consist of neurons connected among themselves by edges. Each neuron is capable of a simple mathematical operation. The NN learns a given example by strengthening or weakening (or even eliminating) the appropriate edges. The actual learning process, known as training, takes place by randomly presenting examples to the network. The proposed method also attempts to overcome a few related obstacles resulting from using NN techniques, such as using adaptive training to eliminate redundant holes (6). NN has the potential for high-quality remeshing, since it converges to the object without any a priori information while preserving the shape and topology of the object. Among the common configurations for NN are: (a) self-organizing map (SOM) (7); (b) growing cell structures (GCS) (8); and (c) growing neural gas (GNG).

The SOM neural network usually employs competitive learning (CL) techniques (9), in which the neurons compete for proximity to sampled data. This method deforms a mesh of a given topology into the geometry represented by the training data set. It has the advantage of rapidly approximating a mesh with predefined topology to a large amount of data with no a priori information about the object. The shortcomings of this method are that it is nondeterministic and that the topology must be known a priori to avoid large distortions.

The GCS neural network can represent faces in the mesh and grow neurons while learning the geometry. However, the initial cell structure of the NN has to be predefined by the user, according to the application, and the topology cannot be changed during the training process.

In the GNG neural network, new neurons are successively added to an initially small network. This addition results from the previous training steps. This method has the advantage of changing its topology during the training process according to the input data of different object regions. The meshing growing neural gas (MGNG) neural network, which is strongly based upon the GNG method, approximates a network consisting only of neurons and edges to the cloud of sampled points. The surfaces are produced later in a post process, thus creating the triangular mesh (10, 11). In our case, the cloud of sampled points is derived from a low-quality mesh.

All these NN methods are variations of artificial neural networks. All try to imitate the way the human mind works. This imitation is, of course, a simplified model of the human brain; nevertheless, a few features were meant to be derived and implemented using this model, namely the ability to learn. Our proposed method for remeshing is an extension of the MGNG (10). In the following section, we present an extension of the MGNG for remeshing of microstructures. In Subheading 3, we demonstrate the MGNG remeshing results on a fragment from a 3D trabecular bone model.

2. Extending the MGNG for Remeshing of Microstructures

Three-dimensional model reconstruction of bone is difficult due to the complexity of the bone's microstructure. The small features lead to a great deal of neighborhood ambiguity near each vertex. Thus, it is imperative to partition the original mesh to smaller sub-meshes using a predefined grid and to redefine the cloud of sampled points in each sub-mesh. Each such sub-mesh will then be remeshed using the MGNG method, and afterwards all sub-meshes will be assembled into the new remeshed model.

The extension of MGNG for microstructure remeshing has been implemented as follows:

- Create a grid-based partition of the original mesh.
- Remesh each cell using the MGNG method.
- Initially, two neurons are connected with an edge. Each neuron represents a 3D coordinate in the Euclidian space and contains an error parameter.
- During the training process, each winning neuron and its neighborhood are updated according to the Euclidean distance between the winner and the current sampled point.
- This neural network can grow until some predefined requirement, such as the number of iterations or number of neurons, is met. Other terminating criteria can be used, for example, the accuracy of the approximation.
- During the training period, only neurons and edges that connect them are created or removed.
- Meshes are reconstructed only after the training process is terminated. The mesh is reconstructed by a local projection algorithm (12).
- Merge all sub-meshes to achieve a global new remeshed model of the microstructure.

The whole process is described in Fig. 3.

The above algorithm was implemented and tested on a typical segment of the microstructure of a trabecular bone. Preliminary results are shown and discussed in the following section.

3. An Example of the Remeshing MGNG Method

We examined the quality of a 3D mesh model (84k triangles) created from μCT images by a typical commercial scanning system (Fig. 4). For the evaluation, we used two characteristic parameters: (a) an aspect ratio that stands for a ratio between the

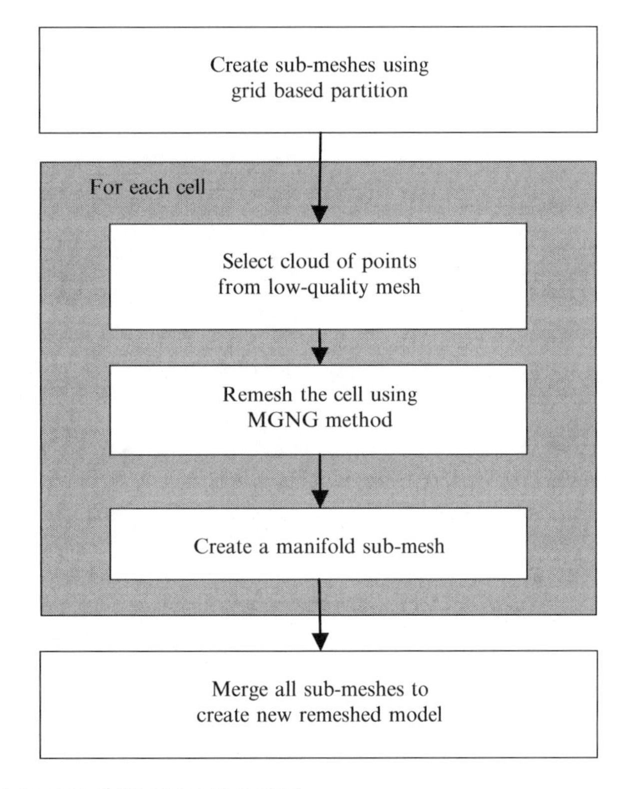

Fig. 3. A diagram of the proposed method.

longest and shortest edge of the triangles, ideally equal to one, and (b) maximum/minimum angles that in an ideal mesh are 60° each. This test resulted in failure rates of 10 and 47%, respectively. Therefore, new reconstruction methods are needed that avoid the above described drawbacks. In response, we have developed the proposed reconstruction method. An additional quality analysis was carried out with respect to three different triangular quality indices (13–15). The improvement in the mesh quality is notable, as seen in Fig. 4c.

4. Summary

This paper describes the MGNG method, a neural network-based method for mesh reconstruction, as it was applied for remeshing a trabecular bone 3D model. The method is suitable for this problem since it can reconstruct freeform 3D objects of genus-n, with no a priori information about the shapes and topologies of the objects. This method has the following benefits:

(a) The algorithm reduces the data significantly.

(b) The shape of the sampled object is learned by the MGNG algorithm. Thus, it is possible to represent a highly detailed

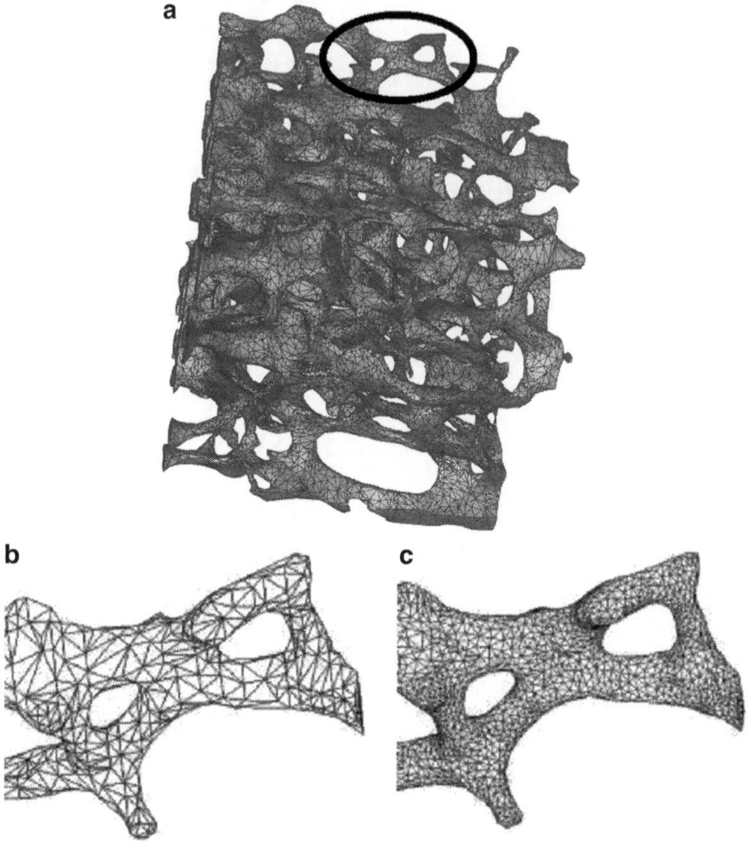

Fig. 4. Microstructures remeshing using neural network: (**a**) macro sample of a bone biopsy; (**b**) original micro-mesh with 47% failure in minimum/maximum angles test; (**c**) remeshed microstructure results with only 22% failure in the same test.

object with a mesh consisting of a relatively small number of vertices.

(c) Though there are no topological constraints on the network during the training process, the topology of the sampled object is preserved.

(d) The MGNG converges towards the geometry of the sampled object with a small error (less than 1%) and low time complexity (less than 50 s for 140k size models represented by 400 neurons).

Applying the proposed method will add the following main benefits for microstructure modeling:

(a) It allows representing highly detailed and porous structures with a mesh consisting of a relatively small number of vertices.

(b) Though there are no topological constraints on the network during the training process, the topology of the remeshed object is preserved.

The disadvantages of the proposed method are:

(a) In most of the cases, the resulting mesh has a finite number of non-manifold edges that cannot be eliminated.

(b) If the density of the sampled points is not uniform or sparse, the resulting mesh usually contains unexpected holes that are difficult to detect and close.

The extended MGNG based method for remeshing of bone microstructure can be used for generating high-quality meshes suitable for finite element analysis, precise medical applications, and more accurate visualization.

Acknowledgment

This research has been supported in part by the Israel Ministry of Science as an infrastructure project on "Advanced Methods for Reconstruction of 3D Synthetic Objects," 2007.

References

1. Bone and Joint Decade. http://www.boneandjointdecade.org. Accessed 1 Feb 2012
2. Arthritis Foundation. http://www.arthritis.org. Accessed 5 Feb 2012
3. Scientific Advisory Board. Osteoporosis Society of Canada (1996) Clinical practice guidelines for the diagnosis and management of osteoporosis. CMAJ 155:1113–1133
4. South-Paul JE (2001) Osteoporosis: part I. Evaluation and assessment. Am Fam Physician 63(5):897–904
5. Haykin S (1998) Neural networks: a comprehensive foundation, 2nd ed. Prentice Hall PTR Upper Saddle River, NJ, USA
6. Barhak J, Fischer A (2001) Parameterization and reconstruction from 3D scattered points based on neural network and PDE techniques. IEEE Trans Vis Comput Graphics 7(1):1–16
7. Kohonen T (1997) Self-organizing maps. Springer-Verlag New York, Inc., Secaucus, NJ, USA
8. Eck M, Hoppe H (1996) Automatic reconstruction of B-spine surfaces of arbitrary topological type. SIGGRAPH 96:325–334
9. Barhak J, Fischer A (2002) Adaptive reconstruction of freeform objects with 3D SOM neural network grids, geometrical modeling and computer graphics. J Comput Graphics 26(5):745–751
10. Holdstein Y, Fischer A (2008) Reconstruction of volumetric freeform objects using neural networks. Vis Comput J 24(4):295–302
11. Fritzke B (1995) A growing neural gas network learns topologies. In: Gerald Tesauro, David S. Touretzky, Todd K. Leen (eds) Advances in neural information processing systems 7. MIT Press, Cambridge, MA, pp 625–632
12. Gopi M, Krishnan S (2002) Fast and efficient projection based approach for surface reconstruction. In: Luiz Marcos Garcia Gonççalves, Soraia Raupp Musse (eds) 15th Brazilian symposium on computer graphics and image processing. SIBGRAPI, Fortaleza-CE, Brazil, pp 7–10
13. Bar-Yoseph PZ, Mereu S, Chippada S, Kalro VJ (2001) Automatic monitoring of element shape quality in 2D and 3D computational mesh dynamics. Comput Mech 27: 378–395
14. Ho-Le K (1988) Finite element mesh generation methods: a review and classification. Comput Aided Design 20(1):27–38
15. Diekmann R, Dralle U, Neugebauer F, Römke T (1996) PadFEM: a portable parallel FEM-tool. In: HPCN, LNCS 1067, Springer-Verlag, pp 580–585

Chapter 10

Using the Taguchi Method to Obtain More Finesse to the Biodegradable Fibers

Ville Ellä, Anne Rajala, Mikko Tukiainen, and Minna Kellomäki

Abstract

The Taguchi method together with Minitab software was used to optimize the melt spun PLLA multifilament fiber finesse. The aim was to minimize the number of spinning experiments to find optimal processing conditions and to maximize the quality of the fibers (thickness, strength, and smoothness). The optimization was performed in two parts. At first, the melt spinning process was optimized considering the drawing that followed and at second step the drawing was optimized. Fine (15 μm) fibers with feasible strength properties (730 MPa) for further processing were produced with the aid of Minitab software.

Key words: Extrusion, Melt spinning, Taguchi method, Fiber, Polylactide, Orientation, Parameter optimizing

1. Introduction

1.1. Fibrous Scaffolds

Biodegradable fibers can be used in several medical applications. Closing the wounds with sutures is probably the best-known and most widely used application (1). Tissue engineering is relatively new concept especially when applied clinically. To perform tissue regeneration in most cases, a porous scaffold is needed together with the cells. Since not all tissue types are equal, there is a demand for different types of scaffolds in tissue engineering regarding different cells, culturing purposes, and target tissues. The scaffolds should direct the growth of the cells to fill in the 3D structure designed and act as a support structure for predescribed period of time (2, 3). There are several different methods for manufacturing porous polymer scaffolds. Methods like salt leaching, phase separation, rapid prototyping, 3D printing, and supercritical carbon dioxide are commonly used (3–8). The availability of the fibers opens a door for scaffold manufacturing using textile technologies, such as knitting, braiding, weaving, and nonwoven technologies (9–12).

Michael A.K. Liebschner (ed.), *Computer-Aided Tissue Engineering*, Methods in Molecular Biology, vol. 868,
DOI 10.1007/978-1-61779-764-4_10, © Springer Science+Business Media, LLC 2012

Polymer fiber in general can be manufactured in several ways, wet-spinning, dry-spinning, electro-spinning, and melt-spinning. Spinning method is selected for example according to polymer features, but methods have also parameters which make methods more or less attractive to specific applications. Being a process not requiring solvents, the melt-spinning is an attractive production method for fibers. Though melt-spinning has been applied for several decades and it is in principal well-known (13), it often requires practical trials to find the optimal processing parameters for each raw polymer. The processing parameters influence the properties of the fibers (strength, modulus, elasticity, thickness, surface smoothness and in the case of biodegradable polymers the hydrolytic degradation rate) which may influence the later processability of the fibers to the textile structures and furthermore the scaffold properties. Out of the requested scaffold properties, pore sizes, degree of porosity, and interconnectivity of the pores have been widely reported (14–16) and it has also been shown that the fiber dimensions should be minimized for the finesse of the scaffold structure (17). Thus, this chapter explains the procedure on how the fiber thickness in melt spinning process is minimized by means of Taguchi method using a Minitab software.

1.2. Taguchi Method

Taguchi et al. (18) developed a method or a strategy on how design and efficiently use minimized test procedures to find solutions for production problems. These are called Taguchi's orthogonal arrays. They are fractional factorial methods used in statistical design of experiments (DOE). The Taguchi experimental method is basically a technique to reduce the number of test iterations required in an experiment. For example, an experiment with nine variables (parameters/factors) which each have two possible levels/values would in a full factorial method generate $2^9 = 512$ test configurations. When using the Taguchi's orthogonal arrays the number of required test configurations can be reduced to 12. With the use of orthogonal arrays, one is trying to find parameters that really do matter for the process. Taguchi has two quality assessment tools to offer: the loss functions and the signal-to-noise ratio (S/N ratio). With the use of loss functions one can define three categories for the product: larger the better, smaller the better, and on-target (nominal value is the best). With the use on S/N-ratio, one tries to find right values for the parameters to minimize the noise in the process so in these tests it is relevant that there are fluctuations between the test results that one can define the S/N-ratio. Taguchi method has its limitations since it cannot handle parameter interactions and some methods like genetic algorithms have been successfully used when optimizing the melt spinning (19, 20). Taguchi method has also been used to optimize the manufacturing of products based on PLLA (21, 22).

2. Materials

1. The material used as a reference in this chapter was commercially available medical-grade poly-L-lactide (PLLA) purchased from Purac Biochem b.v. (Gorinchem, The Netherlands). It is possible to use other polymers according to this method although the parameters will vary according to the polymer properties.
2. The inherent viscosity of 2.3 dl/g measured in Ubbelohde viscometer in 25°C diluted in chloroform, concentration of 0.1 mg/ml.
3. Partially crystalline with melting enthalpy (H_m) 68.8–69.3 J/g, glass transition temperature (T_g) of 59°C, melting range of 162–198°C, peak value of melting temperature (T_m) 189°C (all thermal properties were measured using TA DSC Q1000 equipment under N_2-gas, heating range 10–220°C and rate of 20°C/min, thermal cycle heating-cooling-heating).

3. Methods

3.1. Sample Preparation

1. Raw material was ground and sieved to the size of 0.5–1.4 mm to better fit the gravimetric feeding unit screws (see Note 1).
2. The polymer was dried in the vacuum oven (able to reach 10^{-5} torr) in temperature of 100°C for a period of 16 h and cooled down to room temperature.

3.2. Taguchi Method

In the following chapters, two Taguchi analyses were used. First, the Extrusion process was optimized second the drawing of the fibers was optimized using the extruder settings that were based on the first Taguchi analysis. This is path one follows when using the method.

1. Choose the process to be optimized.
2. Choose the properties to be measured (see Note 2).
3. Choose the number of levels (see Note 3).
4. Choose the parameters (see Note 4).
5. Choose the level values for the parameters (see Note 5).
6. Put the values to the orthogonal matrix and calculate the test matrix based upon the number of the levels and parameters (see Note 6).

3.2.1. Taguchi Method for Extrusion

1. Two levels were chosen for the extruder optimization and L12 orthogonal matrix was chosen.
2. Measured properties of the fibers were: thickness, tensile strength, strain, viscosity (see Note 7), and melt fracture (see Note 8).

Table 1
The L12 matrix suggested by the Minitab software for extrusion optimization

Test	T1 (°C)	T2 (°C)	T3 (°C)	T4 (°C)	T5 (°C)	Screw (rpm)	Mixing unit (yes/no)	Fill up (half/full)	Drawing (yes/no)
1	180	190	200	200	240	5	Yes	Half	No
2	180	190	200	200	240	13	No	Full	Yes
3	180	190	240	240	285	5	Yes	Half	Yes
4	180	230	200	240	285	5	No	Full	No
5	180	230	240	200	285	13	Yes	Full	No
6	180	230	240	240	240	13	No	Half	Yes
7	200	190	240	240	240	5	No	Full	No
8	200	190	240	200	285	13	No	Half	No
9	200	190	200	240	285	13	Yes	Full	Yes
10	200	230	240	200	240	5	Yes	Full	Yes
11	200	230	200	240	240	13	Yes	Half	No
12	200	230	200	200	285	5	No	Half	Yes

3. Choose the parameters and the level values [level 1, level 2] for the extruder.
4. Temperature zones of the extruder (°C), T1 barrel [180, 200], T2 barrel [190, 230], T3 barrel [200, 240], T4 mixing unit [200, 240], and T5 nozzle [240, 285].
5. Screw speed (rpm) [5, 13].
6. Mixing unit [yes, no].
7. Screw fill up state [half = 66%, full = 100%] (see Note 9).
8. Pull out speed (50 m/min) [yes, no] (see Note 10).
9. Calculate the L12 test matrix based on two levels and nine parameters Table 1 (see Notes 11 and 12).

3.2.2. Taguchi Method for Drawing

1. Two levels were chosen for the drawing optimization and L12 orthogonal matrix was chosen.
2. Measured properties of the fibers were: thickness, tensile strength, strain, and spooling value (see Note 13).
3. Choose parameters and level values [level 1, level 2] for the drawing.

Table 2
The L12 matrix suggested by Minitab software for drawing optimization

Test	Cool dist (cm)	Unit dist (cm)	G1 (m/min)	G2 (m/min)	G3 (m/min)	Winding (m/min)	G1 (°C)	G2 (°C)	G3 (°C)
1	13	14	50	90	140	175	70	70	80
2	13	14	50	90	140	213	90	90	90
3	13	14	78	130	170	175	70	70	90
4	13	50	50	130	170	175	90	90	80
5	13	50	78	90	170	213	70	90	80
6	13	50	78	130	140	213	90	70	90
7	50	14	78	130	140	175	90	90	80
8	50	14	78	90	170	213	90	70	80
9	50	14	50	130	170	213	70	90	90
10	50	50	78	90	140	175	70	90	90
11	50	50	50	130	140	213	70	70	80
12	50	50	50	90	170	175	90	70	90

4. Cooling distance (cm) [13, 50] (see Note 14).
5. Distance of the drawing units (cm) [15, 50] (see Note 15).
6. Speed of the godets (heated drawing units) (m/min): G1 [52, 80], G2 [90, 130], G3 [140, 170], and winding unit [175, 213].
7. Temperature of the godets (°C): G1 [70, 90], G2 [70, 90], and G3 [80, 90].
8. Calculate the test matrix Table 2.

3.3. Optimization of the Extrusion Parameters

1. The granules were put to a gravimetric K-Tron twin screw micro feeder (Pitman, USA) and fed to the extruder.
2. A laboratory scale Ø 12 mm single screw extruder (Gimac, Gastronno, Italy) was used along a spinneret/nozzle with 16 orifices each Ø 0.15 mm (see Note 16).
3. During the extrusion, a dry N_2-atmosphere is held in the feeder and in the extruder (see Note 17).
4. Perform the Taguchi L12 matrix trial runs.
5. Gather the samples from the extruder and analyze the selected properties (tensile test, visual smoothness, and diameter measurement) of the fibers.

6. Input the measured data to the Minitab® (Minitab Inc., USA), (or other) software and analyze each of the measured properties together with the data (example of this is given in Chapter 3.4).
7. Strain varied from 0.8 to 450% and tensile strength varied from 11 to 660 MPa.
8. To obtain the best possible result, take the best test of 12 Taguchi runs and estimate this according to the analyzed data (the prediction that the software makes) on the most important measured properties (example of this is given in Chapter 3.4). Evaluate the influence of new or suggested parameters in relation to the best run by making suitable amount of confirmation or check runs. Regarding this chapter, six more test runs were done based on the Taguchi.
9. Each check run combination resulted in different fiber properties. Decision of optimum parameters to use was based on the fiber having highest strain (618%) (higher strain gives the possibility to draw the fiber even more) companied by a fair tensile strength 108 MPa. Processing parameters obtained were: T1: 200°C, T2: 190°C, T3: 200°C, T4: 220°C, and T5: 275°C, screw speed: 5 rpm, mixing unit: no, and fill up: 90%.

3.4. Optimization of the Fiber Drawing Parameters

1. The filaments were gathered from the extruder that was set according to the best parameter settings of the extruder Taguchi check runs and taken through the cooling unit to the first godet G1 (see Note 18) and further to G2, G3, and the winging unit.
2. The test runs were done according to L12 matrix presented in Table 2.
3. Fiber properties were analyzed to receive the data from the test runs Table 3.
4. Data is put to the Minitab and plots for means and loss functions for each of the measured properties are done (see Note 19).
5. In Fig. 1., the *S/N* ratio loss function plot, where smaller the better is applied for fiber diameter. Every property gets its own parameter-dependent plot.
6. The *S/N* decibel ratio in Fig. 1. is at maximum 0.8 dB (strong influence when >1.5 dB) so the relations are not strong, though estimations can be made from the plot to optimize the smallest fiber diameter, example follows.
7. Mild influence when increasing the cooling distance (see Note 20).
8. Mild influence when increasing unit distances (see Note 21).
9. Mild influence when increasing the first godet speed (see Note 22).

Table 3
Measured data of the L12 matrix from the drawing optimization runs (spooling parameter: 1 = bad, 2 = fair, and 3 = good)

Test	Thickness (μm)	Tensile strength (MPa)	Strain (%)	Spooling (1–3)
1	27.0	187	52.1	3
2	23.2	266	28.2	2
3	19.6	208	76.3	3
4	25.3	254	42.2	2
5	21.7	250	59.2	2
6	20.8	231	47.3	2
7	26.3	150	89.4	3
8	19.9	268	49.6	2
9	21.3	286	32.4	1
10	21.6	259	81.0	3
11	20.7	310	32.0	1
12	20.6	337	42.1	3

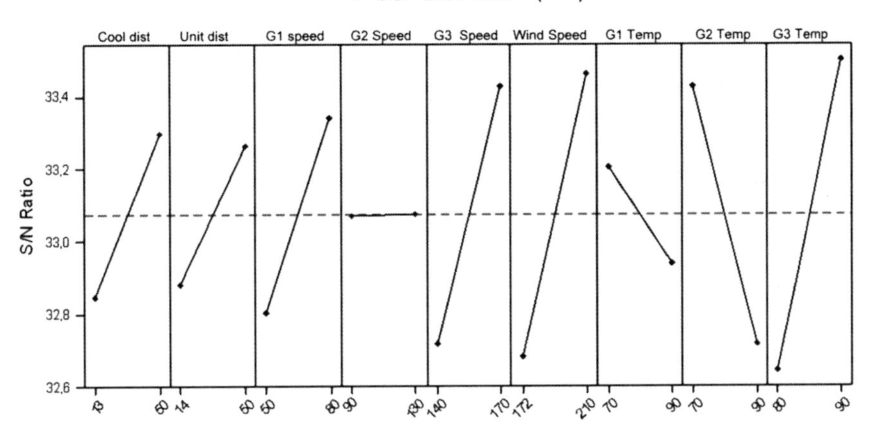

Fig. 1. Minitab *smaller the better* loss function plot shows the relation of the parameters to fiber diameter.

Table 4
Measured data from the drawing optimization confirmation runs

Confirmation	Thickness (μm)	Tensile strength (MPa)	Strain (%)	Spooling (1–3)
1	16.5	482	67.1	2
2	16.6	449	102.9	3
3	17.3	392	82.7	3
4	15.6	731	47.5	3
5	17.3	597	51.2	3
6	15.6	564	95.5	3

10. Second godet speed is irrelevant (see Note 23).
11. Strong influence when increasing the speeds of the third godet and the winder.
12. Mild influence when lowering the first godet temperature.
13. Weak influence when lowering the second godet temperature.
14. Strong influence when increasing the third godet temperature.
15. With similar fashion one can solve all the S/N ratio loss function plots for each measured property and check the relevance of the parameter to the property.
16. With Minitab one can estimate the values of the parameters one is going to use (parameters one did not use but which were in between the levels) but the relevance of this estimation was not found accurate enough to be a useful tool.
17. After processing and evaluating all the data with Minitab the smallest diameter, highest strength revealed further six confirmation runs were made to implement the Taguchi S/N ratio diagrams and evaluate the relevance of the obtained results Table 4.

3.5. The Efficiency of Taguchi Method

1. Taguchi method is not the best optimizing method since there are many parameters that are in interaction with each other so further optimizing by other methods is needed. Some of these methods are available already on Minitab.
2. Before starting the Taguchi optimizing of the melt spinning, it is relevant to know the feasible working limits, raw material, and available equipment since this helps the assessment of the obtained results. This is important when working with for example heat sensitive thermally degrading materials.

3. The method was helpful in the sense of minimizing the number of the tests.
4. The method did not give accurate prediction for offset parameter values.
5. Results were improved by the Taguchi method when Table 3 data together with Taguchi and Minitab was implemented for the confirmation tests shown in Table 4.
6. The sufficient strength level for thinnest possible fibers produced by the available equipment was the goal and it was reached. Thin fibers for further processing that require certain amount of strength, such as knitting and braiding, were manufactured.
7. Taguchi can be used for optimizing the melt spinning procedure to obtain more finesse to the fibers.

4. Notes

1. The grinding and sieving of the polymer might not be necessary although the quality control of the melting process in the extruder becomes easier when the particle size is more homogeneous.
2. Try to decide what it is that you need to optimize in your process and what are the properties that should be measured. These measurements will define the outcome of the Taguchi analysis and guide you to better results.
3. It is possible to choose the number of the levels to be used in the Taguchi according to the levels the parameters are chosen. Typically, one decides the size of the matrix at this point and this outlines the maximum number of levels and parameters.
4. The parameters are chosen according to the desired procedure. Try to choose parameters that really do have an effect on quality (parameters that do not affect the studied properties should be set as a constant) although sometimes it is necessary to choose more parameters than necessary to exactly show what part they actually play in the process.
5. One must carefully choose the level values for chosen parameters according to the polymer and available equipment. This is at the same time the power and the weakness of the Taguchi since if the wrong level values are chosen for the parameters very different optimization results can be obtained.
6. If you use the Taguchi software, it calculates the matrix for you so you must take care that the design test procedures suggested are in-line and suite the process since it is possible that unsuitable combination of parameter values are presented by the software.

7. Viscosity needs to be studied in order to evaluate the thermal degradation. It is recommended to measure also the residual monomer after the melt extrusion.
8. Melt fracture was not measured but only visually graded to give information about the presence of it.
9. To determine whether the different melt length in the screw has an effect to the properties via different rheological behavior in the flow. The state can be adjusted by controlling the speed of the feeding unit.
10. Pull out speed is one of the parameters since the final optimization goal was the oriented fiber pulled from the extruder with a speed of at least 50 m/min.
11. It is possible that the L12 matrix gives test procedures that are impossible or not feasible to perform. If a lot of those combinations appear you should re-evaluate the level values, since the optimization results depend on the reached and measured values of the tests.
12. Extra 12 samples were taken for the samples in Table 1 but opposite drawing levels. Now, two matrices could be formed where the drawing either was or was not a parameter.
13. Spooling value was a numerical value that was depending on the quality of the unwinding from the spool (fiber breakage when unwinding).
14. Cooling distance is the distance measured from the extruder nozzle to the cooling unit.
15. Distance of the drawing units is the distance between the single drawing units, the distance between the godets G1–G2, G2–G3, and G3–winder.
16. Standard extruder is not suitable for bioabsorbable polymer processing; modification for low shear stresses is needed.
17. It is important that granulates are under the N_2-atmosphere already in the feeder.
18. As the filaments are wound to travel around the godets one should always use enough revolutions to transfer the heat and maintain the number of revolutions constant when optimizing the spinning.
19. Before plotting the S/N ratio plots one must decide how the properties will be evaluated by the loss functions of *larger the better*, *smaller the better*, or *on target*.
20. Increasing the cooling distance (distance of the cooling unit from the extruder) will result to different morphology so care must be taken when performing this action.
21. Increasing the unit distance will allow more time for the orientation before coming onto contact with the next godet. This however, may influence the heating profile of the fiber on the next godet.

22. First godet is responsible for the take off from the melt so this speed has the most influence to the fiber diameter.
23. This irrelevance might be a result of the set parameters and level values. It is also a result from the high strain (due to the parameter settings based on the extruder optimization) so this increase in the speed affects more strain and strength than the fiber thickness.

References

1. Chu CC (2001) Textile-based biomaterials for surgical applications. In: Severian D (ed) Polymeric biomaterials, 2nd edn. Marcel Dekker, New York, USA, pp 491–544
2. Patrick CW Jr, Mikos AG, McIntyre LV (eds) (1998) Frontiers in tissue engineering. Pergamon, Oxford, UK
3. Nerem RM, Sambanis A (1995) Tissue engineering: from biology to biological substitutes. Tissue Eng 1:3–13
4. Mikos AG, Thorsen AJ, Czerwonka LA, Bao Y, Langer R (1994) Preparation and characterization of poly (L-lactic acid) foams. Polymer 35(5): 1068–1077
5. Thomson RC, Wake MC, Yaszemski M, Mikos AG (1995) Biodegradable polymer scaffolds to regenerate organs. Adv Polym Sci 122: 247–274
6. Park A, Wu B, Griffith LG (1998) Integration of surface modification and 3D fabrication techniques to prepare patterned poly(L-lactide) substrates allowing regionally selective cell adhesion. J Biomater Sci Polym Edn 9 (2):89–110
7. Landers R, Hübner U, Schmelzeisen R, Mülhaupt R (2002) Rapid prototyping of scaffolds derived from thermoreversible hydrogels and tailored for applications in tissue engineering. Biomaterials 23(23):4437–4447
8. Hile DD, Amirpour ML, Akgerman A, Pishko MV (2000) Active growth factor delivery from poly(D, L-lactide-co-glycolide) foams prepared in supercritical CO_2. J Control Release 66: 177–185
9. Honkanen PB, Kellomäki M, Lehtimäki MY, Törmälä P, Mäkelä OT, Lehto MUK (2003) Bioreconstructive joint scaffold implant arthroplasty in metacarpophlangeal joints: short-term results of a new treatment in rheumatoid arthritis patients. Tissue Eng 9(5): 957–965
10. Laukkarinen J, Sand J, Chow P, Juuti H, Kellomäki M, Kärkkäinen P, Isola J, Yu S, Somanesan S, Kee I, Song IC, Teck HN, Nordback I (2007) A novel biodegradable biliary stent in the normal duct hepaticojejunal anastomosis: an 18-month follow-up in a large animal model. J Gastrointest Surg 11:750–757
11. Gundy S, Manning G, O'Connel E, Ellä V, Sri Harwoko M, Rochev Y, Smith T, Barron V (2008) Human coronary artery smooth muscle cell response to a novel PLA textile/fibrin gel composite scaffold. Acta Biomater 4: 1734–1744
12. Cao Y, Vacanti JP, Paige KT, Upton J, Vacanti CA (1997) Trasplantation of chondrocytes utilizing a polymer-cell construct to produce tissue-engineered cartilage in the shape of a human ear. Plast Reconstr Surg 100:297–302
13. Fourné F (ed) (1999) Synthetic fibers. Hanser Publishers, Munich, Germany
14. Wake NC, Patrick CW, Mikos AG (1994) Pore morphology effects on the fibrovascular tissue growth in porous polymer substrates. Cell Transpl 3:339–343
15. Nehrer S, Breinan HA, Ramappa A, Young G, Shortkroff S, Louie LK, Sledge CB, Yannas IV, Spector M (1997) Matrix collagen type and pore size influence behaviour of seeded canine chondrocytes. Biomaterials 18:769–776
16. Grande DA, Halberstad C, Naughton G, Schwartz R, Manji R (1997) Evaluation of matrix scaffolds for tissue engineering of articular cartilage grafts. J Biomed Mater Res 34:211–220
17. Pulliainen O, Vasara AI, Hyttinen MM, Tiitu V, Valonen P, Kellomäki M, Jurvelin JS, Peterson L, Lindahl A, Kiviranta I, Lammi MJ (2007) Poly-L-D-lactic acid scaffold in the repair of porcine knee cartilage lesions. Tissue Eng 13(6):1347–1355
18. Taguchi G, Chowdhury S, Wu Y (eds) (2004) Taguchi's quality engineering handbook. JohnWiley and Sons, New Jersey, USA
19. Huang CC, Tang TT (2006) Parameter optimization in melt spinning by neural networks and genetic algorithms. Int J Adv Manuf Techol 27:1113–1118

20. Huang CC, Tang TT (2006) Optimizing multiple qualities in as-spun polypropylene yarn by neural networks and genetic algorithms. J Appl Polym Sci 100(3):2532–2541
21. Wang MW, Jeng JH (2009) Optimal molding parameter design of PLA micro lancet needles using the Taguchi method. Polym Plast Technol 48(7):730–735
22. Patra SN, Easteal AJ, Bhattacharyya D (2009) Parametric study of manufacturing poly(lactic) acid nanofibrous mat by electrospinning. J Mater Sci 44(2):647–654

Chapter 11

Numerical Modeling in the Design and Evaluation of Scaffolds for Orthopaedics Applications

Wojciech Swieszkowski and Krzysztof J. Kurzydlowski

Abstract

Numerical modeling becomes a very useful tool for design and preclinical evaluation of scaffold for tissue engineering. This chapter illustrates, how finite element analysis and genetic algorithm maybe applied to predict the mechanical performance of novel scaffolds, with a honeycomb-like pattern, a fully interconnected channel network, and controllable porosity fabricated in layers of directionally aligned microfibers deposited using a computer-controlled extrusion process.

Key words: Scaffolds, Orthopaedics biomechanics, Numerical modeling, Optimization

1. Introduction

1.1. Numerical Methods in Materials Science

The enormous progress made over recent years in the technology of materials was, to a considerable degree, made possible by the progress in the computer modeling of materials' structures and properties. The relevance of modeling to the development of modern materials can be seen in a number of different examples of advanced metallic materials, ceramics, polymers, and particularly composites, which are being widely used in biomedical applications. Modern composites are characterized by complex structures, which differ, not only in the relative volume of each phase, but also in their particle size, shape, and spatial arrangement, as illustrated by the micrographs shown in Fig. 1. As a result, the properties of a composite made of two specific components, for example a ceramic powder and a polymer matrix, can vary over a wide range depending on the size/shape and possible clustering of the powder particles.

In fact, the essence of modern materials engineering is that the properties of materials are governed by their microstructure, which may be tailored for specific applications. However, it should be recognized that the structure of modern materials is often very

Michael A.K. Liebschner (ed.), *Computer-Aided Tissue Engineering*, Methods in Molecular Biology, vol. 868,
DOI 10.1007/978-1-61779-764-4_11,

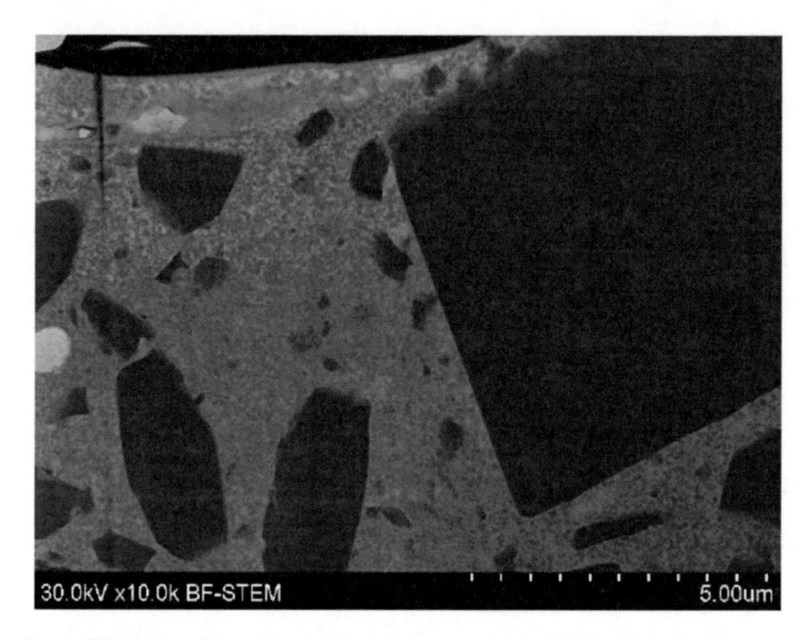

Fig. 1. SEM image of the microstructure of a composite used in dental applications.

complex as the structure will contain a range of point, linear, planar and volumetric elements, such as voids, fibers, interfaces, and grains or particles. Moreover, these features, because of recent advances in the understanding of nanotechnology, frequently have size variations of several orders of magnitude. For example, the particles added to modern composite materials may vary in size from tens of nanometers to hundreds of micrometers and may possess profoundly different geometry as can be seen in Fig. 1. The complex and diverse structures of modern engineering materials make the prediction of their properties a challenging task. Frequently, it is only possible to make rough estimates based on concepts, such as the rule of mixtures. More precise predictions of the properties of modern materials are usually based on extensive and detailed modeling, which is becoming the backbone of modern materials engineering. Modeling is of particular importance in the case of advanced and critical applications, such as products for tissue engineering, which can be designed and optimized only by making use of modern tools for predicting the properties of complex structures.

Modern modeling methods used in the development of new materials can be broadly classified based on the length scale of the description. At the atomistic level, materials are modeled by the so-called ab-initio methods (1). The term ab-initio means "from first principles" and is used to underline the fact that this technique does not require assumptions with regard to the structure of the modeled system. In principle, ab-initio calculations, which take into account the electron structure of atoms, can be used to predict the structure and properties of any combination of the known atoms.

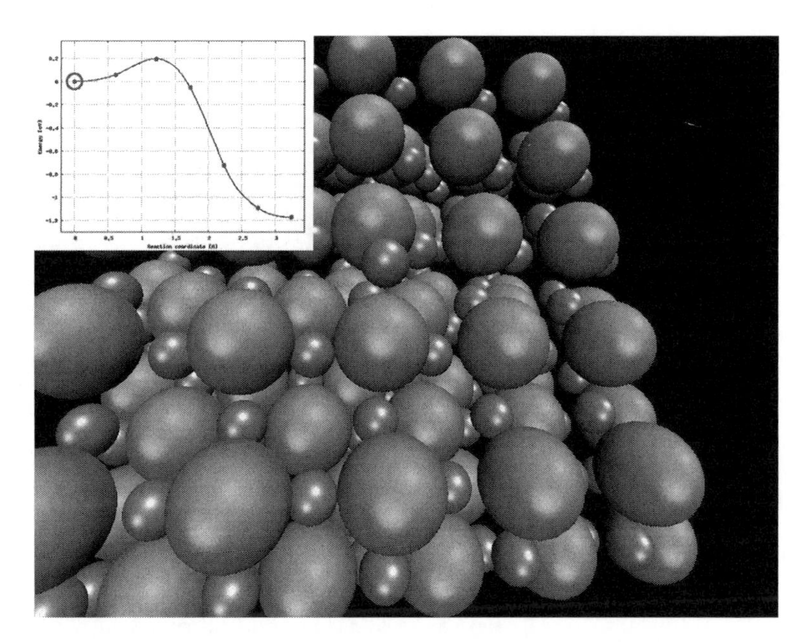

Fig. 2. Structure of the surface of ZrO_2 computed by ab-initio method.

Such predictions disregard whether or not such combinations exist in nature or have been synthesized in the laboratory and are the foundations for combinatorial materials science. There is, however, a price to pay for the versatility of these methods, which currently allow modeling systems of only up to a few hundred atoms. As a result, they are mainly used for predicting the structure of highly uniform materials, such as single-phase metals and ceramics. The analyses can be relatively easily extended to take into account point defects in crystals. Currently, with some additional assumptions they can be used for modeling the properties of surfaces and interfaces (2)—see for example Fig. 2.

The computational tools for ab-initio modeling are commercially available, but their use requires considerable effort by the modeler and it takes significant time to gain familiarity with them. The major application fields for these methods include the prediction of the properties of intermetallic compounds, which combine the properties of metals and ceramics. Papers reporting on the modeling of interfaces, such as interphase boundaries are rare.

For the more complex materials structures, molecular dynamics methods, MD, which can take a significantly larger number of atoms into account, are used. These methods are particularly efficient in modeling structure of large molecules. MD can also be used to simulate such phenomena as deformation and fracture on the atomistic scale (3).

The efficiency of molecular dynamic computations is strongly dependant on the description of the interactions between the atoms forming the material being studied. These interactions are usually described in terms of the interatomic potential, which is approximated

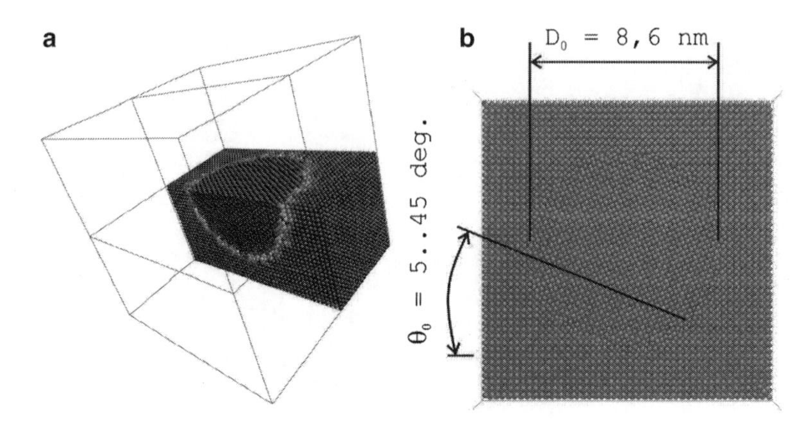

Fig. 3. Molecular dynamic model of nano-metric crystal of iron.

by a number of interatomic potentials (3). Different potentials usually need to be tested in computations and eventually the ones selected are those which offer the best fit to the experimental data. An example of the results of MD computations is shown in Fig. 3, which shows the structure of a nano-metric crystal of iron.

Although molecular dynamics allows relatively large assemblies of atoms to be modeled, they still account for an extremely small volume of material, below 1 mm^3. The interactions and processes occurring in such small volumes are essential for understanding the properties of nano-materials, including nano-laminates and nano-porous structures. However, in many applications, including bio-medical ones, significantly larger volumes of materials are used, which usually possess specific structure and architectures of a length measured in micrometers. Figure 4 shows a scanning electron microscope, SEM, image of a polyurethane scaffold produced by a modified salt leaching technique.

The modeling of such a structure at the atomistic level is neither feasible nor rational, as most of the relevant properties can be explained making an assumption that it is made of continuous phase, in this case a polymeric sponge modified with ceramic particles. The properties of the composite material possessing the structure shown in Fig. 1 can be considered in similar way, except that in this case the porosity is negligible. As a result, both materials/structures can be efficiently modeled by "macroscopic" methods, which apply to relatively large volumes of material, in which the quantum atomistic effects can be ignored. One of the efficient methods for analyzing items of such a size is based on the finite element method, FEM.

The FEM is an efficient numerical technique for obtaining approximate solutions of partial differential equations (PDE) that describe a wide variety of problems occurring in engineering and physics (1). The FEM can be successfully applied to one-, two-, and three-dimensional (3D) problems in solid and fluid mechanics, electrical engineering, and materials science. The basic concept of

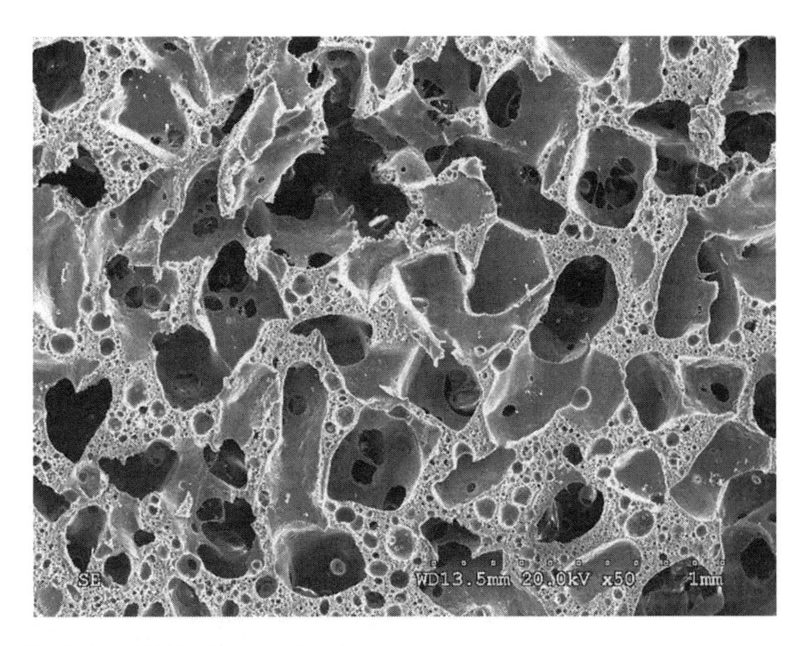

Fig. 4. A scaffold for tissue engineering.

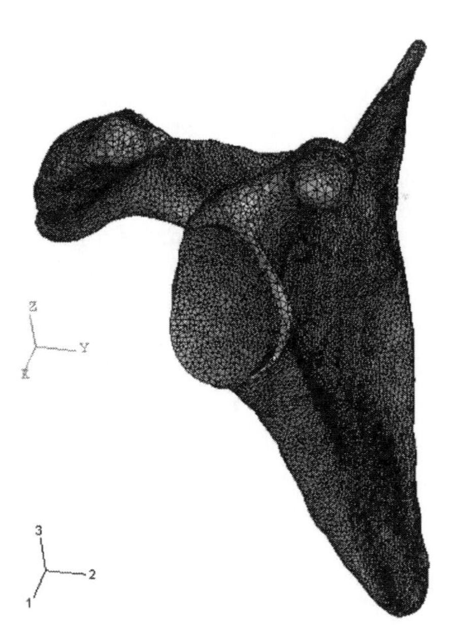

Fig. 5. Finite element model of the scapula with a glenoid component.

the FEM is the subdivision of a complex domain into a finite number of smaller parts or elements (Fig. 5), whose reaction to the applied load, in the general sense of this term, is known and for which approximate solutions to the differential equations can be obtained. Each element may have specific physical properties such, coefficient of thermal expansion, density, Young's modulus, shear modulus, and Poisson's ratio as well as dimensions, etc. Then, by assembling a set of equations for each element, the behavior of the entire domain can be determined. With this approach different domain variables may be calculated, e.g., physical displacements, temperature, heat flux, fluid velocity, etc. (4).

The finite elements method is becoming widely used by the engineering and scientific communities. Commercial codes are now available which can be used to divide the structure into elements and to simulate the properties relatively easily. It should be remembered that FEM provides only approximate solutions, but the precision increases if the element size is decreased.

Although, FEM has been used for the numerical solution of complex problems in structural mechanics, it rapidly became the method of choice for the simulation of complex systems in the bioengineering field. It has proved to be useful to engineer the stresses in implants to prevent fracture and excessive wear (5). FEM has also been used to re-design implants to enhance their performance in clinical applications (6). The method is also very useful for obtaining a better understanding of the properties of natural biomaterials like bone, cartilage and muscles. A better understanding of the natural materials paves the way to bio-mimicking of synthetic biomaterials, made of metals, polymers, ceramics, frequently combined into complex composites.

During the modeling of the properties of materials developed for bio-medical applications, small parts of their volume are considered as the elements for the modeling process. Each element may have specific geometry and properties, which provide the input data for the modeling. Currently available commercial codes allow for modeling the elastic and nonlinear plastic properties as well as the thermal and electrical effects. Combined with computer-aided design and image acquisition, the FEM codes can be used to generate finite element models of real structures visualized by a variety of imaging methods. These models can be used subsequently to compute the properties and to optimize the architecture of the particular structure (7, 8).

Such an application of FEM is illustrated in Fig. 6, which shows an aggregate of crystals in nano-metallic titanium, the FEM model of the imaged structure and the results of modeling of its resistance to plastic deformation.

1.2. Modeling in Tissue Engineering

The tissue engineering (TE) approach to the regeneration of tissue defects can generally be considered as having four steps: (1) isolation of specific cells through a biopsy from the patient, (2) growing the cells on a three-dimensional scaffold under precisely controlled culture conditions, (3) implantation of the tissue engineered product, the scaffold/growth agents + cells, to the damaged site in the patient, and (4) new tissue formation with the possible "dissolution" of the scaffold materials (Fig. 7).

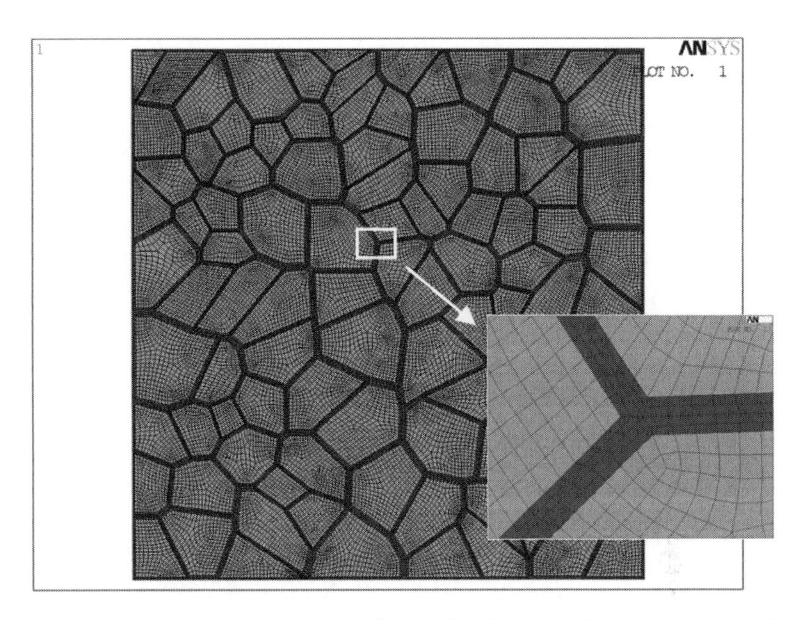

Fig. 6. FEM model of the structure and the results of modeling its resistance to plastic deformation.

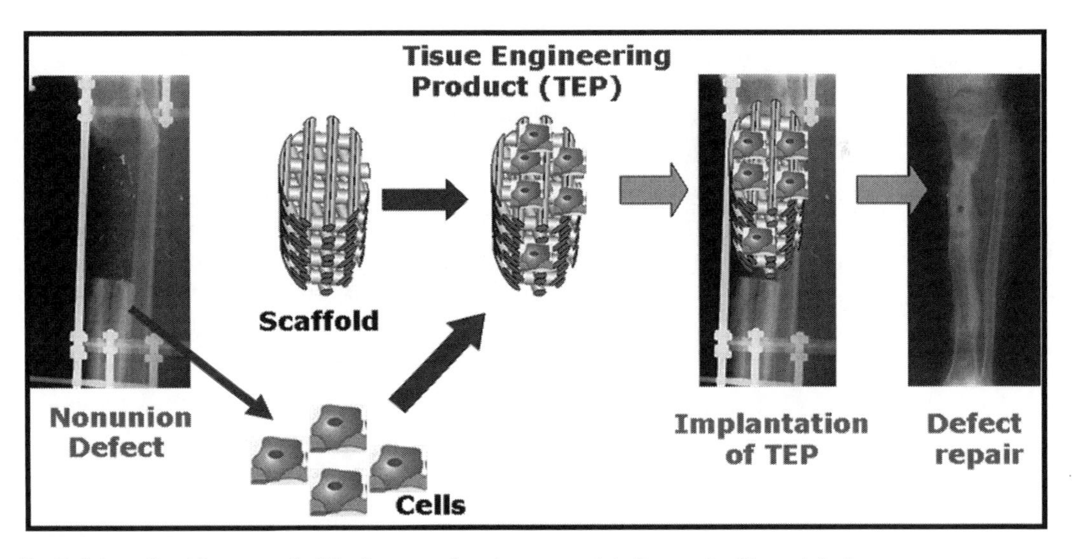

Fig. 7. Schematic of the concept of the tissue engineering approach to the repair of bone defects.

Bone tissue engineering provides an efficient solution for generating new bone tissue having satisfactory functional and mechanical performance, to treat bone defects resulting from high energy open fracture, osteonecrosis, bone neoplasia/tumors, stabilization of spinal segments, and in maxillofacial, craniofacial reconstructive surgery. This approach is of special importance when the size of the bone defects exceeds the healing capacity of the individual patient.

One of the main issues in bone tissue engineering is the design of appropriate scaffolds. Similarly to the natural extra cellular matrix that surrounds cells in the human body is necessary and the scaffold should not only provide physical support for the cells, but also allow for their proliferation, differentiation and the formation of new tissue. To meet all these requirements, a scaffold should be three-dimensional and possess a highly porous architecture with an interconnected pore network. This should ensure that the right conditions for cell growth and the flow of nutrients and metabolic waste exist. The scaffold material should be biocompatible and preferable bio-resorbable with a controllable degradation and absorption rate matching the tissue growth in vivo. Its surface chemistry and topography should promote the attachment of cells, their proliferation, and their differentiation. Last but not least, the mechanical properties of the scaffolds must match those of the tissue at the site of implantation.

Despite the great interest in tissue engineering and a high number of the studies conducted in leading laboratories worldwide, the availability of scaffolds for the regeneration of load-bearing large bones is far from being satisfactory. With the current abundance of materials fabrication routes, the shortage of scaffolds is primarily caused by a deficiency in the understanding of the relationships between scaffold microstructures, cell growth, and nutrient (oxygen) consumption rates in biodegradable scaffolds, which undergo structural changes, such as dissolution, during tissue growth. For the reasons already explained, the progress in the understanding of the phenomena involved requires in-depth experimental studies and robust methods for the computer simulation of the mechanisms occurring in the scaffolds.

Currently, the numerical models presented in literature are focused on the mechanical properties of the structures employed in the formation of new tissues in:

1. Fracture healing (9–15).
2. Callus growth (16), distraction ontogenesis (17), osteochondral defect healing (18).
3. Bone ingrowth into porous implants (19).
4. Bone tissue engineering (20, 21).

Several so-called mechano-regulation theories were explored in these models, with one of the first for tissue differentiation being proposed by Pauwels (9). He hypothesized that the mechanical parameters, like deviatory stresses, stimulate the formation of fibrous connective tissue and that hydrostatic compressive stresses stimulate cartilage formation. Carter et al. (11) proposed that tissue differentiation is effected by principal tensile strains and hydrostatic pressure, or stress. They proved that high compressive hydrostatic stresses result in the formation of cartilage (chondrogenesis), while

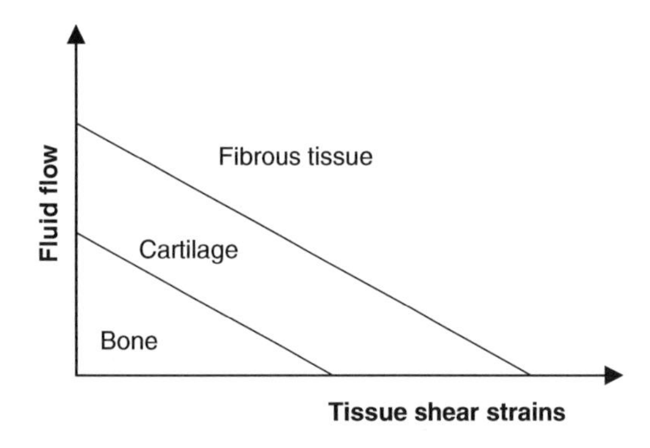

Fig. 8. Mechano-regulation model by Prendergast et al. (10).

low hydrostatic stresses correlate with the bone formation process (osteogenesis). At the same time, high principal strains promote the formation of fibrous connective tissue (fibro cartilage).

Prendergast et al. (10) suggested that two biophysical stimuli: (1) deviatory strain in the solid phase and (2) fluid velocity of the interstitial phase play a dominant role in tissue differentiation and formation as shown in Fig. 8. According to their model, the high values of either of these two, favor the growth of the fibrous tissue and only when both stimuli are low enough, does ossification occurs. This model accurately predicted tissue differentiation at implant-bone interfaces and was further developed and adapted to simulate tissue differentiation during fracture healing (13). Along the same lines Claes and Heigele (12) hypothesized that the level of the local stresses and strains influence the type of the new tissue. A different model was proposed by Ament and Hofer (22), who considered the strain energy density as the main mechanical stimulus for tissue formation.

A comparison of the models relevant to bone formation has been recently provided by Isaksson et al. (15) and Epari et al. (23), who suggested that the model developed by Prendergast and co-workers offers the best correlation with the results of in vivo and in vitro studies. None of the volumetric components, neither pressure nor fluid rate alone, can be used to correctly predict the kinetics of new tissue formation. However, when only deviatory strains were used to stimulate the growth, the fracture healing process has been satisfactorily simulated. This suggests that the deviatory component may be the most significant mechanical parameter to "guide" tissue differentiation during bone formation.

It should be realized, however, that the "cellular mechanisms" are not included in the models discussed so far. This is because it commonly assumed cell migration and proliferation is governed by the diffusion laws up to point of maximum cell density. However, it is clear nowadays that cell biology, in particular cell proliferation, death, and activity as well as the effects of using cytokine (growth factors) and nutrition need to be incorporated in the models for a better understanding of tissue formation.

The effect of two generic growth factors on cell differentiation and fracture healing was investigated by Bailòn-Plaza and van der Meulen (24), but the interactions between the growth factors and the mechanical environments were not taken into account and neovascularization was only implicitly considered. The influence of blood supply on bone formation was included via random migration of the osteoblasts from the existing bone and depended on the initial location of the bone tissue (16). The current understanding is that the role of the blood vessels, which are required for bone formation (25), should be included in a more explicit way in models of tissue growth.

Therefore, it may be concluded that future work on the numeric modeling of tissue formation should address the coupling of the cell migration and differentiation, as developed by Gomez-Benito et al. (16) with the Prendergarst mechano-regulation model (10) and the models which describe the effect of the growth factors on cell differentiation and tissue formation. Additionally, the formation of the blood supply systems and scaffold degradation should be included in the modeling.

1.3. Quantitative Description of Porous Structures

A significant number of engineering applications, including tissue engineering, require materials with a high degree of porosity. Porosity in materials may exist in one of two basic forms:

(a) As isolated "bubbles", either empty or filled with a gas.

(b) As a continuous network of multi-connected "pipes", generally referred to as "open porosity".

Obviously, in some materials both types of pores may be present, with open porosity dominating near the free surface.

It should be appreciated that closed pores generally reduce the specific weight of the materials (e.g., light foams). Open porosity materials are used primarily for various filtration/infiltration techniques, including "infiltration" by the cells involved in the formation of new tissue. This application of open porosity materials justifies a brief note about their quantitative description, which is the basis for modeling their properties and designing their architecture.

The basic parameter defining porosity, either open or close, is the volume fraction, V_V, which describes the volume of pores in a unit volume of the material. This parameter can be measured relatively easily from the images of the porous structure, as explained in Fig. 9. It can be also estimated from the measurements of the specific mass, which decreases linearly with the increasing volume fraction of pores.

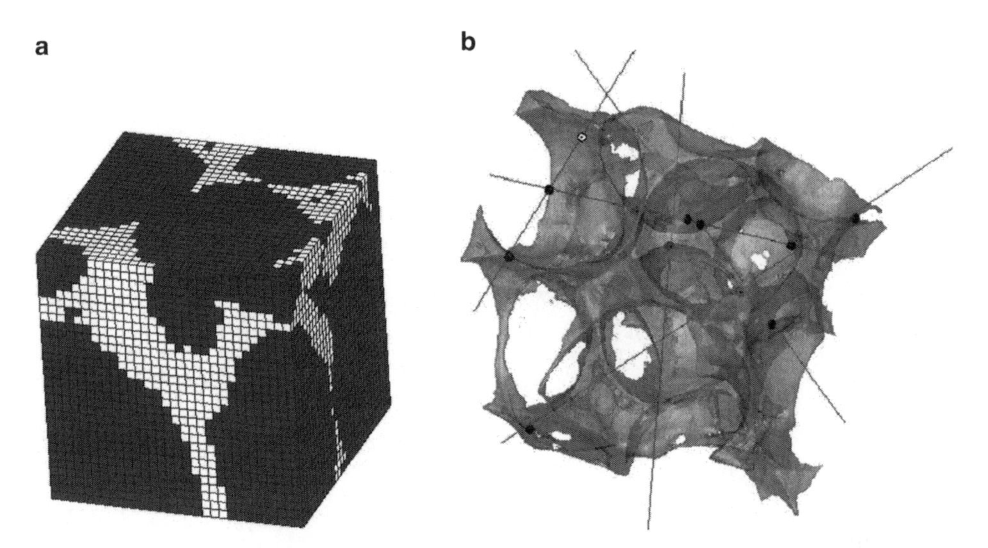

Fig. 9. Explanation of the methods for measuring the volume fraction of pores, V_V, and specific surface area, S_V: (**a**) the volume fraction V_V is equal to the areal fraction, which can be estimated by counting pixels of the image revealed on a cross-section of the studied material; (**b**) the specific surface area S_V can be estimated by counting the number of intersection points of a system of randomly oriented test lines of known length.

As samples of the same material possessing the same volume fraction of pores may contain a large number of small holes or channels or a smaller number of relatively large holes or channels, another relevant parameter is the specific surface of the pores, S_V. This, measured in m^2/m^3, directly defines the surface of the pores in a unit volume of the material and it may be estimated from the images of the structure illustrated in Fig. 9.

In the case of an open porosity network, the two parameters, V_V and S_V, under some general assumption can be used to estimate the dimensions of the volume available for infiltration. Firstly, assume that the pores form a three-dimensional, interconnected system of pipes, channels, in which from any point inside a specific channel it is possible to reach the outer surface of the material by moving along a number of alternative routes. For relatively narrow pore-channels, their total length in the unit volume, L_V, can be estimated by counting the number of pores emerging on a unit area of the surface, N_A.

However, simple geometrical considerations show that:

$L_V \times \pi r^2 = V_V$, where $r =$ the average radius of the porous channels. From this relationship, by measuring L_V and S_V, one may effectively estimate the average radius of the throat of the channel network.

Now, consider another specific situation, in which the material has a very high volume fraction of pores and, in contrast to the situation discussed above, the "solid" part of the material effectively exists as a network of multiconnected rods or wires. In this case, the permeability of the porous material depends on the size of the

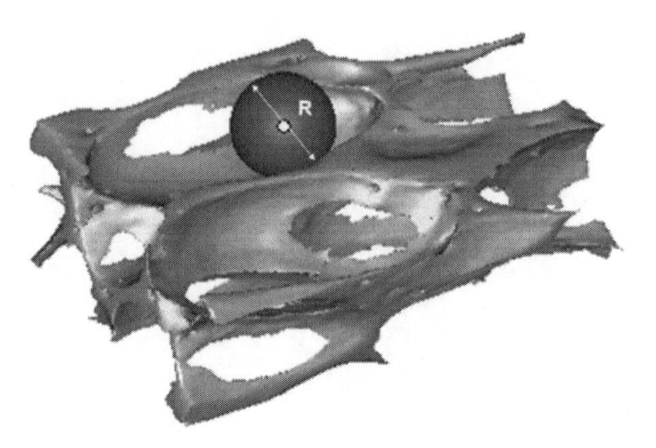

Fig. 10. Schematic of the model used for estimation of the windows in the highly porous materials.

"windows" framed into the system of interconnected elongated pores. It is possible to model such a structure by assuming that the connecting rods/wires are placed on the surfaces of hypothetic spherical cages, of radius R, as illustrated in Fig. 10.

Topological considerations show that the three-dimensional spaces can be filled with elements having on average 14 faces (Thompson octahedrons). This implies that on average, in the system of cages of radius R, the surface area of such faces is $4\pi R^2/14$. The equivalent radius of such a face is approximately 0.54 R and the length of the edges is 3.4 R. If it is assumed that the average radius of connecting rods/wires is r, then as each edge is shared by three adjacent cages, the following estimate can be obtained:

$L_V = 20/R^2$, $V_V = 20\pi r^2/R^2$, values which may be used to determine both r and R. Alternatively, the approximate surface area of the windows between the adjacent cages can be evaluated as $\pi (0.54R)^2$.

An effective method for characterizing porous structures is computerized X-ray tomography (CT). Standard optical and scanning electron microscopes have limitations because they only allow the surface of the samples to be studied. As a result, the three-dimensional geometry of the materials structures can only be assessed from the examination of thin slices of the material. The preparation of the slices for the examination is not easy, particularly when the material contains components of significantly different hardness. Additionally, the preparation of samples is a destructive technique, which precludes the possibility of studying processes taking place in samples subjected to such examinations prior to testing the functionalities of their structures.

Modern CT techniques allow for "virtual" slicing of the samples, which ensures that the sample remains intact. The typical spatial

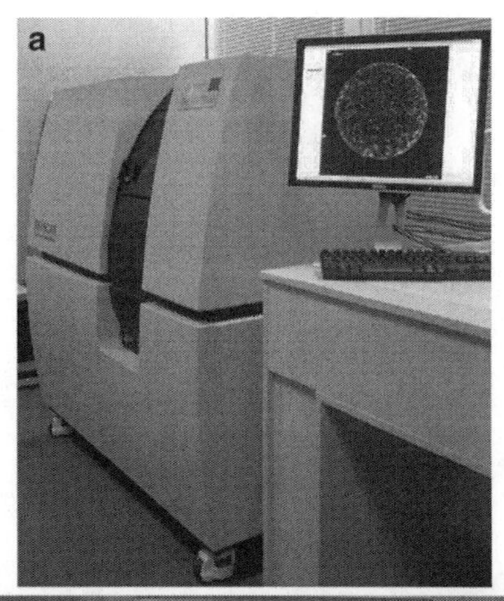

Fig. 11. Examples of: (**a**) nano- and (**b**) micro-CT scanners.

resolution of conventional medical CT-scanners is of the order of 0.75–2.5 mm. Micro- and nano-tomography using the CT scanners shown in Fig. 11 can attain a spatial resolution of about 1 μm and 150 nm, respectively.

During the CT examinations, the investigated samples are rotated incrementally with a fixed rotational step. At each angular position, the X-ray beam is transmitted through the object and the absorption of X-ray is measured. The high-resolution projected image, derived from the degree of absorption of the beam, is produced using a *filtered back projection algorithm*. The projections registered at different angles are then used to compute images of the cross-sections of the object being studied. Subsequently, there are several items of commercial software available, which can import the CT images and generate a 3D numerical model of the scanned objects as illustrated in Fig. 12.

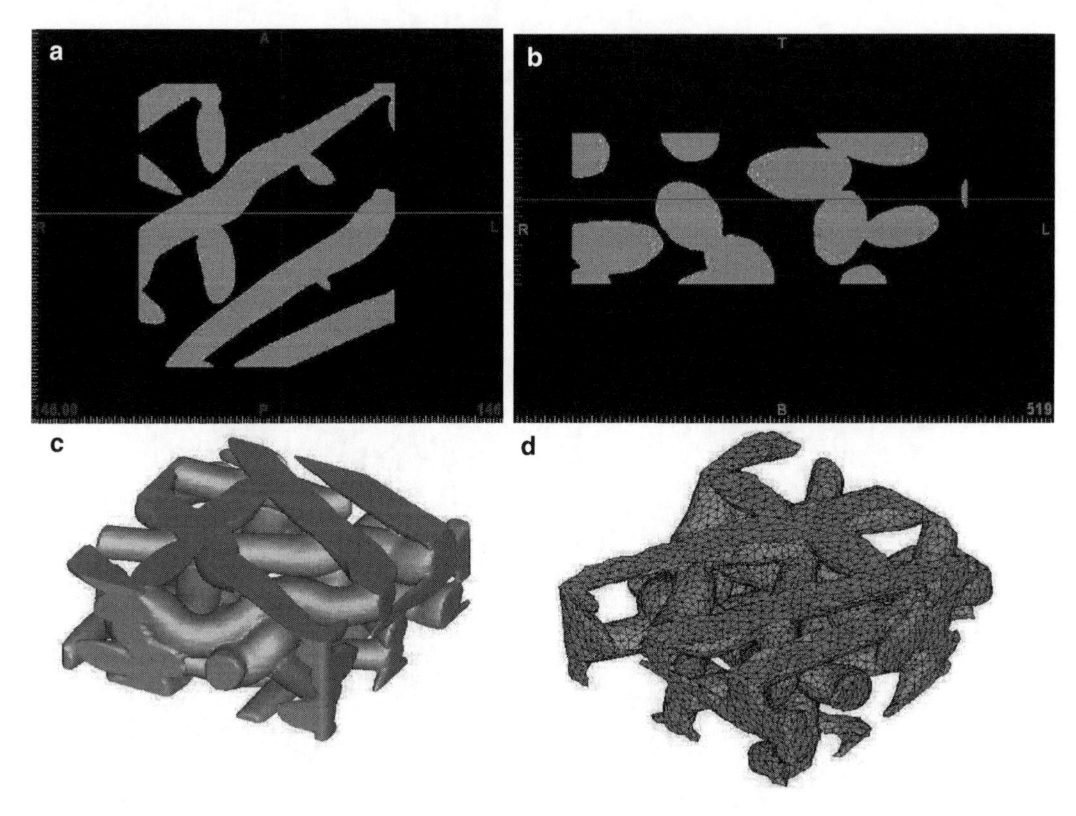

Fig. 12. Images of the cross-sections (axial (**a**), and coronal (**b**)), 3D model- (**c**) - and FEM (**d**) subdivision of a scaffold.

2. Design and Evaluation of the Architecture and Geometric Requirements of a Scaffold Utilizing the FEM

One of the main challenges in bone tissue engineering is scaffold design. A scaffold should mimic the properties of bone while providing a temporary matrix for the growth of cells and the formation of new tissue. In particular, they should be characterized by high mechanical strength and 3D open porosity to provide the space and transportation avenues for tissue regeneration. The performance of scaffolds primarily depends on the materials employed. In addition to mechanical strength, the scaffold materials must not impede cell migration, adherence, proliferation, and formation. These requirements are quite challenging and there are currently only a few materials considered suitable for bone replacement and regeneration. The materials widely investigated in this context are calcium phosphate-based materials, such as synthetic hydroxysapatyt (HAp) and β-tricalcium phosphate (β-TCP) (26). Poly(lactic acid) (PLA), poly(glycolic acid) (PGA), polycaprolactone (PCL) as well as their co-polymers, which have been developed for skeletal repair and regeneration.

The select ion of scaffold materials must also take into account the availability of fabrication methods to produce the required three-dimensional structures with the desired pore architecture. The results published to date show that three-dimensional porous structures of a scaffold maybe efficiently fabricated by: fiber bonding, sintering, solvent casting, particulate leaching, membrane lamination, melt molding, temperature-induced phase separation (TIPS), gas foaming, and rapid prototyping (RP) (27). The RP techniques, such as fused deposition modeling (FDM), selective laser sintering (SLS), three-dimensional printing (3DP), and solid freeform fabrication (SFF) are particularly promising methods for producing customer-designed scaffolds with patient-specific geometry and controlled internal architecture, which can be designed with computer-aided design (CAD) software.

With the availability of a few acceptable biomaterials and the possibility of controlling the architectures of scaffolds, the next challenge is to produce the optimum designs. In this context, the aim is to show that near-optimal architecture of the scaffolds can be obtained by numerical modeling of the structures required for bone tissue engineering. This chapter illustrates, how finite element analysis maybe applied to predict the mechanical performance of novel scaffolds, with a honeycomb-like pattern, a fully interconnected channel network, and controllable porosity fabricated in layers of directionally aligned microfibers deposited using a computer-controlled extrusion process (28). Two forms of these novel scaffolds are analyzed, which were constructed with a different in lay-down pattern of the fibers: 0/90° (type A) and 0/60/120° (type B), which are shown schematically in Fig. 13. Scaffolds made of fibers of different biodegradable biomaterials, PCL and PDLA (poly(D-lactic) acid, are used in this "case study."

The numerical models of the scaffolds were generated using MSC/Patran software. Each scaffold consisted of 24 layers and each layer consisted of four fibers, to form a three-dimensional structure of 96 fibers. Tetrahedral (four node) finite elements were used to generate FE meshes of these structures. The generated meshes for each scaffold were optimized with respected to the number of elements using a strain energy density criterion. After this optimizing process, the scaffold with the 0/90° lay-down of fibers consisted of 211,784 elements.

The structural analysis of the scaffolds architectures was performed using MSC Nastran software. The geometrically optimized scaffolds were compressed by a uniformly distributed load of 0.5 MPa, which corresponds to the load generated by physiological weight. The loading is applied to the scaffold through a rigid plate bonded to the top layer of the scaffold. The bottom part of the scaffold is assumed to stay in a fixed position as shown in Fig. 13. It has been further assumed that the scaffolds are made of polymeric materials with a value of Young's modulus varying from 0.5 to 8.5 GPa, for PCL and PDLA (29, 30), respectively.

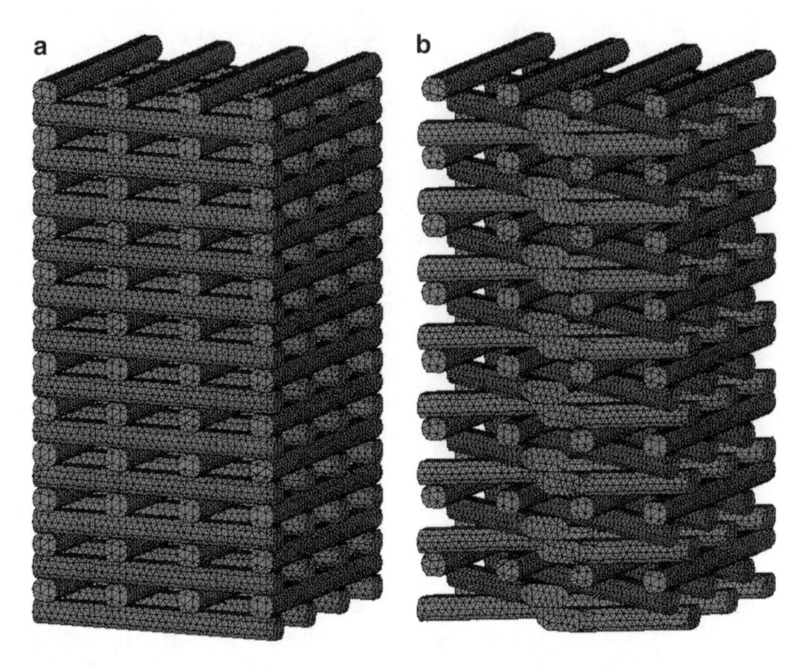

Fig. 13. Schematic of two scaffolds with different fiber patterns: (**a**) 0/90° and (**b**) 0/60/120°.

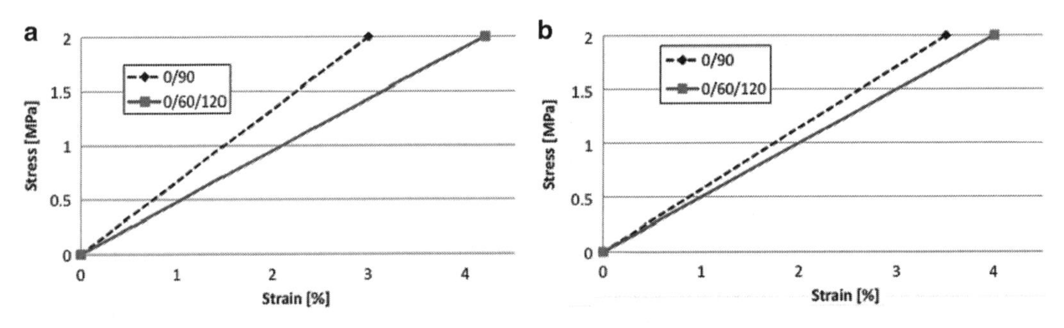

Fig. 14. Comparison of experimental and numerical data for 0/90 and 0/60/120 patterned scaffolds: (**a**) experimental data adopted from Zein et al. (31) and (**b**) numerical results of stress–strain characteristics.

Initially, the FE model of the scaffolds has to be validated. The validation consists of comparing the compression stiffness of the two PCL scaffolds obtained by the numerical simulation and the experiments (31). The results of numerical simulations were consistent with the experimental data, as shown in Fig. 14. Also, Zein et al. (31) showed that scaffolds fabricated with a 0/90 pattern exhibit higher compression stiffness ($E_{average} = 65$ MPa) than those with a 0/60/120 pattern ($E_{average} = 45$ MPa). The compression stiffness obtained by the numerical simulations discussed here was 60.6 and 45.5 MPa for scaffolds having a 0/90 and 0/60/120 patterns, respectively.

After validation of the model and the employed computational codes, three variations of the fiber materials were simulated for each

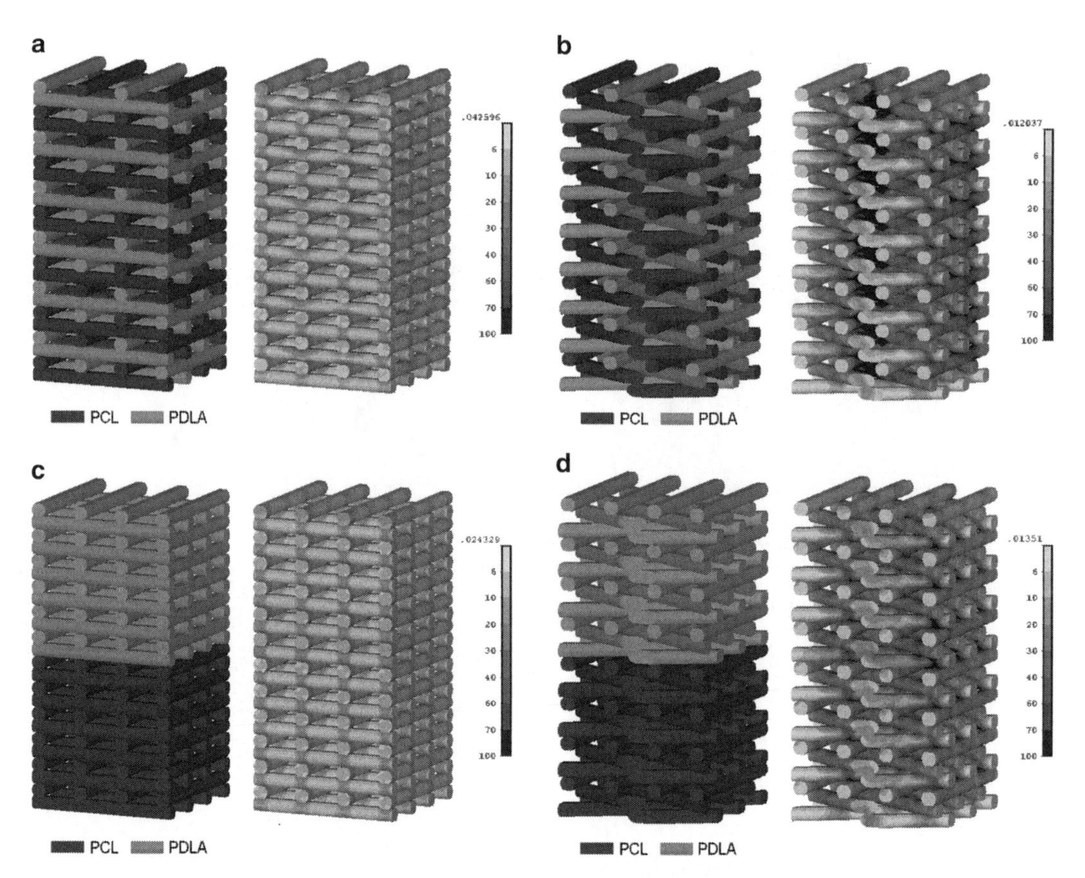

Fig. 15. FE models and Von Mises Stresses for scaffolds with: (**a**) 0/90 lay-down pattern of the fibers made of PCL and PDLA; (**b**) 0/60/120 pattern of the fibers made of PCL and PDLA; (**c**) 0/90 pattern of the fibers made of PCL (*lower*) and PDLA (*upper*); and (**d**) 0/60/120 pattern of the fibers made of PCL (*lower*) and PDLA (*upper*).

type of the scaffold architecture (Fig. 15). Firstly, the scaffolds were made of alternate layers of PCL and PDLA (Fig. 15a, b). This implies that in single layer of the structure, each two-neighbor fibers are made of PCL and PDLA. This combination mimics to some extent nonhomogeneity of the bone tissue. In the second variant (Fig. 15c, d), the scaffolds are divided into two parts, made of two materials. Thus, it shows some similarity to the osteochondral column formed by cartilage and subchondral bone tissues in the natural joint. In order to simulate the structure of a larger volume of the bone, in particular the outer stiffer cortical and the inner trabecular bone, a third variant was proposed which is shown in Fig. 16. In this case, the outer part of the scaffold is made of PDLA and the inner of PCL (PDLA is much stiffer than PCL). The three variants described above are called inhomogeneous, osteochondral, and the bone model, respectively.

The properties of all proposed scaffold models were studied using FE analyses by simulating their reaction to compression and

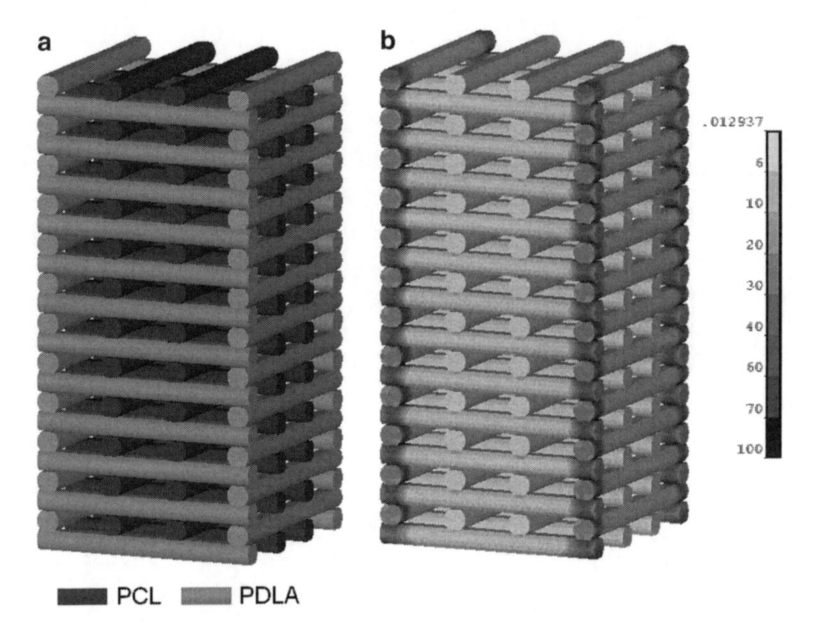

Fig. 16. (**a**) The FE model (**b**) Von Mises Stresses for a scaffold with 0/90 pattern of PCL fibers (*inner part*) and PDLA fibers (*outer part*).

Table 1
Compression stiffness of the analyzed scaffolds

	Inhomogeneous model		Osteochondral model		Bone model
Scaffold (variant/type)	**0/90**	**0/60/120**	**0/90**	**0/60/120**	**0/90**
Compression stiffness (MPa)	130.7	124.2	117.6	95.2	400

the compression stiffness and Von Mises stresses were calculated. The results of the stiffness under compression of the different scaffold architectures are given in Table 1.

It is apparent that the structures with the 0/90 patterns exhibit higher compression stiffness than those with the 0/60/120 arrangement of fibers and the stiffness of the nonhomogeneous model is higher than that for the osteochondral model. The bone model possesses the highest compression stiffness of 400 MPa. However, this value is lower than those measured for the cortical and trabecular bones, for which Young's modulus ranges from 10 to 20 GPa and 500 to 1,000 MPa, respectively (32).

The results of stress analysis (Figs. 15 and 16) show that Von Mises stresses are distributed more homogeneously for the 0/90 pattern than for the 0/60/120 fiber arrangement. However, the maximum values of the stresses are significantly higher for 0/60/120 than for 0/90 pattern. It was found that for the bone model, the highest stresses appear at the outer edges of the scaffolds.

The FEM may also used to predict the permeability of fluids through the scaffold. As already discussed, fluid flow is one of the mechanical stimuli regulating tissue differentiation. Moreover, a high velocity of fluid flow is needed for scaffold cell seeding in vitro in bioreactors (33).

Steady-state Newtonian fluid analyses were performed for the scaffold with the 0/90 fiber pattern. An inlet fluid velocity of 10 mm/s was applied to the top of scaffold (Fig. 17) and the outlet fluid pressure was set as zero at the exit at the bottom of the scaffold. It was assumed that fluid could not flow out through the walls (a confined perfusion system). The simulated fluid density and viscosity were set as similar to the cell culture medium (33). Based on these assumptions, the fluid flow velocity was calculated for the scaffold structure. The results obtained show that the velocity increased during the passage of the medium through the scaffolds due to the porous architecture (Fig. 17). Since the pores form regular channels from the top to the bottom of the modeled scaffold, the increase in the fluid velocity is not as high as for a scaffold with a nonregular architecture, as shown by Sandino et al. (33). It may be expected that, the scaffold with 0/60/120 fiber pattern, having a less regular internal architecture than that with the 0/90 pattern, will exhibit a more significant increase of the velocity of fluid, but this will be verified by further experimentation.

In summary, the results presented, illustrate that FEM can be used to evaluate and design the architecture of novel scaffolds for bone tissue engineering. Although, there are few assumptions in these analyses, such as linearity and idealized elastic properties of the polymeric materials, the results provide useful information on the mechanical performance of the implants under simulated physiological loads. The results showed that the application of different scaffold fibers is an efficient method of tuning the mechanical properties, which have a significant influence on the formation of new tissue. The mechanical performance of the scaffolds is also controlled by the fiber patterns. The fluid flow analysis showed that continuous and nonhomogeneous flow of the fluid through the scaffold stimulates cell seeding and tissue differentiation. Further numerical studies should take into account the process of the materials biodegradation/bioresorption, which cause changes to the mechanical properties of the scaffold with the passage of time after its implantation.

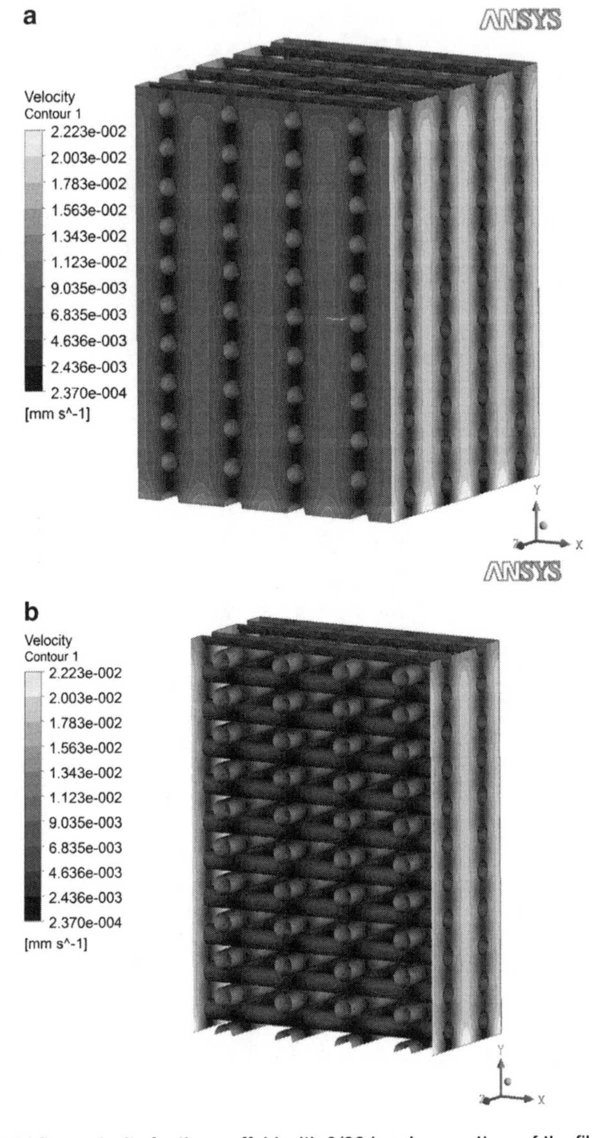

Fig. 17. Fluid flow velocity for the scaffold with 0/90 lay-down pattern of the fibers: (**a**) 3D view and (**b**) cross-section.

3. Optimization of the Scaffold Structure and Materials Using Genetic Algorithms

Bone tissue engineering is a complex procedure, which must take into account dynamic changes in the mechanical functions of the bone–scaffold systems, including coupling of bone regeneration and scaffold degradation. In the first stage of the bone regeneration process, the scaffolds are designed to carry the mechanical load applied to the impaired tissue. Subsequently, when the bone undergoes regeneration and remodeling, the scaffold should degrade in

Table 2
Properties of the materials used in the optimization of the scaffolds architecture (29)

Material	Modulus (GPa)	Elongation (%)	Degradation time (months)
PGA	7.0	15–20	6–12
LPLA	2.7	5–10	>24
DLPLA	1.9	3–10	12–16
PCL	0.4	300–500	>24
DLPLG	2.0	3–10	5–6

the in vivo environment and gradually transfer its mechanical functions to the developing new tissue. Finally, the scaffold should be completely replaced by the newly formed bone. This requires that a scaffold must initially provide conditions for cell and new tissue growth and then decay with the kinetics adjusted to the formation of new bone and the bone remodeling. To fulfill these requirements, optimization of the mechanical strength against the rate of degradation of scaffolds is essential.

Providing suitable stiffness and strength of the scaffold at each stage of bone regeneration is a great challenge. This may require the use of several different materials for the fabrication of a single scaffold. Also, the different types of materials may be arranged within a scaffold in a nonhomogeneous, but specific way, which is now possible because of the advances in rapid prototyping techniques.

The following text gives some insight into the procedures for optimizing the mechanical strength of biodegradable scaffolds against their rate of degradation for complex, nonhomogeneous scaffold architectures. Five different biodegradable materials are considered when designing an optimized scaffold with a 0/90 fiber pattern. These materials differ in mechanical properties and rate of degradation, as indicated in the data shown in Table 2. The geometry of the scaffold used for this study was shown in Fig. 13a.

As no standard information of the optimal rate of degradation of a scaffold has yet been proposed the arbitrary curve shown in Fig. 18 was used for the purpose of this study. As the degradation process of the polymeric scaffold is strongly related with loss of the molecular weight and lowering of Young's modulus, it has been assumed that the degradation curve defines the reduction of Young's modulus with the passage of time. The initial value of the modulus was 400 MPa the same as the PCL material possessing lowest modulus shown in Table 2. The value falls within the range of data for trabecular bone.

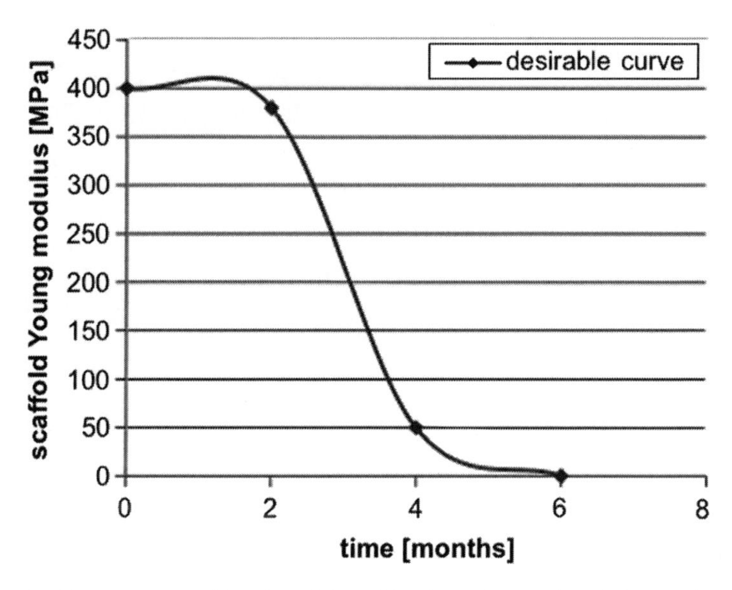

Fig. 18. The curve describing degradation of the Young's modulus of the scaffold as function of the time in the human body.

The objective of the optimization was fitting the degradation curve (Young's modulus reduction over time) to the decay of the polymeric scaffold. The expected result of optimization is selecting a fiber blend, which ensures that the scaffold is as stiff as required at each stage of the new tissue generation. Since the scaffolds considered are subject to compression loading, the optimal selection of the materials for their fabrication can obtained, using the following:

$$\max_{\overline{D}} J(\overline{D}, t), \tag{1}$$

where J is an objective functional, $\overline{D}$ is a vector of the design parameters, which depend on the number of materials considered for the type–type scaffolds

$$J(\overline{D}, t) = \sum_{i=0}^{n} A - \left|E_{\text{obt}}(\overline{D}, t_i) - E_{\text{des}}(\overline{D}, t_i)\right| - (|d_x| + |d_z|), \tag{2}$$

where A is a constant; n is the number of considered stages of degradation; d_x and d_z are displacements in the x and z axis (side displacements) of the scaffold. A requirement, which must be fulfilled by the scaffold, is the uniform distribution of the load to ensure that fiber is overloaded, which could cause the stability of entire structure to be lost. The method, which assures that this requirement is met is the minimization of side displacements d_x and d_z. The parameter E_{obt} describes the scaffold degradation function (function of Young's modulus reduction over time), whereas E_{des} is the desirable degradation function. Young's modulus of scaffold E_{obt} is given by:

$$E_{obt} = \frac{\sigma_{scff}}{\varepsilon_{scff}} \tag{3}$$

$$\sigma_{scff} = \oint_A \frac{F}{dA} \tag{4}$$

$$\varepsilon_{scff} = \frac{d_{top}}{h_{scff}}, \tag{5}$$

where σ_{scff} corresponds to compression stresses, which acts on top of scaffold (F is a force acting on infinitesimal area at the top of scaffold dA), ε_{scff} is scaffold compressive strain, d_{top} is displacement of scaffold top and h_{scff} is height of scaffold.

Optimal assigning of the five different materials to different scaffold fibers, which assure fulfilling the stated objectives, required the use of an efficient optimization method, such as a genetic algorithm (GA). The GA proposed by J. Holland in 1970s is one of the most robust methods for discrete optimization. The GA consists of a stochastic search using the principles of evolution and natural genetics. GAs are applied to populations of individuals (34), with an assumption that each individual has one chromosome. Chromosomes consist of genes, which are equivalent to the design variables in the optimization problem considered in this specific case. In the situation of scaffold optimization, the design variables correspond to the scaffold fibers (struts), whereas the value of each gene corresponds to the material type. There were 96 scaffold fibers and each could be assigned as one of five polymeric materials (Table 2).

The adaptation, the fulfillment of requirements, for a given individual, which corresponds to a particular strut blend, is analyzed using a fitness function (Eq. 2). A flowchart representing the schematic of the standard GA coupled with the structural analysis by FEM adopted from Cappello et al. (32) is shown in Fig. 19. The expected results of the structural analysis are displacements of scaffold top, which can be used to calculate the strain. In the first step, an initial population of individuals was created and the values of the genes of particular individuals were randomly generated. In the next step, the fitness function values of individuals were computed. Then, evolutionary operators (crossover, mutation) change genes of the parent population of individuals (34, 35).

The fitness function was computed for the varied individuals, based on the results of the finite element analyses. The individuals selected for the offspring population become a parent population and the algorithm and the iterations were used until termination conditions were attained, which were defined by a certain maximum number of iterations before reaching the critical value of 160. Taking into account that the crossover probability is 0.8 and

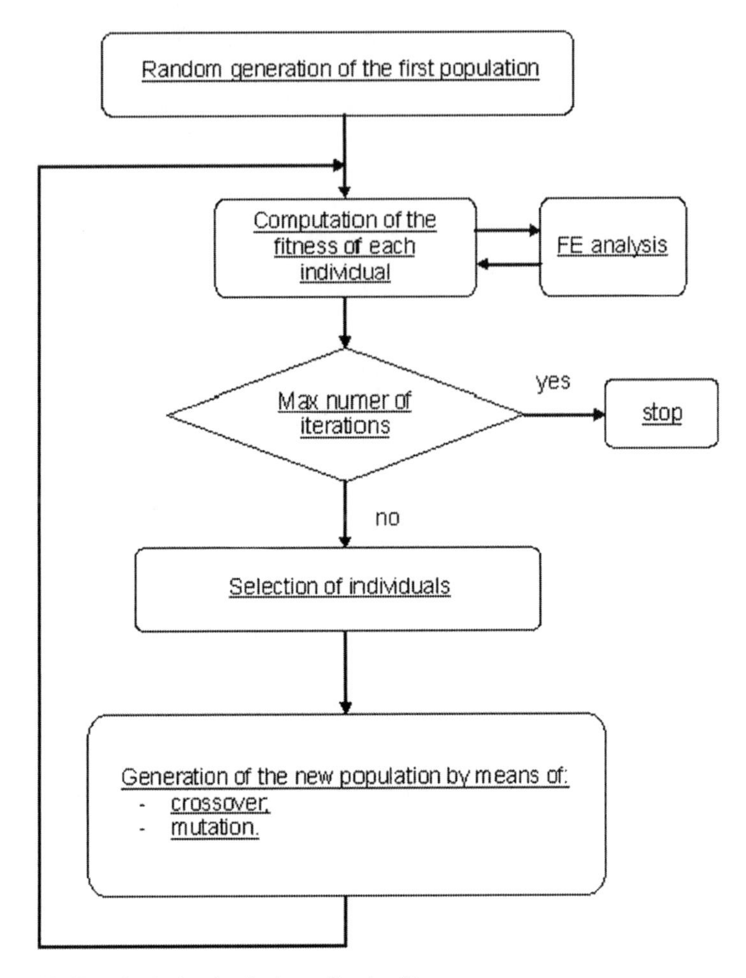

Fig. 19. Flow chart of a standard genetic algorithm.

the mutation probability is 0.15, based on Michalewicz (34), the constant A in Eq. 2 was assumed to be 10,000 as it must be sufficiently large to guarantee that the difference between A and rest of the equation components is a positive number.

It has been assumed that the scaffold is under uniform distributed compressive load. Structural simulations were performed with the use of the MSC/Nastran FEM. The results obtained with the proposed optimization algorithm exhibit convergence to an optimal solution after the hundredth iterations (Fig. 20).

The best properties of the scaffolds are obtained with the algorithm discussed for a combination of five different fiber materials arranged as shown in Fig. 21. The arrangement is quite complex in terms of the polymer distribution. The obtained curve of the scaffold stiffness (curve of scaffold degradation), which is a function of the bone healing time (polymer degradation time) fits well to the desirable curve of degradation (Fig. 22).

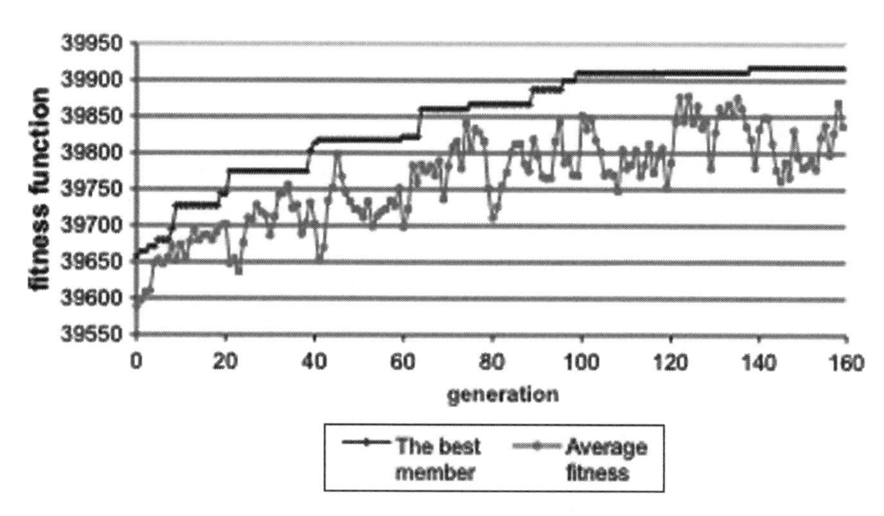

Fig. 20. Fitness function of the best member and averaged number of fitness functions (referring to the entire population).

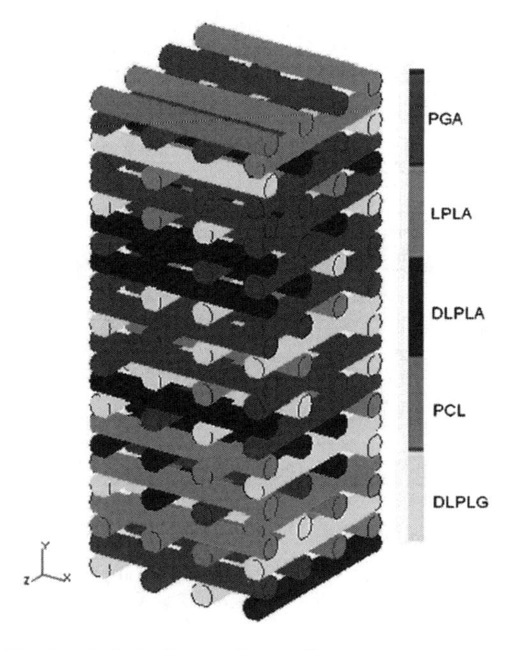

Fig. 21. Fibers blends selected with genetic algorithm.

One of the requirements, which scaffold should fulfill is uniform distribution of the load (none of the struts should be overload) at each stage of the process of degradation. To control the properties of the scaffolds, the displacement in x and z axis direction have been analyzed for the period considered necessary for the process of tissue regeneration. The results are presented in Fig. 23, which clearly show that the stress distribution within the structure is quite uniform. This implies uniform load transfer and ensures stability of the scaffold at each stage of its degradation.

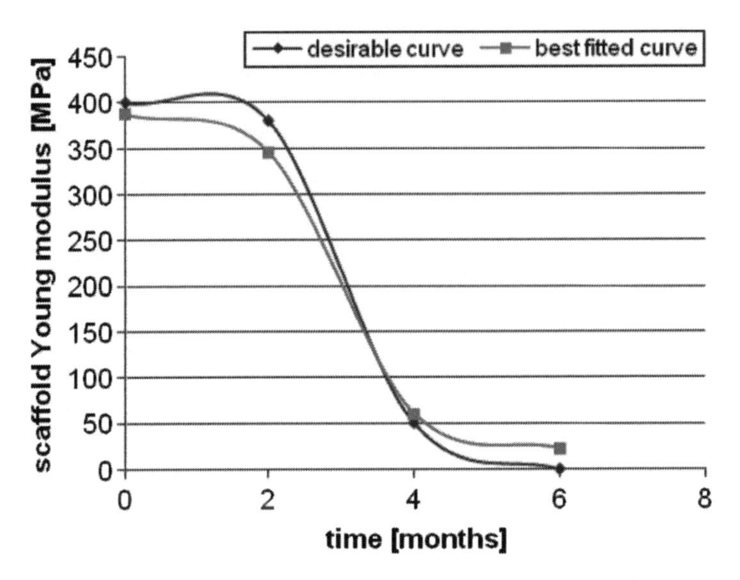

Fig. 22. Desirable curve of degradation and the curve (the best fit) of degradation obtained for the scaffold visualized in Fig. 21.

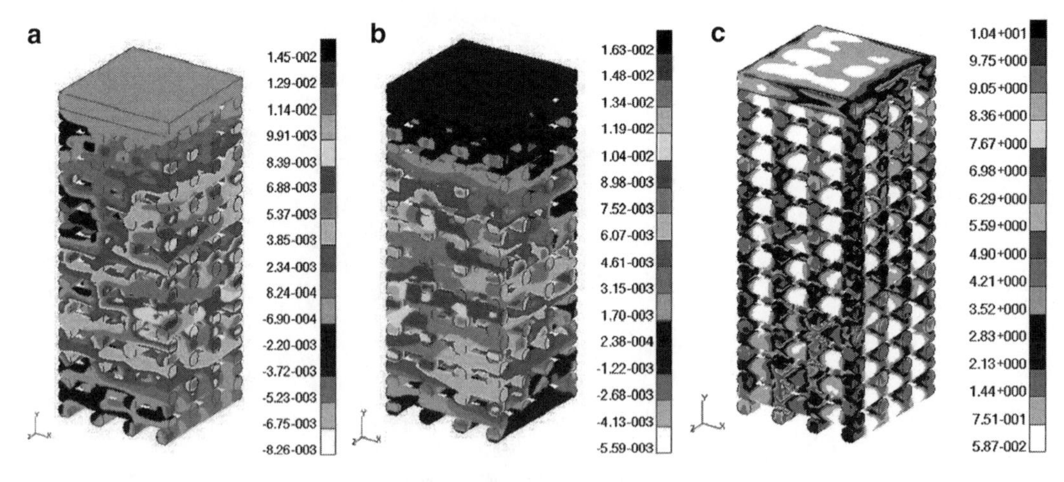

Fig. 23. Displacements (mm) of optimized scaffold (sixth month of degradation) (**a**) in *X* axis, (**b**) in *Z* axis, (**c**) Von Mises stress distribution (MPa) in optimized scaffold (sixth month of degradation).

In conclusion, it should be emphasized that the available methods for the optimization of scaffolds, in particular those based on GA, do achieve the objective. It should be recognized that the optimum structures are complex in terms of the selection of the appropriate fibrous materials and their spatial arrangement. Finally, it should also be appreciated that scaffolds of the optimum design can be fabricated by rapid prototyping methods.

Acknowledgments

The work was partially supported by Polish Ministry of Science and A*STAR under the Polish–Singapore Collaboration. The authors would like to thank Marcin Heljak for his help with the FE simulations.

References

1. Baroni S, Gironcoli S, Dal CA, Giannozzi P (2001) Rev. Mod Phys 73:515
2. Kurzydlowski KJ (2009) Proceedings of LMT, Brisbane, 2009
3. Rapaport DC (2004) The art of molecular dynamics simulation. Cambridge University Press, Cambridge
4. Zienkiewicz OC, Taylor RJ (eds) (2000) Finite element method. Butterworth Heinemann, Oxford
5. Swieszkowski W, Ku D, Bersee H, Kurzydlowski KJ (2006) An elastic material for cartilage replacement in arthritic shoulder joint. Biomaterials 27:1534–1541
6. Rundell SA, Auerbach JD, Balderston RA, Kurtz SM (2008) Total disc replacement positioning affects facet contact forces and vertebral body strains. Spine 33(23):2510–2517
7. del Palomar AP, Calvo B, Doblaré M (2008) An accurate finite element model of the cervical spine under quasi-static loading. J Biomech 41 (3):523–531
8. Balac I, Milovanevic M, Tang Ch, Uskokovic PS, Uskokovic DP (2004) Estimation of elastic properties of a particulate polymer composite using a face-centered cubic FE model. Mater Lett 58(19):2437–2441
9. Pauwels F (1960) A new theory on the influence of mechanical stimuli on the differentiation of supporting tissue. The tenth contribution to the functional anatomy and causal morphology of the supporting structure. Z Anat Entwicklungsgesch 121:478–515
10. Prendergast PJ, Huiskes R, Soballe K (1997) ESB Research Award 1996. Biophysical stimuli on cells during tissue differentiation at implant interfaces. J Biomech 30(6):539–548
11. Carter DR, Beaupre GS, Giori NJ, Helms JA (1998) Mechanobiology of skeletal regeneration. Clin Orthop Relat Res 355(Suppl): S41–S55 (review)
12. Claes LE, Heigele CA (1999) Magnitudes of local stress and strain along bony surfaces predict the course and type of fracture healing. J Biomech 32(3):255–266
13. Lacroix D, Prendergast PJ (2002) A mechano-regulation model for tissue differentiation during fracture healing: analysis of gap size and loading. J Biomech 35(9):1163–1171
14. Shefelbine SJ, Augat P, Claes L, Simon U (2005) Trabecular bone fracture healing simulation with finite element analysis and fuzzy logic. J Biomech 38(12):2440–2450
15. Isaksson H, Wilson W, van Donkelaar CC, Huiskes R, Ito K (2006) Comparison of biophysical stimuli for mechano-regulation of tissue differentiation during fracture healing. J Biomech 39(8):1507–1516
16. Gomez-Benito MJ, Garcia-Aznar JM, Kuiper JH, Doblare M (2005) Influence of fracture gap size on the pattern of long bone healing: a computational study. J Theor Biol 235 (1):105–119
17. Isaksson H, Comas O, van Donkelaar CC, Mediavilla J, Wilson W, Huiskes R, Ito K (2007) Bone regeneration during distraction osteogenesis: mechano-regulation by shear strain and fluid velocity. J Biomech 40 (9):2002–2011
18. Kelly DJ, Prendergast PJ (2005) Mechano-regulation of stem cell differentiation and tissue regeneration in osteochondral defects. J Biomech 38(7):1413–1422
19. Andreykiv A, Prendergast PJ, van Keulen F, Swieszkowski W, Rozing PM (2005) Bone ingrowth simulation for a concept glenoid component design. J Biomech 38(5): 1023–1033
20. Kelly DJ, Prendergast PJ (2006) Prediction of the optimal mechanical properties for a scaffold used in osteochondral defect repair. Tissue Eng 12(9):2509–2519
21. Adachi T, Osako Y, Tanaka M, Hojo M, Hollister SJ (2006) Framework for optimal design of porous scaffold microstructure by computational simulation of bone regeneration. Biomaterials 27(21):3964–3972
22. Ament C, Hofer EP (2000) A fuzzy logic model of fracture healing. J Biomech 33(8): 961–968

23. Epari DR, Taylor WR, Heller MO, Duda GN (2006) Mechanical conditions in the initial phase of bone healing. Clin Biomech (Bristol, Avon) 21(6):646–655
24. Bailon-Plaza A, van der Meulen MC (2001) A mathematical framework to study the effects of growth factor influences on fracture healing. J Theor Biol 212(2):191–209
25. Olsen L, Sherratt JA, Maini PK, Arnold F (1997) A mathematical model for the capillary endothelial cell-extracellular matrix interactions in wound-healing angiogenesis. IMA J Math Appl Med Biol 14(4):261–281
26. Stevens MM (2008) Biomaterials for bone tissue engineering. Mater Today 11(5):18–25
27. Hutmacher DW (2000) Scaffolds in tissue engineering bone and cartilage. Biomaterials 21(24):2529–2543
28. Heljak M, Swieszkowski W, Lam CXF, Hutmacher DW, Kurzydlowski KJ (2008) Numerical analyses of the polymeric scaffolds for bone tissue engineering. Tissue Eng A 14:890
29. Middleton JC, Tipton AJ (2000) Synthetic biodegradable polymers as orthopaedics devices. Biomaterials 21:2335–2346
30. Goetzen N, Lampe F, Nassut R, Morlock MM (2005) Load-shift-numerical evaluation of a new design philosophy for uncemented hip prostheses. J Biomech 38:595–604
31. Zein I et al (2002) Fused deposition modeling of novel scaffold architectures for tissue engineering applications. Biomaterials 23:1169–1185
32. Mow C, Huiskes R (2004) Basic orthopaedics biomechanics and mechano-biology. Lippincott Williams & Wilkins, Philadelphia
33. Sandino C, Planell JA, Lacroix D (2008) A finite element study of mechanical stimuli in scaffolds for bone tissue engineering. J Biomech 41:1005–1014
34. Michalewicz Z (1996) Genetic algorithms + data structures = evolutionary programs. Springer Verlag, Berlin
35. Cappello F, Mancuso A (2003) A genetic algorithm for combined topology and shape optimisations. Comp Aided Design 35:761–769

Chapter 12

Structural and Vascular Analysis of Tissue Engineering Scaffolds, Part 1: Numerical Fluid Analysis

Henrique A. Almeida and Paulo J. Bártolo

Abstract

Rapid prototyping technologies were recently introduced in the medical field, being particularly viable to produce porous scaffolds for tissue engineering. These scaffolds should be biocompatible, biodegradable, with appropriate porosity, pore structure, and pore distribution on top of presenting both surface and structural compatibility. This chapter presents the state-of-the-art in tissue engineering and scaffold design using numerical fluid analysis for optimal vascular design. The vascularization of scaffolds is an important aspect due to its influence regarding the normal flow of biofluids within the human body. This computational tool also allows to design either a scaffold offering less resistance to the normal flow of biofluids or reducing the possibility for blood coagulation through forcing the flow toward a specific direction.

Key words: Tissue engineering, Scaffolds, Biofabrication, Scaffold vascularization, Computer simulation

1. Introduction

The loss or failure of an organ or tissue is a frequent, devastating, and costly problem in health care. Currently, this problem is treated by transplanting organs from one individual to another or performing surgical reconstructions by transferring tissue from one location in the human body into the diseased site. The need for substitutes to replace or repair tissues or organs due to disease, trauma, or congenital problems is overwhelming. In order to overcome this organ deficiency, a new field named tissue engineering emerged and has been growing and gaining more and more importance in the scientific community. The disparity between the need and availability of donor tissues (Fig. 1) has motivated the development of tissue engineering approaches, aimed at creating cell-based substitutes of native tissues (1, 2).

Michael A.K. Liebschner (ed.), *Computer-Aided Tissue Engineering*, Methods in Molecular Biology, vol. 868,
DOI 10.1007/978-1-61779-764-4_12,

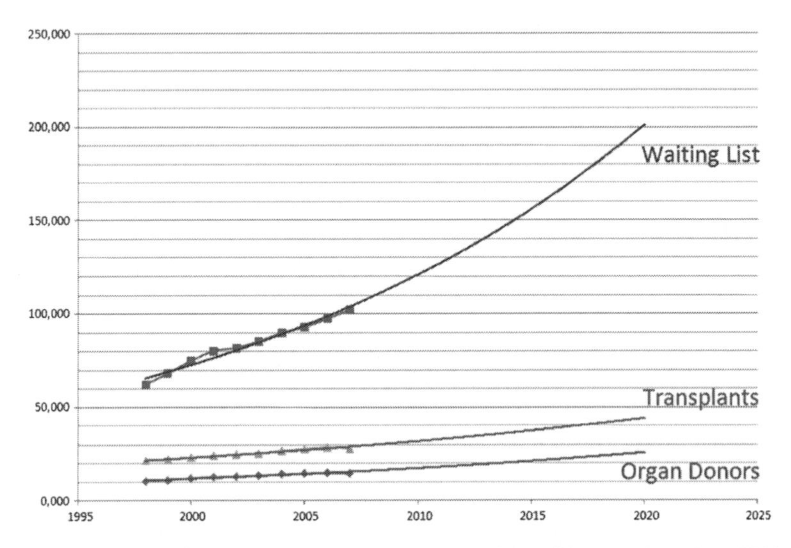

Fig. 1. Organ donations, waiting lists and transplants of the USA population, since 1998 until 2007 (U.S. Organ Procurement and Transplantation Network) with a prediction curve until the year 2020.

Tissue engineering is a multidisciplinary field that combines the principles of biology, engineering, and medicine to create biological substitutes for lost or defective native tissues (1, 3–8). According to Skalak and Fox (9), tissue engineering can be described as "the application of the principles and methods of engineering and life sciences toward the fundamental understanding of structure–function relationships in normal and pathological mammalian tissues and the development of biological substitutes to restore, maintain, or improve function" (5).

Three general strategies have been adopted for the creation of new tissues. These strategies are (2, 10, 11):

1. Cell self-assembly, which corresponds to the direct in vivo implantation of isolated cells or cell substitutes and it is based on cells synthesizing their own matrix. This approach avoids the complications of surgery and allows replacement of only those cells that supply the needed function. The main limitations include immunological rejection and failure of the encapsulated cells.
2. Acellular scaffold, which is based on the ingrowth of tissue cells into a porous material, loaded with growth factors or any other therapeutic agent.
3. Cell-seeded temporary scaffolds, which is based on the use of a temporary scaffold that provides a substrate for the implanted cells and a physical support to organize the formation of the new tissue. In this approach, transplanted cells adhere to the scaffold, proliferate, secrete their own extracellular matrices and stimulate new tissue formation. This is the most commonly used strategy in tissue engineering.

Cells used in tissue engineering may be allogenic, xenogenic, syngeneic, or autologous (10). They should be nonimmunogenic, highly proliferate, easy to harvest and with high capacity to differentiate into a variety of cell types with specialized functions (10, 12). Cell attachment to materials is correlated to many factors, such as the stiffness and attachment area. Skeletal muscle satellite cells, cardiomyocytes, and endothelial cells have been used in many tissue engineering applications. An ideal cell source for tissue engineering should have the capacity to proliferate and then differentiate in vitro, in a manner that can be reproducibly controlled. The choice of cell type can affect both in vitro culture requirements and in vivo function of scaffolds.

The growth and differentiation of many cell types is regulated by four major sources of external signalling:

1. Soluble growth and differentiation factors.
2. Nature and organization of insoluble and soluble extracellular matrix (ECM) constituents.
3. Intercellular interactions.
4. Environmental stress induced by fluid flow and/or mechanical stimuli, as well as other physical cues (oxygen, tension, and pH effects).

The cell-seeded temporary scaffolds approach is the most widely used. In this case, living cells are obtained from a tissue harvest, either from the patient (autograft) or from a different person (allograft), and cultured in vitro on a three-dimensional scaffold to obtain a tissue construct suitable for transplantation (10). Scaffolds provide an initial biochemical substrate so that the novel tissue may grow until the cells produce their own extracellular matrix (ECM). Therefore, scaffolds not only define the 3D space for the formation of new tissues, but also serve to provide tissues with appropriate functions.

The primary roles of a scaffold in tissue engineering are (2, 13–17):

- To serve as an adhesion substrate for the cell.
- To provide temporary mechanical support to the newly grown tissue.
- To guide the development of new tissues with the appropriate function.

An ideal scaffold must satisfy some biological and mechanical requirements (5, 18–21):

(a) *Biological requirements*

- Biocompatibility—the scaffold material must be nontoxic allowing cell attachment, proliferation, and differentiation (22).

- Biodegradability—the scaffold material must degrade into nontoxic products.
- Controlled degradation rate—the degradation rate of the scaffold must be adjustable in order to match the rate of tissue regeneration.
- Appropriate macro and microstructure of the pores and shape, highly interconnected pore structure and large surface area to allow high seeded cells and to promote neovascularization (23–37). Large number of pores may be able to enhance vascularization while smaller pore diameter is preferable to provide large surface per volume ratio.
- Ability to carry biomolecular signals, such as growth factors. Numerous growth factors have been identified, such as fibroblast growth factor (FGF), platelet-derived growth factor, bone morphogenic protein (BMP), insulin growth factor, transforming growth factor, epidermal growth factor, vascular endothelial growth factor, etc. (38, 39).

(b) *Mechanical and physical requirements*

- Sufficient strength and stiffness to withstand stresses in the host tissue environment (40–43).
- Adequate surface finish to guarantee a good biomechanical coupling achieved between the scaffold and the tissue (44–47). Cell response is affected by the physicochemical parameters of the biomaterial surface, such as surface energy, surface charges, or chemical composition (Fig. 2). New efforts to encourage cell attachment focusing on mimicking the surface chemistry of autogenous ECM has been also reported (48). Surface properties, such as surface charge and surface topography, can influence biocompatibility (49).
- Easy sterilization by either exposure to high temperatures or immersing in a sterilization agent, remaining unaffected by either of these processes.

Scaffolds must be processed into open porous structure that fills the tissue defect enabling tissue ingrowth. The techniques for 3D scaffold fabrication can be divided into two broad categories: conventional techniques and additive technologies (3, 5). Conventional techniques include fibre bonding, solvent casting and particulate leaching, melt moulding, and gas foaming (19, 41, 50). Each of these techniques presents several limitations as they usually do not enable to properly control pore size, pore geometry, and spatial distribution of pores, besides being almost unable to construct internal channels within the scaffold (51). Beyond these limitations, these techniques usually involve the use of toxic organic solvents,

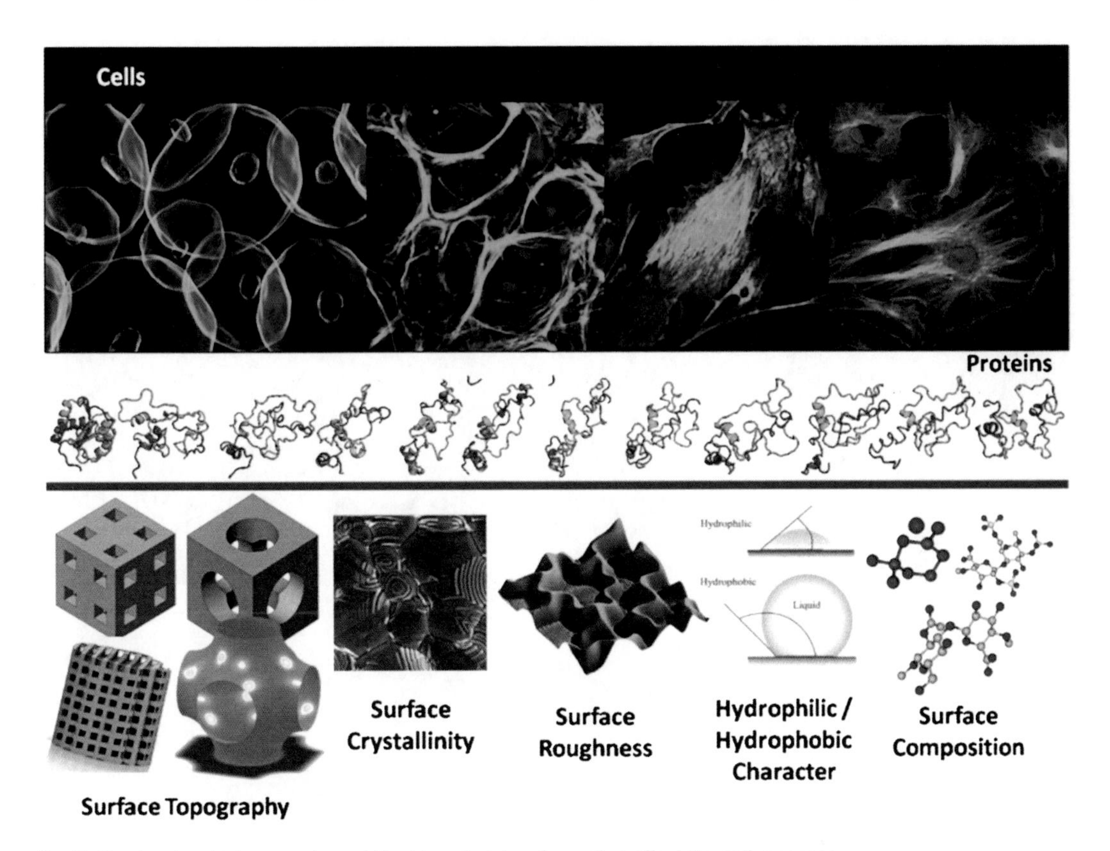

Fig. 2. Physicochemical parameters of the biomaterial surfaces that affect the cell response.

long fabrication times on top of being labour-intensive processes. Through the use of conventional techniques it is not possible to produce scaffolds that will enable the formation of thick 3D tissues as cells proliferate only at the surface of the produced matrices (Fig. 3). Besides this important limitation due to the lack of pore interconnectivity that promotes the so-called M&M effect (cells died due to the lack of nutrients supply and vascularization), conventional techniques are limited in terms of reproducibility and repeatability.

Additive technologies, also designated as biomanufacturing technologies, are considered a viable alternative to fabricate scaffolds for tissue engineering applications. Biomanufacturing was initially defined in 2005 during the Biomanufacturing Workshop hosted by Tsinghua University in China as "the use of additive technologies, biodegradable and biocompatible materials, cells, growth factors, etc., to produce biological structures for tissue engineering applications" (52). More recently, during the Spring 2008 Meeting sponsored by the American National Science Foundation, biomanufacturing was defined as "the design, fabrication, assembly and measurement of bio-elements into structures, devices, and systems, and their interfacing and integration into/with larger scale structures in vivo or in vitro environment such

Fig. 3. Illustration of the cell growth and proliferation of the outer layer of the scaffold, inducing the M&M effect.

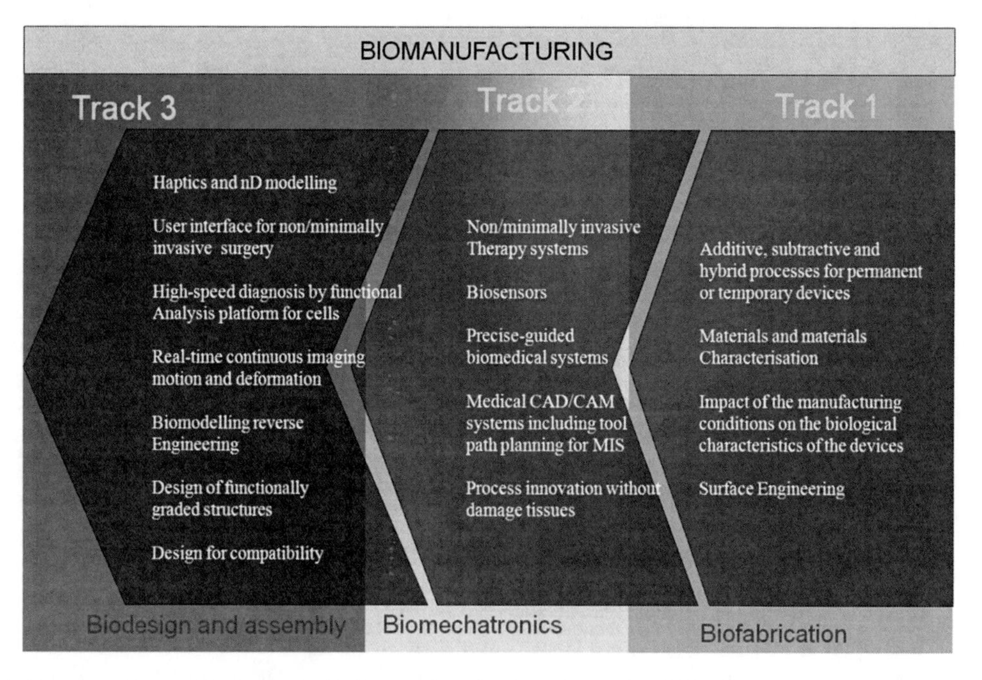

Fig. 4. The main pillars and topics related to biomanufacturing according to the CIRP CWG on Biomanufacturing.

that heterogeneity, scalability and sustainability are possible." In 2009, during the 59th General Assembly of the International Academy for Production Engineering (CIRP) in Boston a collaborative working group on biomanufacturing that is based on three main pillars was established: biofabrication, biomechatronics, and biodesign and assembly (Fig. 4).

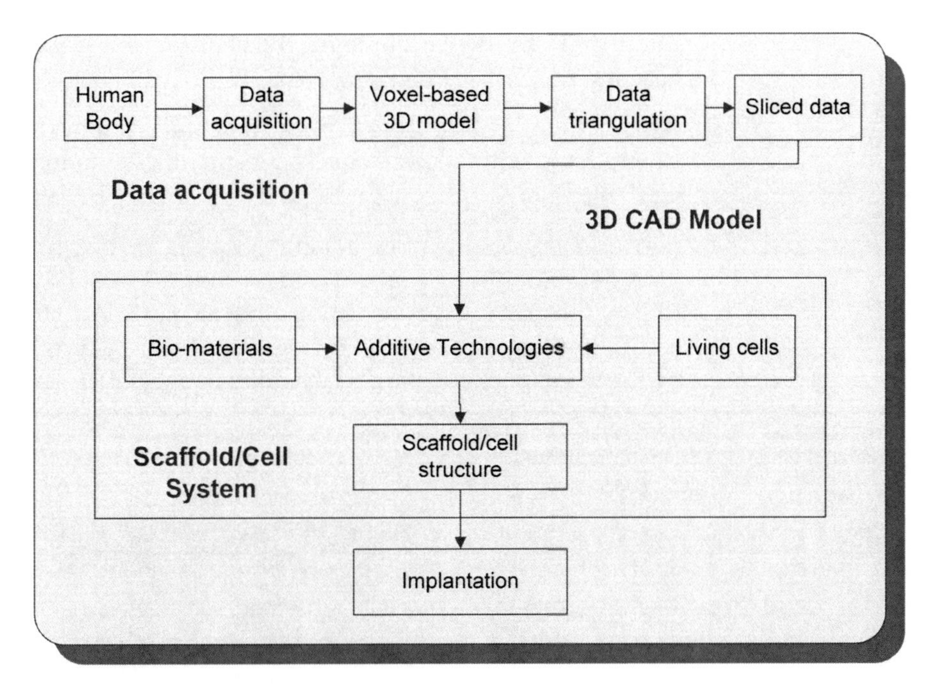

Fig. 5. Steps of biomanufacturing in tissue engineering.

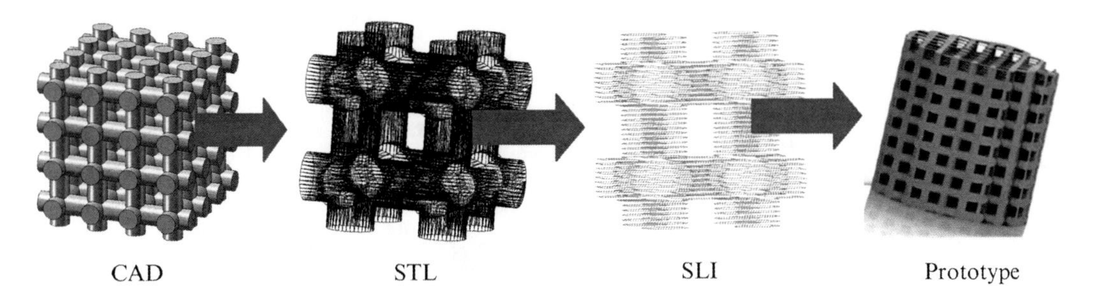

Fig. 6. Flowchart of information to produce a 3D scaffold.

Figure 5 provides a general overview of the necessary steps to produce scaffolds for tissue engineering through biomanufacturing technologies (53, 54). The first step is the generation of the corresponding computer solid model through one of the currently available medical imaging techniques, such as computer tomography, magnetic resonance imaging, etc. These imaging methods produce continuous volumetric data (voxel-based data), which provide the input data for the digital model generation (53). The model is then tessellated as an STL file (Fig. 6). The STL model is then mathematically sliced into thin layers (sliced model). This data is then sent to a biomanufacturing device in order to produce the biocompatible and biodegradable construct containing or not cells and growth factors. Once produced the scaffold model, it is then ready for implantation.

Biomanufacturing technologies include several fabrication approaches, such as (see Fig. 7):

- Stereolithographic processes: MicroStereolithography (55, 56), stereolithography, (57–60) and stereo-thermal-litography (61).
- Laser sintering processes (8, 62).
- Extrusion-based processes: 3D Fibre deposition (63), bioextruder (64), bioplotting (65, 66), cell printing (53, 67, 68), direct cell writing (69), fused deposition modelling (70–74), precision extruding deposition (75, 76), pressure-assisted microsyringe (6), and rapid prototyping robotic dispersing (77).

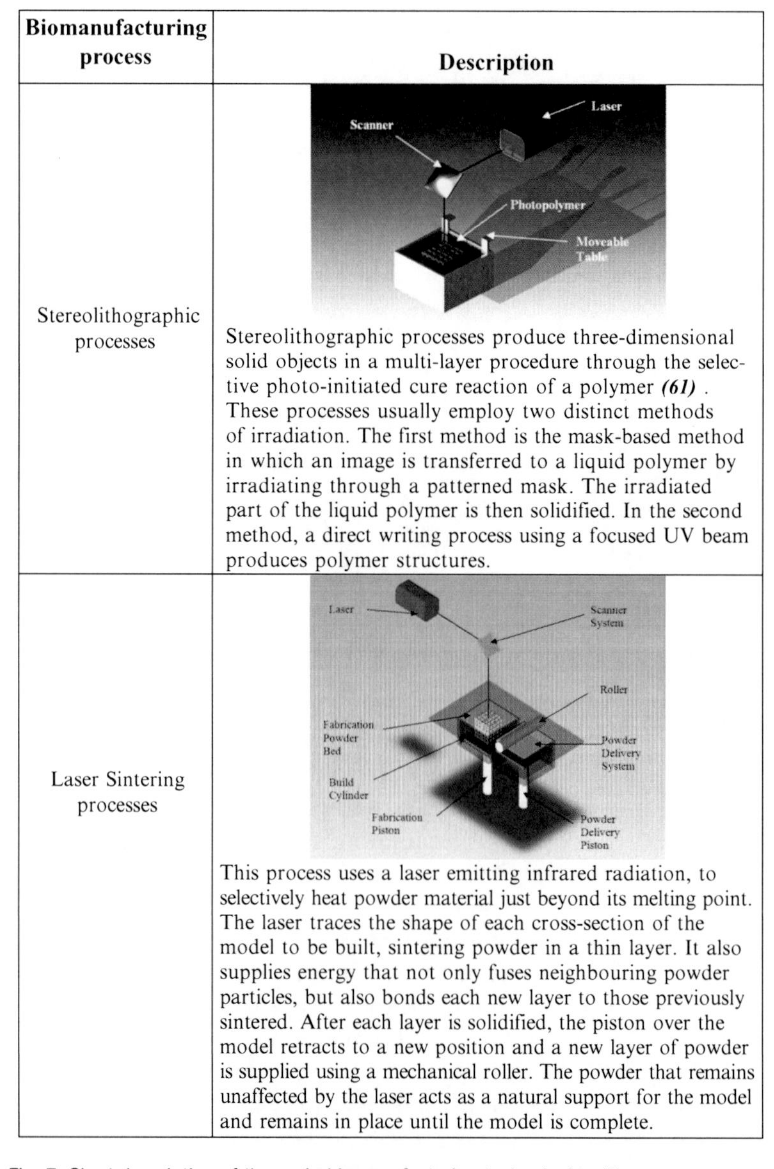

Biomanufacturing process	Description
Stereolithographic processes	Stereolithographic processes produce three-dimensional solid objects in a multi-layer procedure through the selective photo-initiated cure reaction of a polymer ***(61)*** . These processes usually employ two distinct methods of irradiation. The first method is the mask-based method in which an image is transferred to a liquid polymer by irradiating through a patterned mask. The irradiated part of the liquid polymer is then solidified. In the second method, a direct writing process using a focused UV beam produces polymer structures.
Laser Sintering processes	This process uses a laser emitting infrared radiation, to selectively heat powder material just beyond its melting point. The laser traces the shape of each cross-section of the model to be built, sintering powder in a thin layer. It also supplies energy that not only fuses neighbouring powder particles, but also bonds each new layer to those previously sintered. After each layer is solidified, the piston over the model retracts to a new position and a new layer of powder is supplied using a mechanical roller. The powder that remains unaffected by the laser acts as a natural support for the model and remains in place until the model is complete.

Fig. 7. Short description of the main biomanufacturing technologies (5).

Extrusion-based processes	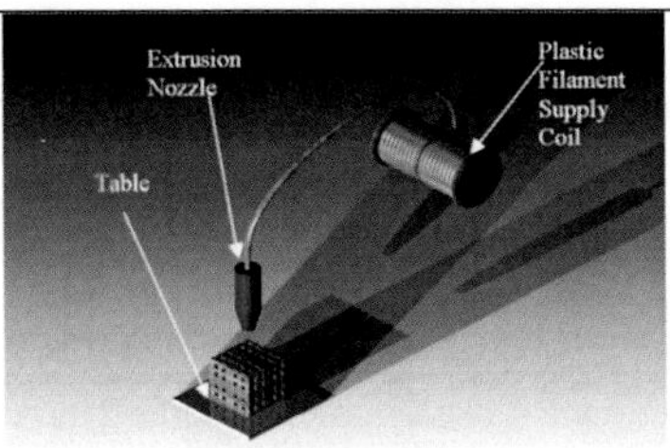 By this process, thin thermoplastic filaments are melted by heating and guided by a robotic device controlled by a computer, to form the object. The material leaves the extruder in a liquid form and hardens immediately. The previously formed layer, which is the substrate for the next layer, must be maintained at a temperature just below the solidification point of the thermoplastic material to assure good interlayer adhesion
Three-dimensional printing	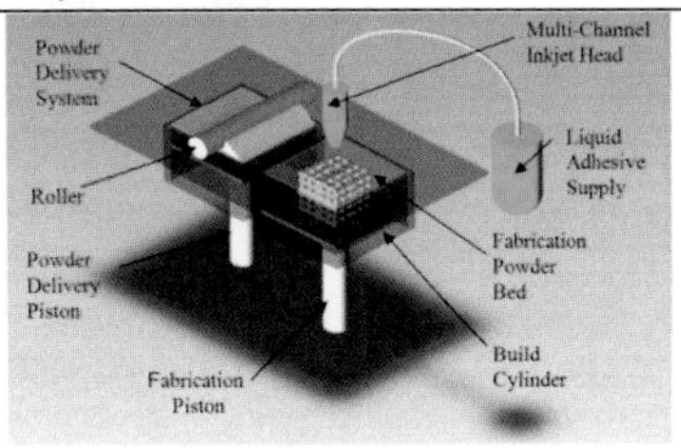 This process deposits a stream of microparticles of a binder material over the surface of a powder bed, joining particles together where the object is to be formed. A piston lowers the powder bed so that a new layer of powder can be spread over the surface of the previous layer and then selectively joined to it.

Fig. 7. (continued)

- 3D printing processes: 3D Printing (35, 78–81) and ModelMaker II (82, 83).

The main advantages of biomanufacturing technologies are both the capacity to rapidly produce very complex 3D models and the ability to use various raw materials. In the field of tissue engineering, biomanufacturing technologies are used to produce scaffolds with customized external shape and predefined internal morphology, allowing good control of pore distribution and size (5, 18, 41, 51, 53, 84–86). Reviews of currently used biomanufacturing technologies for tissue engineering applications can be found at (3–5, 18).

2. Smart Materials Selector

In the past decades, several tools have been developed to support the design of scaffolds for tissue engineering applications. Examples of computation tools are: computer-aided system for tissue scaffolds (CASTS) developed by Naing (87) and computer-aided tissue engineering (CATE) developed by Sun (30, 88). These computation tools integrate several tools ranging from design to engineering to manufacturing, all with the objective of aiding the design of scaffolds for tissue engineering applications. Another computational tool to aid the design of scaffolds for tissue engineering applications is discussed in the next section.

To aid this process, a novel computational tool, designated expert system for medical applications (ESYSMA) (89) has been developed to aid the selection of biomaterials for a given medical application. The expert system is structured in three main levels:

- Level 1: automatic identification of the appropriated biomaterial for a certain application.
- Level 2: automatic definition of the appropriated geometry, porosity, and pore topology of scaffolds for a certain application.
- Level 3: automatic definition of the growth factor to be used in a certain application.

According to the hierarchical knowledge-based structure combining case-based and ruled-based reasoning, the material definition for a specific medical application is performed on specifications of the desired biomaterial's application. The identification of the most adequate material is carried out though a set of questions and corresponding answers concerning issues, such as: the biodegradability of the material, its permeability to water, its absorption, if it will be in contact with organic fluids, if it will need to sustain a structural application, if it has to be environmentally responsive to changes of pH and/or temperature, etc. An extensive database of biomaterials was also developed an integrated to the proposed system. The existence of a computational tool to support the selection process of appropriate biomaterials for medical applications is extremely important.

3. Vascular Analysis of Scaffolds in Tissue Engineering

Most tissues in the body rely on blood vessels to supply the individual cells with nutrients and oxygen. For a tissue to grow beyond 100–200 μm (the diffusion limit of oxygen), new blood-vessel formation is required (90), and this is also true for tissue-engineered constructs. During in vitro culture, larger tissue-engineered constructs can be supplied with nutrients, for instance in perfusion bioreactors (91, 92). However, after implantation of tissue constructs, the supply of oxygen and nutrients to the implant is often limited by diffusion processes that can only supply cells in a proximity of 100–200 μm from the next capillary. In order for implanted tissues of greater size to survive, the tissue has to be vascularized, which means that a capillary network capable of delivering nutrients to the cells is formed within the tissue. After implantation, blood vessels from the host generally invade the tissue to form such a network, in part in response to signals that are secreted by the implanted cells as a reaction to hypoxia. However, this spontaneous vascular ingrowth is often limited to several tenths of micrometres per day (93), meaning that the time needed for complete vascularization of an implant of several millimetres is in the order of weeks. During this time, insufficient vascularization can lead to nutrient deficiencies and/or hypoxia deeper in the tissue. Moreover, nutrient and oxygen gradients will be present in the outer regions of the tissue, which could result in nonuniform cell differentiation and integration and thus decreased tissue function (94).

Because the speed of vascularization after implantation is a major problem in tissue engineering, the successful use of tissue-engineered constructs is currently limited to thin or avascular tissues, such as skin or cartilage, for which postimplantation neovascularization from the host is sufficient to meet the demand for oxygen and nutrients (95). To succeed in the application of tissue engineering for bigger tissues, such as bone and muscle, the problem of vascularization has to be solved (96).

3.1. Blood Characteristics

Blood is a specialized bodily fluid (technically a tissue) composed of a liquid called blood plasma and blood cells suspended within the plasma. The blood cells present in blood are red blood cells (also called erythrocytes), white blood cells (including both leukocytes and lymphocytes), and platelets (also called thrombocytes). Plasma is predominantly water containing dissolved proteins, salts, and many other substances, which amounts up to 55% of blood by volume (97).

The most abundant cells in blood are by far red blood cells. These contain haemoglobin, an iron-containing protein, which facilitates transportation of oxygen by reversibly binding to this respiratory gas and greatly increasing its solubility in blood. In contrast, carbon dioxide is almost entirely transported extracellularly dissolved in plasma as bicarbonate ion. White blood cells help to resist infections and parasites, and platelets are important in blood clotting.

Blood is circulated around the body through blood vessels by the pumping action of the heart. Arterial blood carries oxygen from inhaled air to the body tissues and venous blood carries carbon dioxide, a waste product of metabolism produced by cells, from the tissues to the lungs to be exhaled.

Blood pressure refers to the force exerted by circulating blood within the walls of blood vessels. The pressure of the circulating blood decreases as blood moves through arteries, arterioles, capillaries, and veins. Usually, blood pressure refers to arterial pressure, i.e., the pressure in the larger arteries, where arteries are the blood vessels taking blood away from the heart. Arterial pressure is most commonly measured via a sphygmomanometer, which uses the height of a column of mercury to reflect the circulating pressure. Although many modern vascular pressure devices no longer use mercury, vascular pressure values are still universally reported in millimetres of mercury (mmHg).

The systolic arterial pressure is defined as the peak pressure in the arteries, which occurs near the beginning of the cardiac cycle, while the diastolic arterial pressure is the lowest pressure at the resting phase of the cardiac cycle. The average pressure throughout the cardiac cycle is a mean arterial pressure, while the pulse pressure reflects the difference between the maximum and minimum pressures measured.

Typical values for a resting, healthy adult human are approximately 120 mmHg (16 kPa) systolic and 80 mmHg (11 kPa) diastolic with large individual variations. These measures of arterial pressure are not static, but undergo natural variations from one heartbeat to another, as well throughout the day in a circadian rhythm. They also change in response to stress, nutritional factors, drugs, or disease (97).

Whole blood has a pseudoplastic flow behaviour. A pseudoplastic behaviour is characterized by a viscosity (resistance to flow) that is lower at higher shear rates than at lower shear rates. A pseudoplastic behaviour is also called shear thinning, mainly due to particles within the fluid aggregating at low shear rates but "breaking" at higher shear rates, which lowers the viscosity. A pseudoplastic

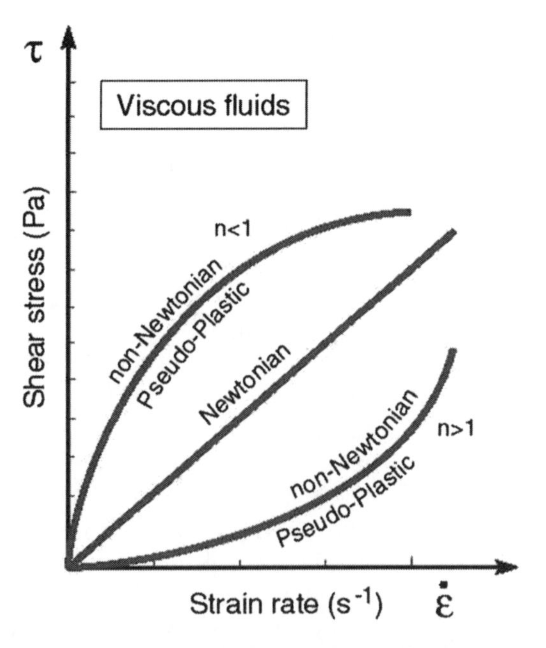

Fig. 8. Illustration of Newtonian and non-Newtonian fluid behaviour.

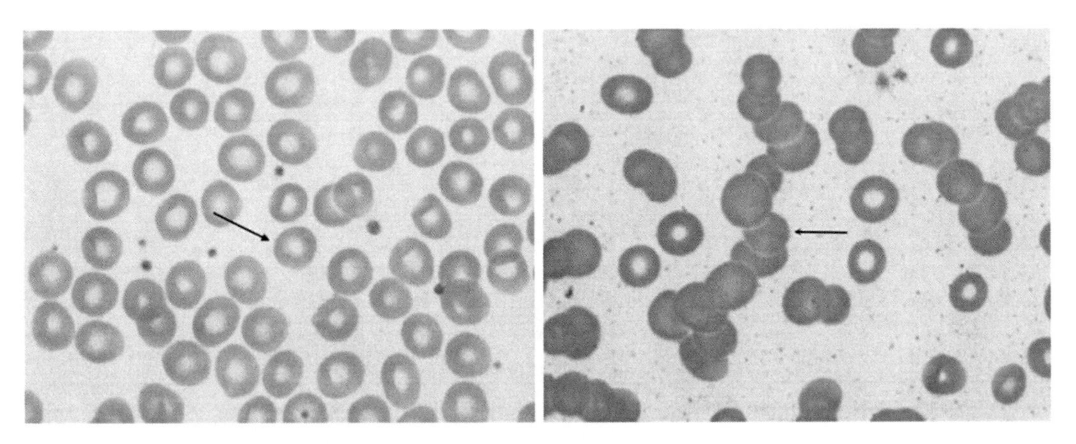

Fig. 9. Illustration of single red blood cells and an aggregate of red blood cells, rouleaux phenomenon.

fluid is a fluid in which its viscosity decreases with an increasing shear rate (Fig. 8).

Red blood cells tend to aggregate at lower shear rates, a phenomenon known as rouleaux (Fig. 9), which depends on the presence of fibrinogen and globulins. In the limit, as the shear rate goes to zero, the blood will tend to further aggregate, eventually leading to a process known as clotting, which involves additional mechanisms, including platelet activation and the conversion of fibrinogen to fibrin, an essential component of a clot. As the shear rate increases from low, but nonzero values, the rouleaux break up and blood behaves like a Newtonian fluid. The latter is often assumed in large arteries with a viscosity $\mu \sim$ constant (often cited to be ~3.5 cP, or centiPoise) and the Navier–Poisson equations to describe most blood flows. Plasma (whole blood minus cells) always behaves as a Newtonian fluid, with a viscosity $\mu \sim 1.2$ cP. In capillaries, which are ~5–8 μm in diameter, the red blood cells go through one at a time, with plasma in between. In this case, the blood should be treated as a two-phase flow—a solid and a fluid mixture (98).

3.2. Constitutive Equation of Blood Flow

As mentioned before, blood is a mixture of blood plasma and blood cells. The mixture, when tested in viscometers whose characteristic dimension is much larger than the characteristic size of the blood cells present a homogeneous fluid behaviour. This assumption of isotropy is motivated by the idea that when the shear stress and shear strain rate are zero, the blood cells have no preferred orientation. This assumption of incompressibility is based on the fact that, in the range of pressures concerned in physiology, the mass densities of the plasma, the cells, and the mixture as a whole are unaffected by the pressure. The rheology of blood differs from that of a Newtonian fluid only in the fact that the coefficient of viscosity is not constant. The constitutive equation of an isotropic incompressible Newtonian fluid is (99):

$$\sigma_{ij} = -p\delta_{ij} + 2\mu V_{ij}, \tag{1}$$

where

$$V_{ij} = \frac{1}{2}\left(\frac{\partial u_i}{\partial x_j} + \frac{\partial u_j}{\partial x_i}\right) \text{ and } V_{ii} = V_{11} + V_{22} + V_{33} = 0. \tag{2}$$

Here, σ_{ij} is the stress tensor, V_{ij} the strain-rate tensor, u_i the velocity component, δ_{ij} the isotropic tensor or Kronecker delta, p the hydrostatic pressure, and μ is a constant called the coefficient of viscosity. The indices i and j range over 1, 2, and 3, and the components of the tensors and vectors are referred to a set of rectangular Cartesian coordinates x_1, x_2, and x_3.

Blood does not obey Eq. 1 because μ is not constant due to its variation with the strain rate. Thus, blood is said to be non-Newtonian. Similar experiments do verify, however, that normal plasma alone is Newtonian. Therefore, the non-Newtonian feature of whole blood comes from the cellular bodies in the blood.

One of the most important principles in continuum mechanics is that any equation must be tensorially correct, i.e., every term in an equation must be a tensor of the same level. Thus, if one tries to adapt this principle to Eq. 1 for blood, under the assumption that its mechanical behaviour is isotropic, and then μ is a scalar, i.e., μ must be a scalar function of the strain rate tensor V_{ij}. Now, V_{ij} is a symmetric tensor of level 2 in three dimensions. It has three invariants which are scalars. Hence, the coefficient of viscosity μ must be a function of the invariants of V_{ij}. These invariants are (98, 99):

$$\begin{aligned} I_1 &= V_{11} + V_{22} + V_{33} \\ I_2 &= \begin{vmatrix} V_{11} & V_{12} \\ V_{21} & V_{22} \end{vmatrix} + \begin{vmatrix} V_{22} & V_{23} \\ V_{32} & V_{23} \end{vmatrix} + \begin{vmatrix} V_{33} & V_{31} \\ V_{13} & V_{11} \end{vmatrix} \\ I_3 &= \begin{vmatrix} V_{11} & V_{12} & V_{13} \\ V_{21} & V_{22} & V_{23} \\ V_{31} & V_{32} & V_{33} \end{vmatrix}. \end{aligned} \tag{3}$$

Since blood is assumed to be incompressible; I_1 vanishes. But when $I_1 = 0$, I_2 becomes negatively and it is more convenient to use a positively valued invariant J_2 defined as:

$$J_2 = \tfrac{1}{3}I_1^2 - I_2 = \tfrac{1}{2}V_{ij}V_{ij}. \tag{4}$$

Hence, μ must be a function of J_2 and I_3. From Eq. 4, it seems that J_2 is directly related to the shear strain rate. For instance, if V_{12} is the only nonvanishing component of the shear rate, then $J_2 = V_{12}^2$. Blood viscosity depends on the shear rate, hence on J_2. Whether it depends on I_3 or not, is unknown because in most viscometric flows (e.g., Couette, Poiseuille, and cone-plate flows) I_3 is zero. Subsequently, μ is a function of J_2 and the following constitutive equation for blood flow is obtained (99):

$$\sigma_{ij} = -p\delta_{ij} + 2\mu(J_2)V_{ij}. \tag{5}$$

According to experimental results, the shear rate is defined as:

$$\dot{\gamma} = \frac{v}{h} = 2 \times \frac{1}{2}\left(\frac{\partial v_1}{\partial x_2} + \frac{\partial v_2}{\partial x_1}\right) = 2V_{12}. \tag{6}$$

In this case, all other components V_{ij} vanish, and:

$$J_2 = V_{12}^2, \tag{7}$$

so that from Eq. 6:

$$|\dot{\gamma}| = 2\sqrt{J_2}, \tag{8}$$

whereas from Eq. 5, the shear stress is defined as:

$$\sigma_{12} = 2\mu(J_2)V_{12} = \mu\dot{\gamma}. \tag{9}$$

It is shown that in steady flow in larger vessels, the experimental results may be expressed in Casson's equation as follows:

$$\sigma_{12} = \left[\sqrt{\tau_y} + \sqrt{\mu|\dot{\gamma}|}\right]^2, \tag{10}$$

in a wide range of $\dot{\gamma}$. Comparing Eqs. 9 and 10:

$$\mu = \frac{\left[\sqrt{\tau_y} + \sqrt{\mu|\dot{\gamma}|}\right]^2}{|\dot{\gamma}|}. \tag{11}$$

Thus, the constitutive equation for blood is (in the range of $\dot{\gamma}$ when blood flows) (99):

$$\sigma_{ij} = -p\delta_{ij} + 2\mu(J_2)V_{ij}, \tag{12}$$

where

$$\mu(J_2) = \left[\left(\eta^2 J_2\right)^{1/4} + 2^{-1/2}\tau_y^{1/2}\right]^2 J_2^{-1/2}. \tag{13}$$

In a simplified form according to the Cartesian coordinates, Eq. 12 is described as follows:

$$\begin{aligned} \sigma_{xx} &= -p + 2\mu(J_2)V_{xx} & \sigma_{xy} &= 2\mu(J_2)V_{xy} \\ \sigma_{yy} &= -p + 2\mu(J_2)V_{yy} & \sigma_{yz} &= 2\mu(J_2)V_{yz} \\ \sigma_{zz} &= -p + 2\mu(J_2)V_{zz} & \sigma_{xz} &= 2\mu(J_2)V_{xz}. \end{aligned} \tag{14}$$

The previous equations are valid when $J_2 \neq 0$ is sufficiently small and μ tends to become large. On the other hand, when J_2 is sufficiently large, the experimental results reduce to the simple statement that μ = constant and the yield stress becomes negligible, which in this case Eq. 1 may be applied. Hence, Fung's relation accounts for the pseudoplastic character as illustrated in Fig. 8, including Newtonian behaviour at high shear rates (98, 99).

Because the pseudoplastic character of blood is due largely to the red blood cells, the transition point of both Eqs. 12 and 13 to the Newtonian equation, with μ = constant, depends on the hematocrit H (the volume fraction of red blood cells in whole blood). For normal blood with a low hematocrit, $H = 8.25\%$, μ was found to be constant over the entire range of shear rate from 0.1 to 1,000/s. When $H = 18\%$, the blood appears to be Newtonian when $\dot{\gamma} > 600/\text{s}$, but obeys both previous equations for smaller values of $\dot{\gamma}$. For higher values of H, the transition point increases to $\dot{\gamma} \cong 700/\text{s}$. Normal shear rates range from 100 to 2,000/s in large arteries and from 20 to 200/s in large to small veins. Normal values of the hematocrit are about 42% and 47% in women and men, respectively, hence the Newtonian response only at high shear rates. As noted earlier, the value of $\mu(J_2)$ at high shear rates is on the order of 2–3.5 cP for whole blood.

In the absence of blood flow, when $V_{ij} = 0$, a blood flow rule must be supplemented by another stress–strain relation. Little is known about the behaviour of blood in these conditions, however, a hypothetical constitutive equation may be proposed. For instance, Hooke's law may be suggested when there is no blood flow, because the stress and strain values are both very small. But for a complete formulation, another condition is needed, the "yielding condition" to define at which stress level, blood flow must occur. In plasticity, the yielding condition is often stated in terms of the second invariant of the stress deviation, as follows (99):

$$J_2' = \frac{1}{2}\sigma_{ij}'\sigma_{ij}', \tag{15}$$

where

$$\sigma_{ij}' = \sigma_{ij} - \frac{1}{3}\sigma_{kk}\delta_{ij}. \tag{16}$$

Flow or yielding occurs when J_2' exceeds a certain number k. Therefore:

$$V_{ij} = \left\{ \begin{array}{ll} 0 & \text{if } J_2' < k \\ \frac{1}{2\mu}\sigma_{ij}' & \text{if } J_2' \geq k \end{array} \right\}. \tag{17}$$

The stress–strain law defines the mechanical behaviour of blood in the absence of blood flow. The yielding condition and the blood flow law combined describe the mechanical behaviour of blood flow.

3.3. Scaffold Vascularization

The architecture and design of a scaffold has a profound effect on the rate of vascularization after implantation (97, 100). First, the

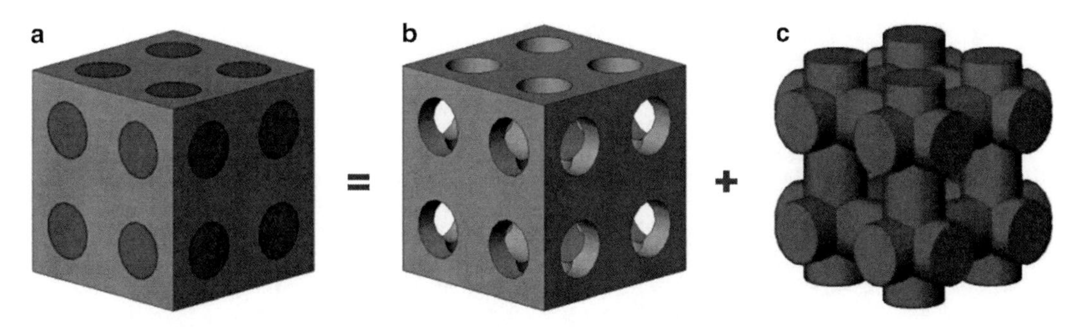

Fig. 10. Illustration of a scaffold hosted within a biofluid environment and the resulting volumes (**a**) scaffold with biofluid, (**b**) scaffold, and (**c**) biofluid geometry.

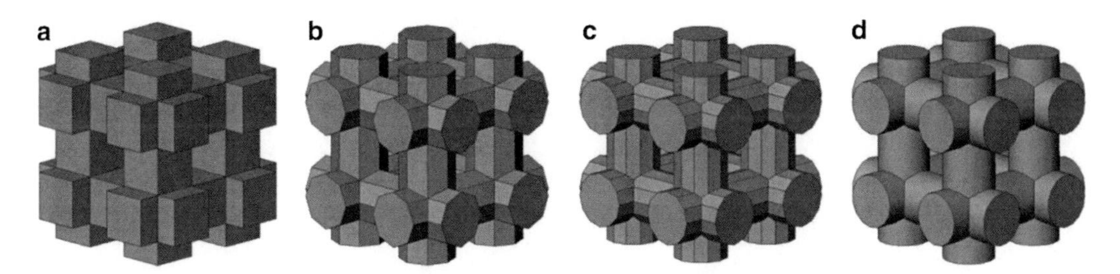

Fig. 11. The family of scaffolds representing the volume related with the flow of biofluids inside the human body, which is classified according to the number of faces per pore. (**a**) 4f unit, (**b**) 8f unit, (**c**) 12f unit, and (**d**) cf unit.

pore size of the scaffold is a critical determinant of blood-vessel ingrowth. Druecke et al. (101) showed that vessel ingrowth was significantly faster in scaffolds with pores greater than 250 μm than in those with smaller pores. However, it is not only the pore size that is important for vascularization: the interconnectivity of the pores is also significant because cell migration, and thus vascularization, will be inhibited if pores are not interconnected, even if the scaffold porosity is high (26, 50, 102).

In computational simulations, computer fluid dynamic analysis is used to evaluate the scaffold's vascular performance. These types of simulations are specially indicated to analyze the flow of fluids, either for gases or liquids. The first step is to obtain the geometry representing the existing fluid. Taking into account "n face per pore" geometries, the empty space within the scaffold unit represents the biofluid volume, if the scaffold is hosted within a biofluid rich environment (Fig. 10). Figure 11 illustrates the unfilled volumes of the scaffold units, which represent the biofluid volume. Presently, the biofluid considered for simulation purposes is blood and the fluid geometries were obtained from "n face per pore" scaffold units with 25, 50, and 75% porosity.

3.4. Blood Flow Simulation

Flow simulation enables to study blood mass flow, pressure, and blood velocity for different scaffold topologies. For simulation

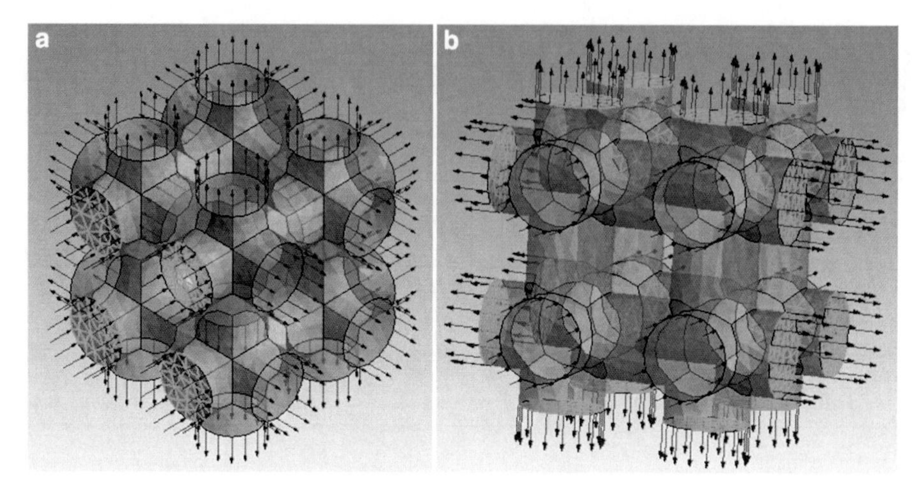

Fig. 12. Blood flow within the scaffold (**a**) inlet and (**b**) outlet.

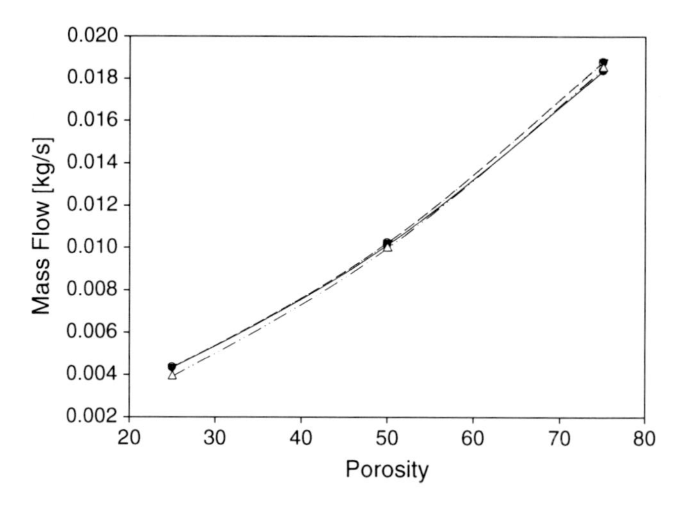

Fig. 13. Variation of the mass flow with porosity for each type of pore.

purposes, it was considered for the blood a density of 1,080 kg/m^3 and a dynamic viscosity of 0.0035 Pas. The inlet value corresponds to the blood pressure, 100 mmHg, and the outlet value is a velocity of 30 cm/s. This velocity corresponds to the velocity of the blood flow at the given pressure. The inlet and outlet of the blood flow within a scaffold is illustrated in Fig. 12a, b.

Figure 13 illustrates the variation of the blood mass flow with porosity for the different pore architectures. As expected, increasing the porosity level, increases the blood flow mass. The pore architecture has also a significant impact on the blood flow mass as shown in Figs. 14–16. It is also possible to observe that, from a topological point of view, there is a critical value from which the increase of the number of faces per pore reduces the blood mass

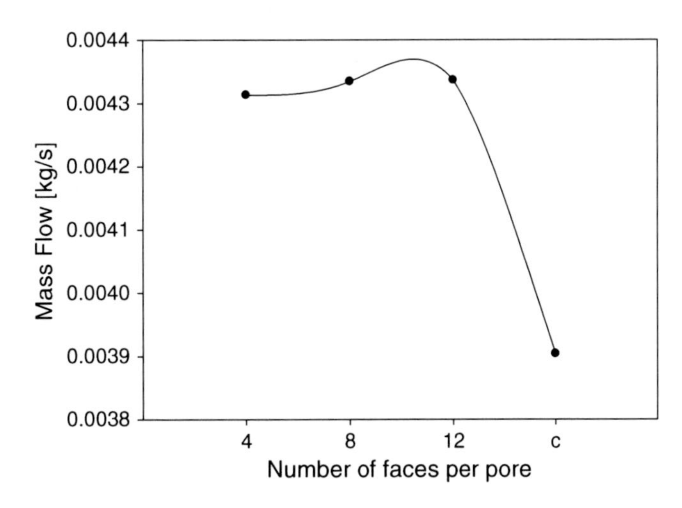

Fig. 14. Variation of the mass flow with the number of faces per pore with 25% porosity.

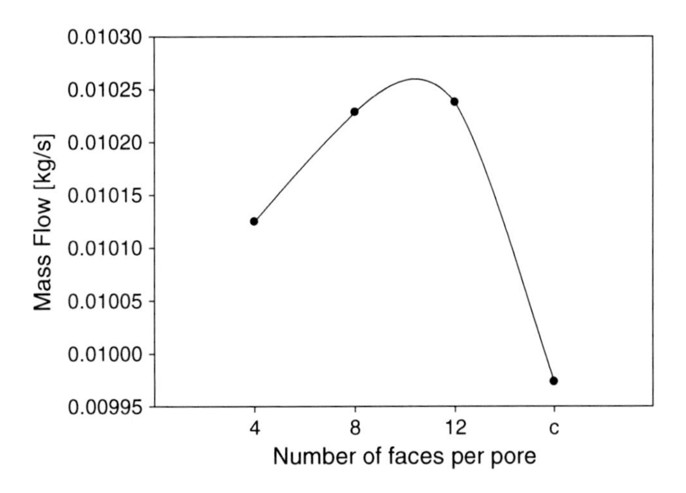

Fig. 15. Variation of the mass flow with the number of faces per pore with 50% porosity.

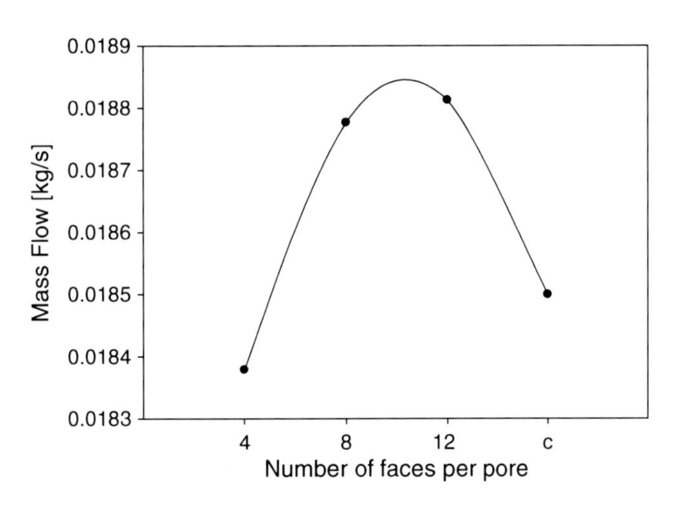

Fig. 16. Variation of the mass flow with the number of faces per pore with 75% porosity.

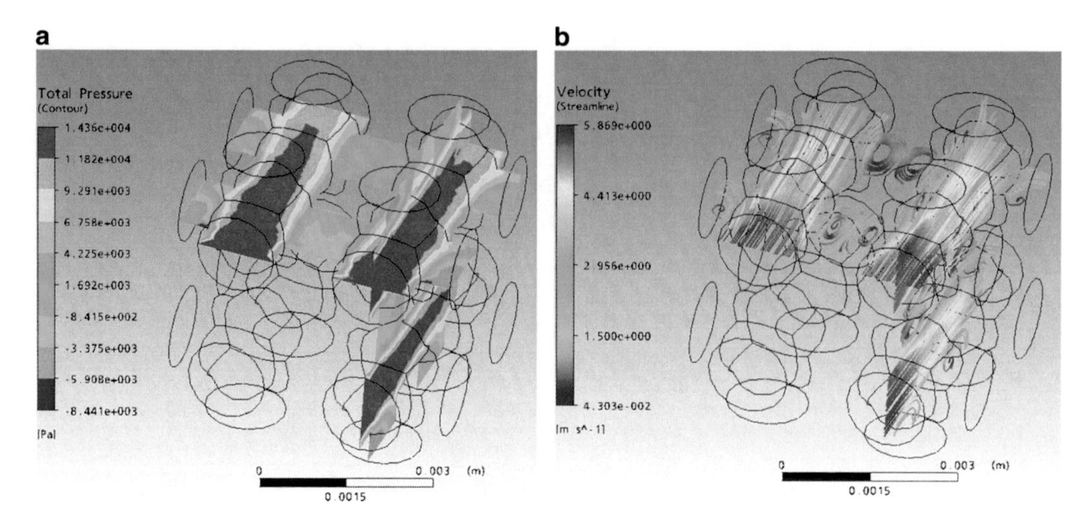

Fig. 17. Fluid (**a**) pressure and (**b**) velocity variation within the scaffold.

flow. The better results were observed for pores having a number of faces between 8 and 12.

The pressure and fluid velocity within a scaffold is illustrated in Fig. 17a, b. Highest values of pressure and velocity are aligned with the direction of the inlet, where we can observe that lateral directions will have less blood flow. If an equal blood flow is needed for all directions, the internal scaffold design needs to be changed in order to force an equal blood flow. The blood has almost no circulation within the scaffold, which may result in blood coagulation. This is illustrated in Fig. 17b with blue velocity streamlines.

4. Conclusions

Rapid prototyping technologies were recently introduced in the medical field, being particularly viable to produce porous scaffolds for tissue engineering. These scaffolds should be biocompatible, biodegradable, with appropriate porosity, pore structure, and pore distribution on top of presenting both surface and structural compatibility. Surface compatibility means a chemical, biological, and physical suitability with the host tissue. Structural compatibility corresponds to an optimal adaptation to the mechanical behaviour of the host tissue. Thus, structural compatibility refers to the mechanical properties and deformation capabilities of the scaffold.

This chapter presents the state-of-the-art in tissue engineering and scaffold design using numerical fluid analysis for optimal vascular design. The vascularization of scaffolds is an important aspect due to its influence regarding the normal flow of biofluids within the human body. This computational tool also allows to design either a scaffold offering less resistance to the normal flow of biofluids or reducing the possibility for blood coagulation through forcing the flow toward a specific direction.

Acknowledgements

This research is supported by the Portuguese Foundation of Science and Technology through a PhD grant of Henrique Almeida (SFRH/BD/37604/2007). The authors also wish to thank the sponsorship given by CYTED through a Biomanufacturing Network "Rede Iberoamericana de Biofabricação".

References

1. Risbud M (2001) Tissue engineering: implications in the treatment of organ and tissue defects. Biogerontology 2:117–125
2. Langer R, Vacanti JP (1993) Tissue engineering. Science 260:920–926
3. Bártolo PJ, Chua CK, Almeida HA, Chou SM, Lim ASC (2009) Biomanufacturing for tissue engineering: present and future trends. Virt Phys Prototyping 4(4):203–216
4. Bártolo PJ, Almeida H, Laoui T (2009) Rapid prototyping and manufacturing for tissue engineering scaffolds. Int J Comput Appl Technol 36(1):1–9
5. Bártolo PJ, Almeida HA, Rezende RA, Laoui T, Bidanda B (2008) Advanced processes to fabricate scaffolds for tissue engineering. In: Bidanda B, Bártolo PJ (eds) Virtual prototyping and bio-manufacturing in medical applications. Springer, New York
6. Vozzi G, Flaim C, Ahluwalia A, Bhatia S (2003) Fabrication of PLGA scaffolds using soft lithography and microsyringe deposition. Biomaterials 24:2533–2540
7. Gibson LJ (2005) Biomechanics of cellular solids. J Biomech 38:377–399
8. Tan KH, Chua CK, Leong KF, Cheah CM, Gui WS, Tan WS, Wiria FE (2005) Selective laser sintering of biocompatible polymers for applications in tissue engineering. BioMed Mater Eng 15:113–124
9. Skalak R, Fox CF (1988) Tissue engineering. Alan R, Liss, New York
10. Fuchs JR, Nasseri BA, Vacanti JP (2001) Tissue engineering: a 21st century solution to surgical reconstruction. Ann Thorac Surg 72:577–581
11. Langer R (1997) Tissue engineering: a new field and its challenges. Pharm Res 14:840–841
12. Marler JJ, Upton J, Langer R, Vacanti JP (1998) Transplantation of cells in matrices for tissue regeneration. Adv Drug Del Rev 33:165–182
13. Gross KA, Rodríguez-Lorenzo LM (2004) Biodegradable composite scaffolds with an interconnected spherical network for bone tissue engineering. Biomaterials 25:4955–4962
14. Kim BS, Mooney DJ (2001) Development of biocompatible synthetic extracellular matrices for tissue engineering. Trends Biotechnol 16:224–230
15. Tan PS, Teoh SH (2007) Effect of stiffness of polycaprolactone (PCL) membrane on cell proliferation. Mater Sci Eng C 27:304–308
16. Kreke MR, Huckle WR, Goldstein AS (2005) Fluid flow stimulates expression of osteopontin and bone sialoprotein by bone marrow stromal cells in a temporally dependent manner. Bone 36:1047–1055
17. Kreeger PK, Shea LD (2002) Biomimetic materials and design: biointerfacial strategies, tissue engineering and targeted drug delivery. In: Dillow AK, Lowman AM (eds) Scaffolds for directing cellular responses and tissue formation in biomimetic materials and design. Marcel Dekker, Inc, New York, pp 283–309
18. Leong KF, Chua CK, Sudarmadjia N, Yeong WY (2008) Engineering functionally graded tissue engineering scaffolds. J Mech Behav Biomed Mater 1:140–152
19. Ma PX (2004) Scaffolds for tissue fabrication. Mater Today 7:30–40

20. Hutmacher DW (2000) Scaffolds in tissue engineering bone and cartilage. Biomaterials 21:2529–2543
21. Mikos AG, Temenoff JS (2000) Formation of highly porous biodegradable scaffolds for tissue engineering. Electron J Biotechnol 3:114–119
22. Kim TK, Sharma B, Williams CG, Ruffner MA, Malik A, McFarland EG, Elisseeff JH (2003) Experimental model for cartilage tissue engineering to regenerate the zonal organization of articular cartilage. Osteoarthr Cartil 11:653–664
23. Sun T, Norton D, Ryan AJ, MacNeil S, Haycock JW (2007) Investigation of fibroblast and keratinocyte cellscaffold interactions using a novel 3D cell culture system. J Mater Sci Mater Med 18:321–328
24. Chong AKS, Chang J (2006) Tissue engineering for the hand surgeon: a clinical perspective. J Hand Surg 31A:349–358
25. Beckstead BL, Pan S, BrattLeal AM, Ratner BD, Giachelli CM, Bhrany AD (2005) Esophageal epithelial cell interaction with synthetic and natural scaffolds for tissue engineering. Biomaterials 26:6217–6228
26. Karageorgiou V, Kaplan D (2005) Porosity of 3D biomaterial scaffolds and osteogenesis. Biomaterials 26:5474–5491
27. Miot S, Woodfield T, Daniels AU, Suetterlin R, Peterschmitt I, Heberer M, Blitterswijk CAv, Riesle J, Martin I (2005) Effects of scaffold composition and architecture on human nasal chondrocyte redifferentiation and cartilaginous matrix deposition. Biomaterials 26:2479–2489
28. Woodfield TBF, Van Blitterswijk CA, Riesle J, De Wijn J, Sims TJ, Hollander AP (2005) Polymer scaffolds fabricated with poresize gradients as a model for studying the zonal organization within tissue engineered cartilage constructs. Tissue Eng 11:1297–1311
29. Martin I, Wendt D, Heberer M (2004) The role of bioreactors in tissue engineering. Trends Biotechnol 22:80–86
30. Sun W, Darling A, Starly B, Nam J (2004) Review – computer aided tissue engineering: overview, scope and challenges. Biotechnol Appl Biochem 39:29–47
31. Taguchi T, Sawabe Y, Kobayashi H, Moriyoshi Y, Kataoka K, Tanaka J (2004) Preparation and characterization of osteochondral scaffold. Mater Sci Eng C 24:881–885
32. Salem AK, Stevens R, Pearson RG, Davies MC, Tendler SJB, Roberts CJ, Williams PM, Shakesheff KM (2002) Interactions of 3T3 fibroblasts and endothelial cells with defined pore features. J Biomed Mater Res 61:212–217
33. Sherwood JK, Riley SL, Palazzolo R, Brown SC, Monkhouse DC, Coates M, Griffith LG, Landeen LK, Ratcliffe A (2002) A threedimensional osteochondral composite scaffold for articular cartilage repair. Biomaterials 23:4739–4751
34. Freyman TM, Yannas IV, Gibson LJ (2001) Cellular materials as porous scaffolds for tissue engineering. Progress in Mate Sci 46:273–282
35. Zeltinger J, Sherwood JK, Graham DA, Müeller R, Griffith LG (2001) Effect of pore size and void fraction on cellular adhesion, proliferation, and matrix deposition. Tissue Eng 7:557–572
36. Nehrer S, Breinan HA, Ramappa A, Young G, Shortkroff S, Louie LK, Sledge CB, Yannas IV, Spector M (1997) Matrix collagen type and pore size influence behaviour of seeded canine chondrocytes. Biomaterials 18:769–776
37. Whang K, Thomas CH, Healy KE, Nuber G (1995) A novel method to fabricate bioabsorbable scaffolds. Polymer 36:837–842
38. Wei G, Jin Q, Giannobile WV, Ma PX (2007) The enhancement of osteogenesis by nanofibrous scaffolds incorporating rhBMP-7 nanospheres. Biomaterials 28:2087–2096
39. Nathan C, Sporn M (1991) Cytokines in context. J Cell Biol 113:981–986
40. Bignon A, Chouteau J, Chevalier J, Fantozzi G, Carret JP, Chavassieux P, Boivin G, Melin M, Hartmann D (2003) Effect of microand macroporosity of bone substitutes on their mechanical properties and cellular response. J Mater Sci Mater Med 14:1089–1097
41. Leong KF, Cheah CM, Chua CK (2003) Solid freeform fabrication of three-dimensional scaffolds for engineering replacement tissues and organs. Biomaterials 24:2363–2378
42. Prendergast PJ & van der Meulen MCH (2001) Mechanics of bone regeneration. In: Cowin SC (ed) Bone mechanics handbook. CRC Press LLC, Boca Raton 32:32.1–32.19
43. Temenoff JS, Lu L, Mikos AG (2000) Bone-tissue engineering using synthetic biodegradable scaffolds. In: Davies JE (ed) Bone engineering. Toronto, Em Squared Incoporated, 454–461
44. Mustafa K, Oden A, Wennerberg A, Hultenby K, Arvidson K (2005) The influence of surface topography of ceramic abutments on the attachment and proliferation of human oral fibroblasts. Biomaterials 26:373–381

45. Cheng Z, Teoh SH (2004) Surface modification of ultra thin poly (caprolactone) films using acrylic acid and collagen. Biomaterials 25:1991–2001
46. Zhao K, Deng Y, Chen GQ (2003) Effects of surface morphology on the biocompatibility of polyhydroxyalkanoates. Biochem Eng J 16:115–123
47. Singhvi R, Stephanopoulos G, Wang DIC (1994) Effects of substratum morphology on cell physiology – review. Biotechnol Bioeng 43:764–771
48. Hynes RO (1992) Integrins, versatility, modulation, and signalling in cell adhesion. Cell 69:11–25
49. Yildirim E, Ayan H, Vasilets V, Fridman A, Guceri S, Sun W (2008) Effect of dielectric barrier discharge plasma on the attachment and proliferation of osteoblasts cultured over poly (e-caprolactone) scaffolds. J Plasma Process Polym 5(1):58–66
50. Yang S, Leong KF, Du Z, Chua CK (2001) The design of scaffolds for use in tissue engineering. Part I. Traditional factors. Tissue Eng 7:679–689
51. Yeong WY, Chua CK, Leong KF, Chandrasekaran M (2004) Rapid prototyping in tissue engineering: challenges and potential. Trends Biotechnol 22:643–652
52. Bártolo PJ, Chua CK (2008) Editorial: celebrating the 70th anniversary of Professor Yongnian Yan: a life dedicated to science and technology. Virt Phys Prototyping 3 (4):189–191
53. Bártolo PJ (2006) State of the art of solid freeform fabrication for soft and hard tissue engineering. In: Brebbia CA (ed) Design and nature III: comparing design in nature with science and engineering. WIT Press, Wessex Institute of Technology, UK, 233–243
54. Bártolo PJ, Lagoa R, Mendes A (2003) Rapid prototyping system for tissue engineering. In: Bártolo PJ et al (eds) Proceedings of the international conference on advanced research in virtual and physical prototyping. Leiria, pp 419–426
55. Gaspar J, Bártolo PJ, Duarte FM (2008) Cure and rheological analysis of reinforced resins for stereolithography. Mater Sci Forum 587/588:563
56. Bertsch A, Jiguet S, Bernhards P, Renaud P (2003) Microstereolithography: a review. In: Pique A, Holmes AS, Dimos DB (eds) Rapid prototyping technologies. 758:3–15
57. Cooke MN, Fisher JP, Dean D, Rimnac C, Mikos AG (2002) Use of stereolithography to manufacture critical-sized 3D biodegradable scaffolds for bone ingrowth. J Biomed Mater Res B Appl Biomater 64B:65–69
58. Matsuda T, Mizutani M (2002) Liquid acrylate-endcapped poly(ε-caprolactone-cotrimethylene carbonate). II. Computer-aided stereolithographic microarchitectural surface photoconstructs. J Biomed Mater Res 62:395–403
59. Chu TMG, Halloran JW, Hollister SJ, Feinberg SE (2001) Hydroxyapatite implants with designed internal architecture. J Mater Sci Mater Med 12:471–478
60. Griffith ML, Halloran JW (1996) Freeform fabrication of ceramics via stereolithography. J Am Ceram Soc 79:2601–2608
61. Bártolo PJ, Mitchell G (2003) Stereo-thermal-lithography. Rapid Prototyping J 9:150–156
62. Williams JM, Adewunmi A, Schek RM, Flanagan CL, Krebsbach PH, Feinberg SE, Hollister SJ, Das S (2005) Bone tissue engineering using polycaprolactone scaffolds fabricated via selective laser sintering. Biomaterials 26(23):4817–4827
63. Woodfield TBF, Malda J, de Wijn J, Péters F, Riesle J, van Blitterswijk CA (2004) Design of porous scaffolds for cartilage tissue engineering using a three-dimensional fiber-deposition technique. Biomaterials 25:4149–4161
64. Mateus AJ, Almeida HA, Ferreira NM, Alves NM, Bártolo PJ, Mota C, Sousa JP (2008) Bioextrusion for tissue engineering applications. In: Bártolo PJ et al (ed) Virtual and rapid manufacturing. Taylor&Francis, London
65. Landers R, Hubner U, Schmelzeisen R, Mulhaupt R (2002) Rapid prototyping of scaffolds derived from thermoreversible hydrogels and tailored for applications in tissue engineering. Biomaterials 23:4437–4447
66. Landers R, Pfister A, Hubner U, John H, Schmelzeisen R, Mulhaupt R (2002) Fabrication of soft tissue engineering scaffolds by means of rapid prototyping techniques. J Mater Sci 37:3107–3116
67. Mironov V, Boland T, Trusk T, Forgacs G, Markwald RR (2003) Organ printing: computer-aided jet-based 3D tissue engineering. Trends Biotechnol 21:157–161
68. Yan Y, Zhang R, Lin F (2003) Research and applications on bio-manufacturing. In: Bártolo PJ et al (eds) Proceedings of the international conference on advanced research in virtual and physical prototyping. Leiria, pp 23–29
69. Chang R, Nam J, Sun W (2008) Effects of dispensing pressure and Nozzle diameter on cell survival from solid freeform fabrication-based direct cell writing. Tissue Eng 14 (1):41–48

70. Hoque ME, Feng W, Wong YS, Hutmacher DW, Li S, Huang MH, Vert M, Bártolo PJ (2008) Scaffolds designed and fabricated with elastic biomaterials applying cad-cam technique. Tissue Eng A 14:907
71. Koh Y-H, Jun I-K, Kim H-E (2006) Fabrication of poly(ε-caprolactone)/hydroxyapatite scaffold using rapid direct deposition. Mater Lett 60:184–1187
72. Zein I, Hutmacher DW, Tan KC, Teoh SH (2002) Fused deposition modeling of novel scaffold architectures for tissue engineering applications. Biomaterials 23:1169–1185
73. Too MH, Leong KF, Chua CK, Du ZH, Yang SF, Cheah CM, Ho SL (2002) Investigation of 3D nonrandom porous structures by fused deposition modeling. Int J Adv Manuf Technol 19:217–223
74. Hutmacher DW, Schantz T, Zein I, Ng KW, Teoh SH, Tan KC (2001) Mechanical properties and cell cultural response of polycaprolactone scaffolds designed and fabricated via fused deposition modelling. J Biomed Mater Res 55:203–216
75. Shor L, Guceri S, Wen X, Gandhi M, Sun W (2007) Fabrication of three-dimensional polycaprolactone/hydroxyapatite tissue scaffolds and osteoblast-scaffold interactions *in vitro*. Biomaterials 28(35):5291–5297
76. Wang F, Shor L, Darling A, Khalil S, Güçeri S, Lau A (2004) Precision deposition and characterization of cellular poly-ε-caprolactone tissue scaffolds. Rapid Prototyping J 10:42–49
77. Ang TH, Sultana FSA, Hutmacher DW, Wong YS, Fuh JYH, Mo XM, Loh HT, Burdet E, Teoh SH (2002) Fabrication of 3D chitosan-hydroxyapatite scaffolds using a robotic dispersing system. Mater Sci Eng C20:35–42
78. Seitz H, Rieder W, Irsen S, Leukers B, Tille C (2005) Threedimensional printing of porous ceramic scaffolds for bone tissue engineering. J Biomed Mater Res B Appl Biomater 74B:782–788
79. Sachlos E, Reis N, Ainsley C, Derby B, Czernuszka JT (2003) Novel collagen scaffolds with predefined internal morphology made by solid freeform fabrication. Biomaterials 24:1487–1497
80. Lam CXF, Mo XM, Teoh SH, Hutmacher DW (2002) Scaffold development using 3D printing with a starch-based polymer. Mater Sci Eng C Biomimet Supramol Syst 20:49–56
81. Kim SS, Utsunomiya H, Koski JA, Wu BM, Cima MJ, Sohn J, Mukai K, Griffith LG, Vacanti JP (1998) Survival and function of hepatocytes o a novel three-dimensional synthetic biodegradable polymer scaffolds with an intrinsic network of channels. Ann Surg 228:8–13
82. Manjubala I, Woesz A, Pilz C, Rumpler M, Fratzl-Zelman N, Roschger P, Stampfl J, Fratzl P (2005) Biomimetic mineral-organic composite scaffolds with controlled internal architecture. J Mater Sci Mater Med 16:1111–1119
83. Yeong WY, Chua CK, Leong KF, Chandrasekaran M, Lee MW (2006) Indirect fabrication of collagen scaffold based on inkjet printing technique. Rapid Prototyping J 12:229–237
84. Chua CK, Yeong WY, Leong KF (2005) Development of scaffolds for tissue engineering using a 3D inkjet model maker. In: Bártolo PJ et al (ed) Virtual modelling and rapid manufacturing. Taylor&Francis, London
85. Hutmacher DW, Sittinger M, Risbud MV (2004) Scaffold-based tissue engineering: rationale for computer-aided design and solid free-form fabrication systems. Trends Biotechnol 22:354–362
86. Tsang VL, Bhatia SN (2004) Three-dimensional tissue fabrication. Adv Drug Deliv Rev 56:1635–1647
87. Naing MW, Chua CK, Leong KF, Wang Y (2005) Fabrication of customised scaffolds using computer aided design and rapid prototyping techniques. Rapid Prototyping J 11:249–259
88. Sun W, Starly B, Nam J, Darling A (2005) BioCAD modeling and its applications in computeraided tissue engineering. Comput Aided Design BioCAD 37:1097–1114
89. Moura CS, Bártolo PJ, Almeida HA (2010) Intelligent biopolymer selector system for medical applications. In: Bártolo PJ et al (eds) Innovative developments in design and manufacturing. Taylor&Francis, London, pp 81–86
90. Carmeliet P, Jain RK (2000) Angiogenesis in cancer and other diseases. Nature 407:249–257
91. Janssen FW, Oostra J, Oorschot A, van Blitterswijk CA (2006) A perfusion bioreactor system capable of producing clinically relevant volumes of tissue-engineered bone: in vivo bone formation showing proof of concept. Biomaterials 27:315–323
92. Pörtner R, Nagel-Heyer S, Goepfert C, Adamietz P, Meenen NM (2005) Bioreactor design for tissue engineering. Biosci Bioeng 100:235–245
93. Clark ER, Clark EL (2005) Microscopic observations on the growth of blood

capillaries in the living mammal. Am J Anat 64:251–301

94. Malda J, Rouwkema J, Martens DE, le Comte EP, Kooy FK, Tramper J, van Blitterswijk CA, Riesle J (2004) Oxygen gradients in tissue-engineered PEGT/PBT cartilaginous constructs: measurement and modeling. Biotechnol Bioeng 86:9–18
95. Jain RK, Au P, Tam J, Duda DG, Fukumura D (2005) Engineering vascularized tissue. Nat Biotechnol 23:821–823
96. Johnson PC, Mikos AG, Fisher JP, Jansen JA (2007) Strategic directions in tissue engineering. Tissue Eng 13:2827–2837
97. Almeida HA, Bártolo PJ (2008) Computer simulation and optimisation of tissue engineering scaffolds: mechanical and vascular behaviour. In: Halevi Y, Fischer A (eds) 9th Biennial ASME conference on engineering systems design and analysis (ESDA2008). Haifa Isreal
98. Humphrey JD, Delange SL (2003) An introduction to biomechanics – solids and fluids, analysis and design. Springer, New York
99. Fung YC (1990) Biomechanics: motion, flow, stress and growth. Springer-Verlag, New York
100. Almeida HA, Bártolo PJ, Ferreira J (2007) Mechanical behaviour and vascularisation analysis of tissue engineering scaffolds. In: Bártolo PJ et al (eds) Virtual and rapid manufacturing. Taylor&Francis, London, pp 73–80
101. Druecke D, Langer S, Lamme E, Pieper J, Ugarkovic M, Steinau HU, Homann HH (2004) Neovascularization of poly(ether ester) block-copolymer scaffolds in vivo: long-term investigations using intravital fluorescent microscopy. Biomed Mater Res A 68:10–18
102. Laschke MW et al (2006) Angiogenesis in tissue engineering: breathing life into constructed tissue substitutes. Tissue Eng 12:2093–2104

Chapter 13

Structural and Vascular Analysis of Tissue Engineering Scaffolds, Part 2: Topology Optimisation

Henrique A. Almeida and Paulo J. Bártolo

Abstract

Rapid prototyping technologies were recently introduced in the medical field, being particularly viable to produce porous scaffolds for tissue engineering. These scaffolds should be biocompatible, biodegradable, with appropriate porosity, pore structure, and pore distribution, on top of presenting both surface and structural compatibility. Surface compatibility means a chemical, biological, and physical suitability with the host tissue. Structural compatibility corresponds to an optimal adaptation to the mechanical behaviour of the host tissue. This chapter presents a computer tool to support the design of scaffolds to be produced by rapid prototyping technologies. The software enables to evaluate scaffold mechanical properties as a function of porosity and pore topology and distribution, for a wide rage of materials, suitable for both hard and soft tissue engineering.

Key words: Tissue engineering, Scaffolds, Computer simulation, Structural analysis, Topological optimisation

1. Computer-Aided Design of Scaffolds

The prediction of the effective optimal properties of tissue scaffolds is very important for tissue engineering applications, either regarding mechanical, vascular, or topological (1–3) properties. To aid this specific issue of the design process of scaffolds, a computational tool, *Computer-Aided Design of Scaffolds* (*CADS*) (4, 5), has been developed, which enables to quantify the structural heterogeneity and mechanical and vascular properties of a scaffold with a designed macro and microstructure (6–8). Besides these topics, this computational tool already contemplates other issues such as degradation. In this case, it was possible to predict the mechanical properties of alginate scaffolds considering the influence of the water loss/shrinkage phenomenon (9).

CADS makes the bridge between different computer applications by managing material database tools (knowledge-based

Michael A.K. Liebschner (ed.), *Computer-Aided Tissue Engineering*, Methods in Molecular Biology, vol. 868, DOI 10.1007/978-1-61779-764-4_13, © Springer Science+Business Media, LLC 2012

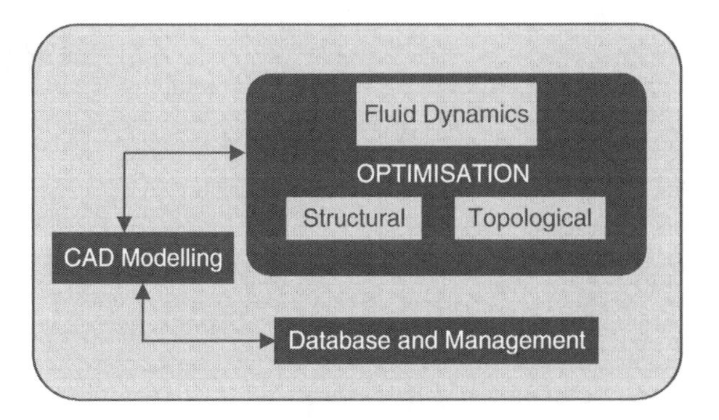

Fig. 1. Structure of CADS.

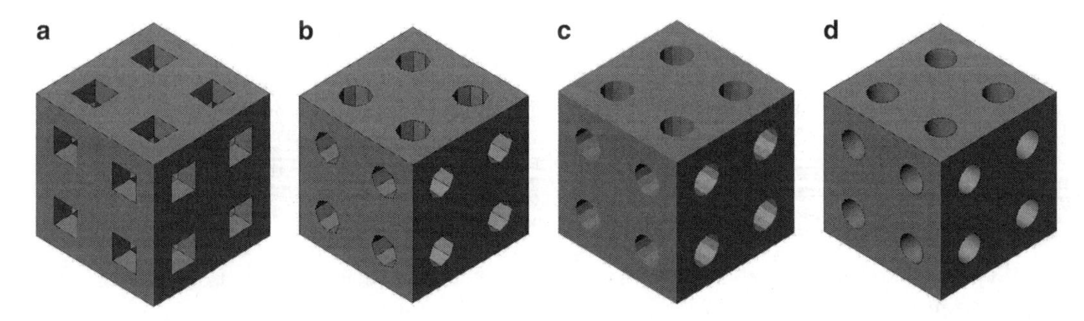

Fig. 2. The *n faces pore* blocks family classified according to the number of faces per pore. (**a**) 4 face (4f) pore unit, (**b**) 8f unit, (**c**) 12f unit, and (**d**) circular (cf) unit.

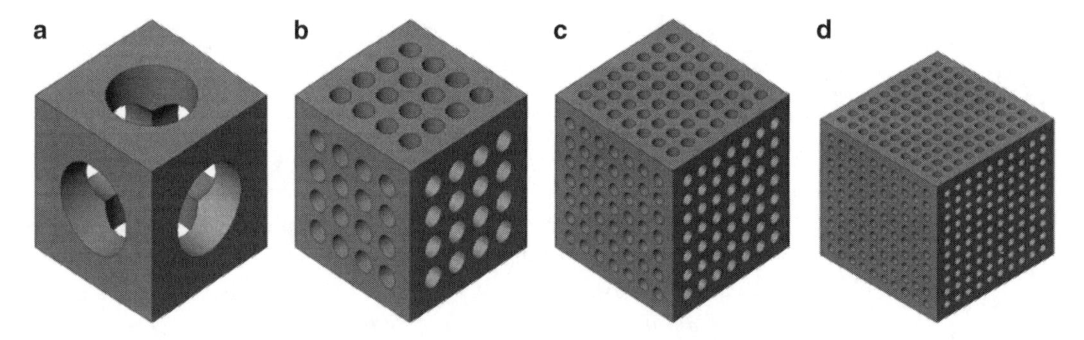

Fig. 3. The *m pores per face* blocks family classified according to the number of pores per face. Only four of the ten geometries with circular pores are represented and all with 50% porosity. (**a**) 1-pore (1p) unit, (**b**) 16p unit, (**c**) 49p unit, and (**d**) 100p unit.

systems for materials selection), CAD modelling systems, and numerical simulation tools for structural, topological, and fluid dynamics optimisation based on the finite element method (Fig. 1).

Two types of elementary building blocks were considered, as shown in Figs. 2 and 3. Figure 2 illustrates elementary scaffolds of the so-called "*n faces per pore*" block family and Fig. 3 represents elementary scaffolds of the so-called "*m pores per face*" block family.

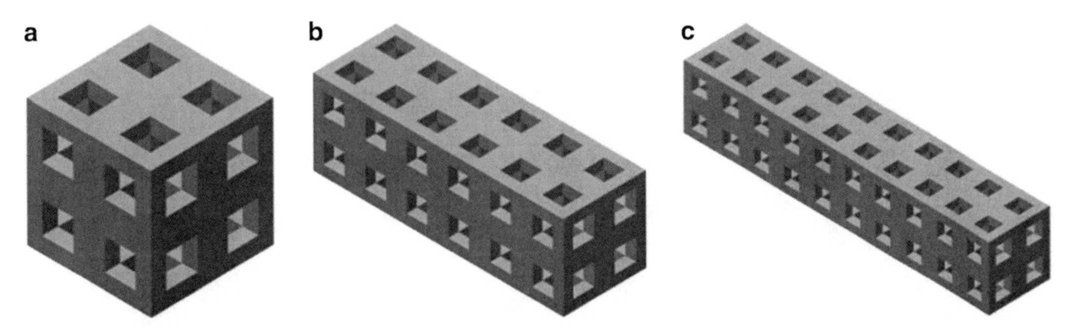

Fig. 4. Association of primary units for the numerical analysis from (**a**) single unit up to (**b**) three units and (**c**) five units in association.

The "*n faces per pore*" block family are scaffolds consisting of 4 pores per face in which the number of faces per pores varies from 4 faces per pore to 8 faces, 12 faces, and infinite faces (circular pores). The "*m pores per face*" block family are scaffolds consisting of multiple pores per face within each unit, while maintaining a specific scaffold's porosity, ranging from 1 to 100 pores per face. In this case, two types of topological architectures were considered, 4 face and circular pores. All elementary family scaffolds are cubic scaffold units of 5 mm in size.

Scaffolds are considered here as a LEGO structure formed by an association of small elementary units or blocks. This association influences the scaffolds mechanical behaviour. In order to understand its influence, a set of primary scaffold units have been associated, varying from 1 to 5 as illustrated in Fig. 4. The structural analysis was performed in both directions, along the single unit direction (Y direction) and along the repeated unit's direction (X direction). Only units with 4 face pores and circular pores were considered.

Another class of elementary blocks are the scaffolds in which its design is based on the triply periodic minimal surfaces (TPMS). Hyperbolic surfaces have attracted the attention of physicists, chemists, and biologists as they commonly exist in natural structures. Amongst various hyperbolic surfaces, minimal surfaces (those with mean curvature of zero) are the most studied. If a minimal surface has space group symmetry, it is periodic in three independent directions. These surfaces are known as TPMS (Fig. 5).

In nature, TPMS are found in lyotropic liquid crystals, zeolite sodalite crystal structures, diblock polymers, soluble proteins in lipid–protein–water phases, and in certain cell membranes (10–12). TPMS allow very high surface-to-volume ratios and provide good analytic description of highly porous structures.

A periodic surface can be generally defined as:

$$\phi(\mathbf{r}) = \sum_{k=1}^{K} A_k \cos[2\pi(\mathbf{h}_k \cdot \mathbf{r})/\lambda_k + p_k] = C, \qquad (1)$$

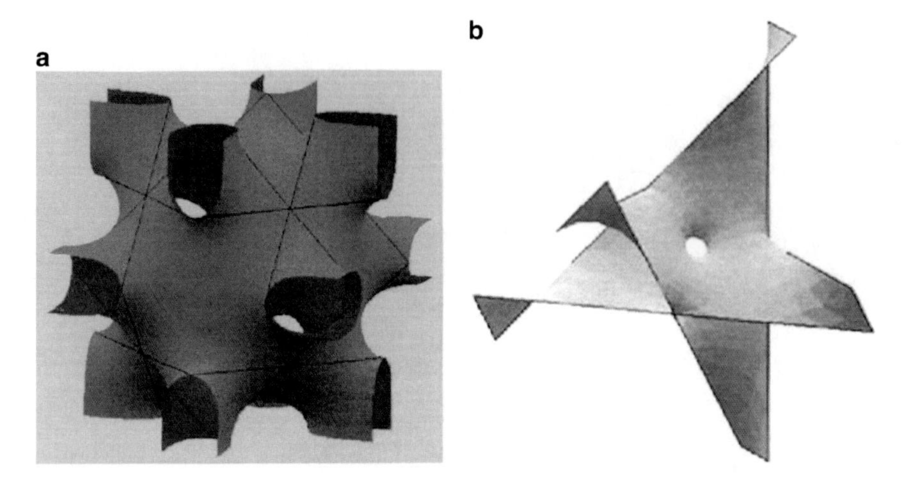

Fig. 5. Examples of triply periodic minimal surfaces (TPMS): (**a**) Neovius' surface and (**b**) Fischer's surface (13).

where r is the location vector in Euclidean space, $\mathbf{h}_k$ is the kth lattice vector in reciprocal space, A_k is the magnitude factor, λ_k is the wavelength of periods, p_k is the phase shift, and C is a constant. Specific periodic structures and phases can be constructed based on this implicit form (14).

In the case of TPMS, the Weierstrass formula describes their parametric form as follows:

$$\begin{cases} x = \mathrm{Re}\displaystyle\int_{\omega_0}^{\omega_1} \mathrm{e}^{i\theta}(1-\omega^2)R(\omega)\mathrm{d}\omega \\ y = \mathrm{Im}\displaystyle\int_{\omega_0}^{\omega_1} e^{i\theta}(1+\omega^2)R(\omega)\mathrm{d}\omega\,, \\ z = -\mathrm{Re}\displaystyle\int_{\omega_0}^{\omega_1} e^{i\theta}(2\omega)R(\omega)\mathrm{d}\omega \end{cases} \tag{2}$$

where ω is a complex variable, θ is the so-called Bonnet angle, and $R(\omega)$ is the function which varies for different surfaces.

From a multi-dimensional control parameter space point of view, the geometric shape of a periodic surface is specified by a periodic vector, such as (14):

$$\mathbf{V} = \langle \mathbf{A}, \mathbf{H}, \mathbf{P}, \Lambda \rangle_{K\times 6}, \tag{3}$$

where

$$\begin{aligned} \mathbf{A} &= [A_k]_{K\times 1} \\ \mathbf{H} &= [\mathbf{h}_k]_{K\times 3} \\ \mathbf{P} &= [p_k]_{K\times 1} \\ \mathbf{\Lambda} &= [\lambda_k]_{K\times 1} \end{aligned}$$

are row concatenations of magnitudes, reciprocal lattice matrix, phases, and period lengths respectively.

Fig. 6. Schwartz TPMS primitive.

An important subclass of TPMS are those that partition space into two disjoint but intertwining regions that are bi-continuous. An example of such surfaces includes the so-called Schwartz primitives (Fig. 6) for which each disjoint region has a volume fraction equal to ½.

The periodic Schwartz primitive surface is given by (14):

$$\phi(\mathbf{r}) = A_{\mathrm{P}}[\cos(2\pi x/\lambda_x) + \cos(2\pi y/\lambda_y) + \cos(2\pi z/\lambda_z)] = 0. \quad (4)$$

The concatenations of magnitude vector, reciprocal lattice matrix, and phase vector are given by (14):

$$\mathbf{A}^{\mathrm{T}} = [1 \quad 1 \quad 1]$$

$$\mathbf{H}^{\mathrm{T}} = \begin{bmatrix} 1 & 0 & 0 \\ 0 & 1 & 0 \\ 0 & 0 & 1 \end{bmatrix} \quad (5)$$

$$\mathbf{P}^{\mathrm{T}} = [0 \quad 0 \quad 0]$$

Schwartz primitives can be easily manipulated from a computational point of view through operations such as union, difference, intersection, modulation, convulation, etc. (Fig. 7).

Two important parameters can be used as modelling control constraints: thickness and radius. Figures 8 and 9 illustrate the effect of these parameters on the Schwartz primitives (P-minimal surfaces) obtained structures.

Fig. 7. Addition operation with Schwartz periodic primitives.

Fig. 8. P-minimal surfaces obtained through thickness variation with constant surface radius.

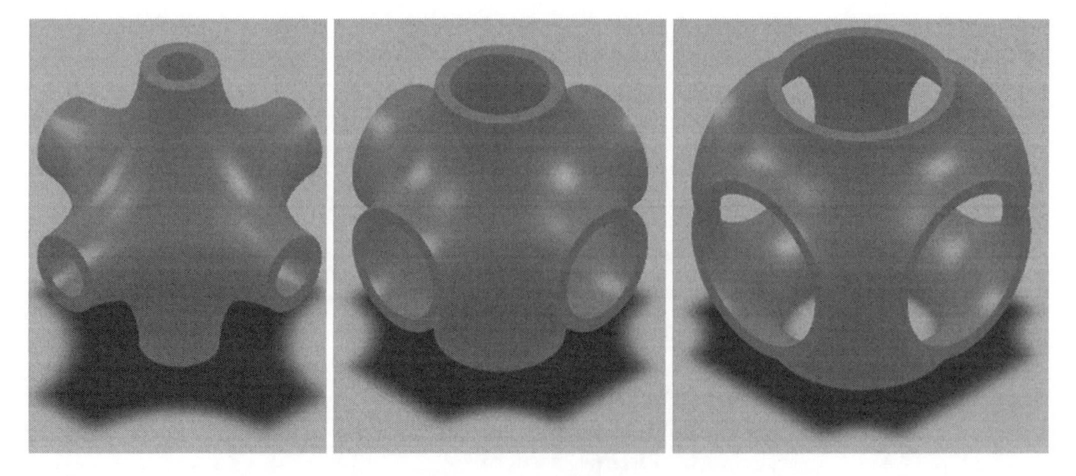

Fig. 9. P-minimal surfaces obtained through radius variation with constant surface thickness.

2. Structural Analysis of Scaffolds for Tissue Engineering

Once the scaffold is produced and placed, formation of tissues with desirable properties relies on the scaffold's mechanical properties, both at a macroscopic and microscopic level. Macroscopically, the scaffold must bear loads to provide stability to tissues as it forms and to fulfil its volume maintenance function. At the microscopic level, cell growth and differentiation and ultimate tissue formation are dependent on mechanical input to cells. Consequently, the scaffold must be able to both withstand specific loads and transmit them in an appropriate manner to the growing and surrounding cells and tissues. Specific mechanical properties of scaffolds include elasticity, compressibility, viscoelastic behaviour, tensile strength, failure strain, and their time-dependent fatigue.

2.1. Constitutive Equation for Structural Analysis

The properties of materials are specified by constitutive equations, and since there are a wide variety of materials, one must not be surprised that there are many great constitutive equations describing an almost infinite variety of materials. What should be surprising, therefore, is the fact that three simple, idealised stress–strain relationships, namely, the nonviscous fluid, the Newtonian viscous fluid, and the Hookean elastic solid, give a good description of the mechanical properties of many materials around us. Within certain limits of strain and strain rate, air, water, and many engineering structural materials can be described by these idealised equations. Most biological materials, however, cannot be described so simply.

A constitutive equation describes a physical property of a material. Hence, it must be independent of any particular set of coordinates of reference with respect to which the components of various physical quantities are resolved. Hence, a constitutive equation must be a tensor equation and every term in the equation must be a tensor of the same rank (15, 16).

A Hookean elastic solid is a solid that obeys Hooke's law, which states that the stress tensor is linearly proportional to the strain tensor; i.e. (15, 16):

$$\sigma_{ij} = C_{ijkl} e_{kl}, \tag{6}$$

where σ_{ij} is the stress tensor, e_{kl} is the strain tensor, and C_{ijkl} is a tensor of elastic constants, or moduli, which are independent of stress or strain.

A great reduction in the number of elastic constants is obtained when the material is isotropic, i.e., when the constitutive equation is isotropic, and the array of elastic constants C_{ijkl} remains unchanged with respect to rotation and reflection of coordinates. Considering that an isotropic material has exactly two independent elastic constants, Hooke's law is described as follows:

$$\sigma_{ij} = \lambda e_{xx} \delta_{ij} + 2\mu e_{ij}. \tag{7}$$

The constants λ and μ are called the Lamé constants. In the engineering literature, the second Lamé constant μ is practically always written as G and identified as the shear modulus.

By extending out the previous equation with x, y, and z as rectangular cartesian coordinates, we have Hooke's law for an isotropic elastic solid:

$$\begin{aligned} \sigma_{xx} &= \lambda(e_{xx} + e_{yy} + e_{zz}) + 2Ge_{xx} & \sigma_{xy} &= 2Ge_{xy} \\ \sigma_{yy} &= \lambda(e_{xx} + e_{yy} + e_{zz}) + 2Ge_{yy} & \sigma_{yz} &= 2Ge_{yz} \\ \sigma_{zz} &= \lambda(e_{xx} + e_{yy} + e_{zz}) + 2Ge_{zz} & \sigma_{zx} &= 2Ge_{zx} \end{aligned} \quad (8)$$

These equations may be solved for e_{ij}. Usually the inverted form is described as follows:

$$e_{ij} = \frac{1+v}{E}\sigma_{ij} - \frac{v}{E}\sigma_{kk}\delta_{ij} \quad (9)$$

or

$$\begin{aligned} e_{xx} &= \tfrac{1}{E}[\sigma_{xx} - v(\sigma_{yy} + \sigma_{zz})] & e_{xy} &= \tfrac{1+v}{E}\sigma_{xy} = \tfrac{1}{2G}\sigma_{xy} \\ e_{yy} &= \tfrac{1}{E}[\sigma_{yy} - v(\sigma_{xx} + \sigma_{zz})] & e_{yz} &= \tfrac{1+v}{E}\sigma_{yz} = \tfrac{1}{2G}\sigma_{yz} \\ e_{zz} &= \tfrac{1}{E}[\sigma_{zz} - v(\sigma_{xx} + \sigma_{yy})] & e_{zx} &= \tfrac{1+v}{E}\sigma_{zx} = \tfrac{1}{2G}\sigma_{zx} \end{aligned} \quad (10)$$

The constants E, v, and G are related to the Lamé constants λ and G (or μ), E is called Young's modulus, v is called Poisson's ratio, and G is called the modulus of elasticity in shear, or shear modulus. They can be written out as follows (15, 16):

$$\begin{aligned} \lambda &= \frac{2Gv}{1-2v} = \frac{G(E-2G)}{3G-E} = \frac{Ev}{(1+v)(1-2v)} \\ G &= \frac{\lambda(1-2v)}{2v} = \frac{E}{2(1+v)} \\ v &= \frac{\lambda}{2(\lambda+G)} = \frac{\lambda}{(3K-\lambda)} = \frac{E}{2G} - 1 \\ E &= \frac{G(3\lambda+2G)}{\lambda+G} = \frac{\lambda(1+v)(1-2v)}{v} = 2G(1+v). \end{aligned} \quad (11)$$

Under a compressive solicitation in the z direction, the strain is described as follows:

$$e_{zz} = \frac{1}{E}\sigma_{zz}. \quad (12)$$

But simultaneously the lateral sides of the compressed part will bulge out somewhat. For a linear material, the bulging strain is proportional to σ_{zz} and is in sense opposite to the stress: A compression induces lateral bulging; a tension induces lateral shrinking. Hence, the following formulations are described:

$$e_{xx} = -\frac{v}{E}\sigma_{zz}, \quad e_{yy} = -\frac{v}{E}\sigma_{zz}. \quad (13)$$

This is the case in which σ_{zz} is the only nonvanishing stress. If the part is also subjected to σ_{xx} and σ_{yy}, and if the material is isotropic and linear (in which the causes and effects are linearly superposable), then the influence of σ_{xx} on e_{yy}, e_{zz} and σ_{yy} on e_{xx}, e_{zz} must be the same as the influence of σ_{zz} on e_{xx}, e_{yy}. Hence, the following equation is described:

$$e_{zz} = \frac{1}{E}\sigma_{zz} - \frac{v}{E}\sigma_{xx} - \frac{v}{E}\sigma_{yy}. \tag{14}$$

For shear, the stress σ_{ij} and the strain $e_{ij}(i \neq j)$ are directly proportional.

2.2. Mechanical Simulations

To study the mechanical behaviour of scaffolds, two different mechanical behaviours were considered:

- A linear elastic behaviour for hard tissue applications.
- An hyperelastic behaviour for soft tissues, described through a polynomial form according to the Neo-Hookean model, as follows:

$$U = C_{10}(\overline{I_1} - 3) + \frac{1}{D_1}(J^{el} - 1)^2. \tag{15}$$

The main goal for simulating the scaffold mechanical behaviour is to evaluate the porosity dependence on both elastic and shear modulus. For a given unit block with a specific open pore architecture, boundary and loading conditions considered for evaluating mechanical properties are shown in Fig. 10. For the numerical computation of the elastic modulus, a uniform displacement in a single direction is considered (the X direction), which is equivalent to the strain on the same direction (ε_x), imposed to a face of the block (Face A). The opposite face (Face B) of the scaffold unit is constrained and unable to have any displacement (Fig. 10a).

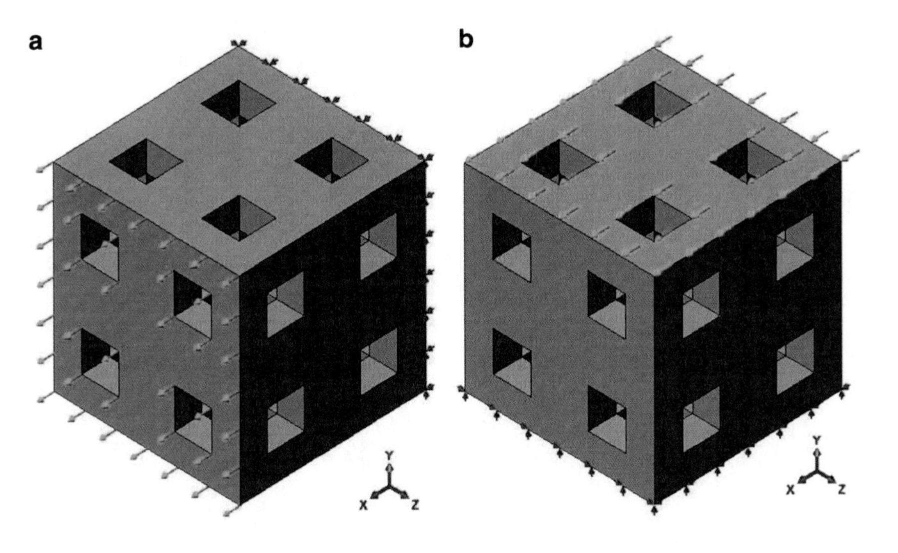

Fig. 10. Loads and constraints for the numerical analysis: (**a**) tensile solicitation and (**b**) shear solicitation.

The average reaction force produced on Face B is used to determine the elastic modulus, due to the imposed displacement.

For the numerical evaluation of the shear modulus, a uniform displacement is applied on the top of a surface, being the opposite face also unable to have any displacement (Fig. 10b). Each unit is considered isotropic.

It is possible, through computer simulation, to establish a cubic mathematical law relating the scaffold porosity and the material's modulus as follows:

$$\begin{aligned} E(\phi) &= E_0 + C_1\phi + C_2\phi^2 + C_3\phi^3 \\ G(\phi) &= G_0 + C_4\phi + C_5\phi^2 + C_6\phi^3. \end{aligned} \tag{16}$$

where E corresponds to the scaffold elastic modulus, E_0 corresponds to the material elastic modulus, G corresponds to the scaffold shear modulus, G_0 corresponds to the material shear modulus, ϕ corresponds to the scaffold porosity, and C_1, C_2 ... C_6 are material and geometric-dependent constants.

3. Results

The effect of the pore architecture is illustrated in Fig. 11a, b, which shows the decrease of both the elastic and shear modulus for poly (caprolactone) with the increase of porosity. For tensile loads, the findings show that a 4f unit is the unit with the worst performance below 50% porosity, becoming the best scaffold unit for higher percentages of porosity. Below 50% porosity, the circular pore scaffolds are the units with better material modulus. For shear solicitations, the results show that the 4f unit is the unit with the worst performance, being the circular pore scaffolds the unit with better performance.

As expected, the influence of the deformation value on the mechanical behaviour of scaffolds is significant. This can be observed in Fig. 12, which indicates the variation of the elastic modulus of circular alginate pore scaffold units submitted to different deformation values. The material modulus decreases as the deformation increases.

Another aspect of extreme importance is the effect of the number of pores per face on the mechanical behaviour of the scaffold. As illustrated in Figs. 13–15, it is possible to observe that the scaffold's modulus increases with the number of pores per face, for each unit.

Regarding the structural analysis of the association of scaffold units, for the *X* direction, the variation of the scaffold's modulus has little variation, but in the *Y* direction, as the number of association increases, so does the scaffold's modulus (Fig. 16).

Regarding the P-minimal surface thickness, as illustrated in Fig. 17, porosity decreases with the P-minimal surface thickness.

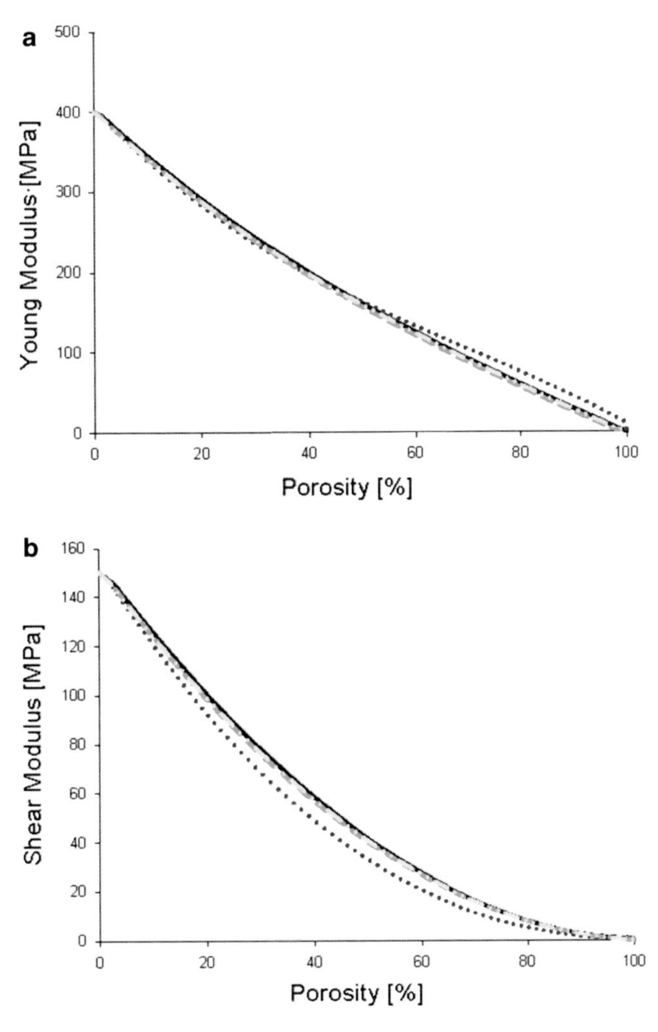

Fig. 11. The variation of the material modulus of poly(caprolactone) material, according to the scaffold porosity for all scaffold units: (**a**) elastic modulus and (**b**) shear modulus.

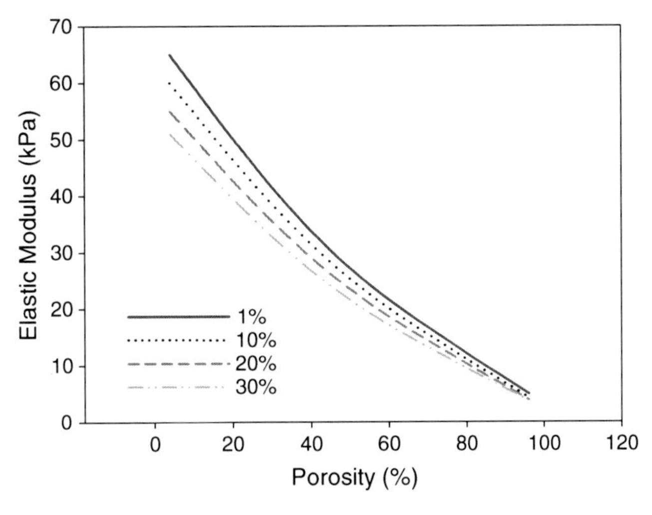

Fig. 12. The variation of the elastic modulus versus porosity for circular scaffold units of alginate, under different deformation values.

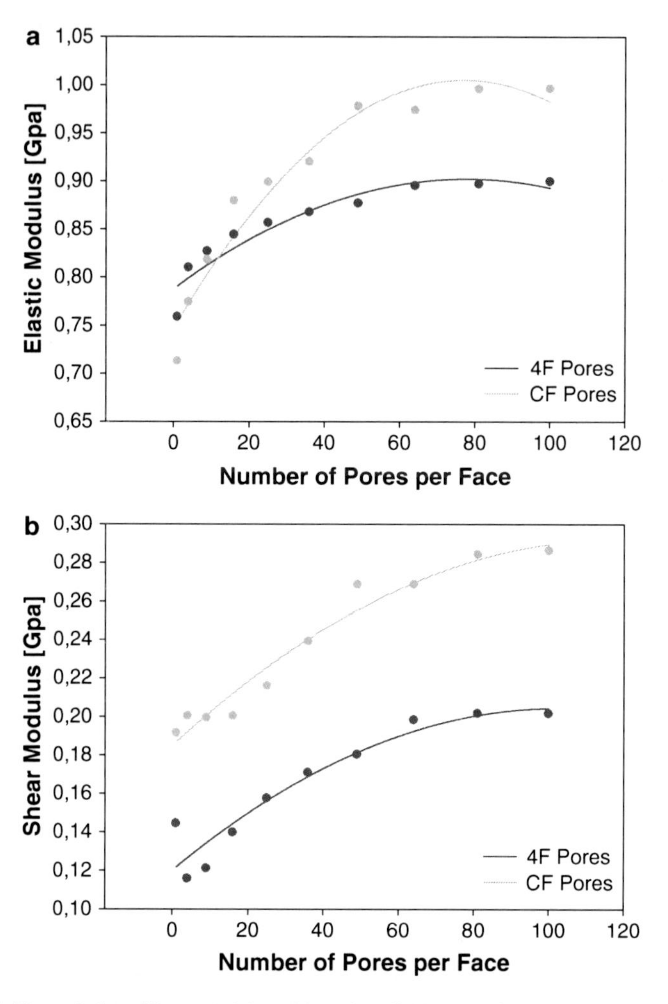

Fig. 13. The variation of the material modulus of scaffold units in function of the number of pores per face with cf and 4f pores: (**a**) elastic modulus and (**b**) shear modulus.

Figure 18 shows that the elastic modulus increases with thickness. A linear dependence between the surface thickness and the elastic modulus was obtained as observed in Fig. 18. Figure 19 illustrates that the elastic modulus decreases with the increase of the scaffold's porosity, as demonstrated before.

Regarding the effect of the P-minimal surface radius variations, Fig. 20 shows that porosity decreases till a threshold value for the surface radius from which starts to increase. The elastic modulus decreases by increasing the P-minimal surface radius as shown in Fig. 21.

As the porosity and the radius have a hyperbolic behaviour, the same was observed for the elastic modulus (Fig. 22). Therefore, we may decrease or increase the elastic modulus of the scaffold while maintaining high porosity values.

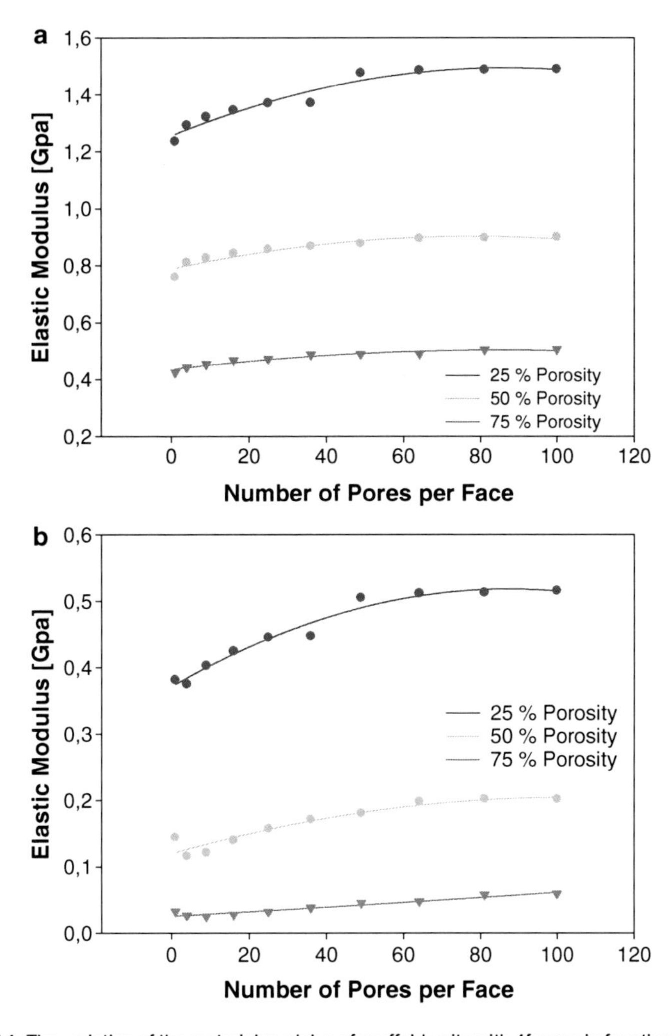

Fig. 14. The variation of the material modulus of scaffold units with 4f pores in function of the number of pores per face and level of porosity: (**a**) elastic modulus and (**b**) shear modulus.

4. Topological Optimisation of Scaffolds in Tissue Engineering

Topological optimisation, aiming to find the best use of material according to a "maximum-stiffness" design, requires neither parameters nor the explicit definition of optimisation variables. The objective function is predefined, as are the state variables (constrained-dependent variables) and the design variables (independent variables to be optimised). The topological optimisation problem requires the problem definition (material properties, model, and loads), the objective function (the function to be minimised or maximised), and the state variables corresponding to the percentage of material to be removed (17–24).

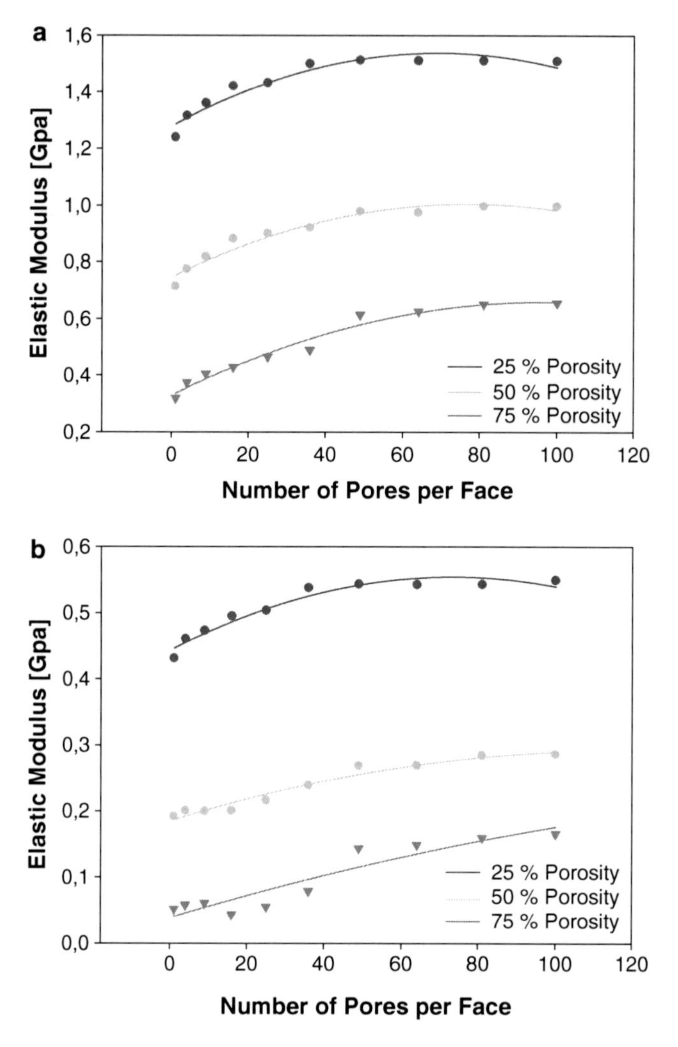

Fig. 15. The variation of the material modulus of scaffold units with cf pores in function of the number of pores per face and level of porosity: (**a**) elastic modulus and (**b**) shear modulus.

From a mechanical point of view, the goal of topological optimisation is to minimise the total compliance, which is proportional to the strain energy. Figure 23 illustrates the general topology optimisation scheme considered in this work.

The design variables are internal, pseudo-densities that are assigned to each finite element in the topological problem. The pseudo-density for each element varies from 0 to 1, where $\eta_i \approx 0$ represents material to be removed, and $\eta_i \approx 1$ represents material that should be kept (Fig. 24).

For a given domain $\Omega \subseteq \Re^2(\Re^3)$, regions $\Omega(\Gamma_t)$, and fixed boundaries, the optimisation goal is to find the optimal elasticity tensor $E_{ijkl}(x)$, which takes the form (25, 26):

$$E_{ijkl}(x) = \eta(x)\overline{E}_{ijkl}, \tag{17}$$

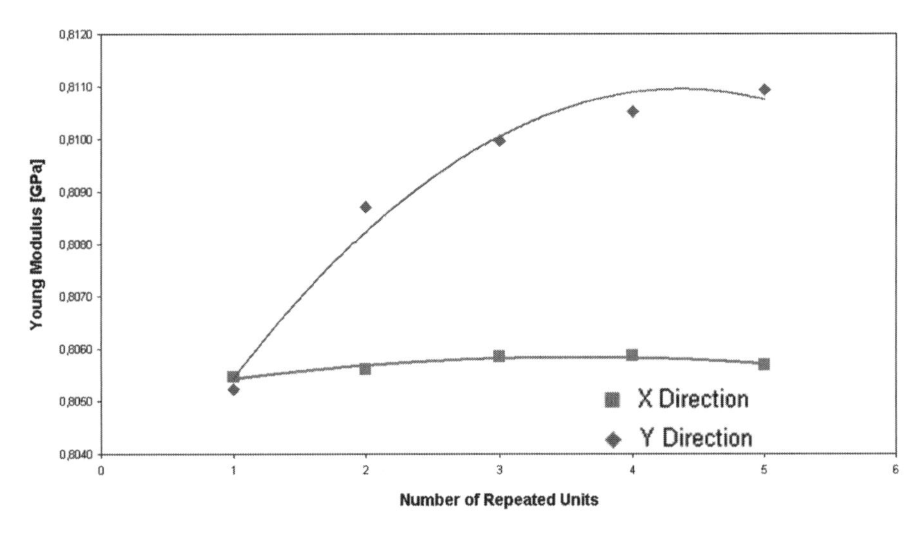

Fig. 16. The variation of the material modulus of scaffold units with 4f pores in function of the number of repeated units in the X direction (*red line*) and in the Y direction (*green line*).

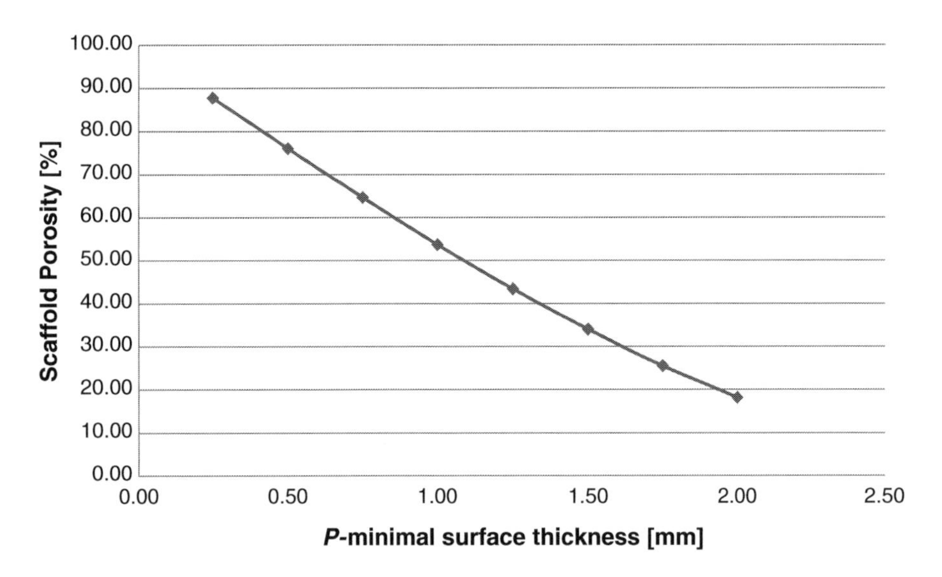

Fig. 17. Variation of the scaffold porosity with the P-minimal surface thickness.

where $\overline{E}_{ijkl}$ is the constant rigidity tensor for the considered material and $\eta(x)$ is an indicator function for a region $\Omega^* \subseteq \Omega$ that is occupied by material:

$$\eta(x) = \begin{cases} 1, & \text{if} \quad x \in \Omega^* \\ 0, & \text{if} \quad x \notin \Omega^* \end{cases}. \tag{18}$$

Considering the energy bilinear form

$$a(u, v) = \int_{\Omega} \sum_{i,j,k,l=1}^{3} E_{ijkl}\varepsilon_{ij}(u)\varepsilon_{kl}(v)\mathrm{d}x \tag{19}$$

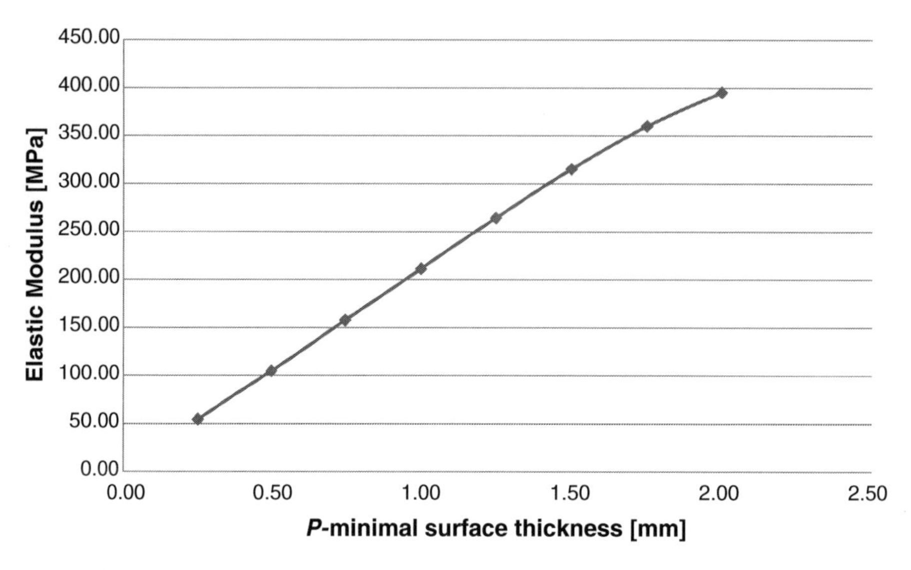

Fig. 18. Variation of the elastic modulus with the P-minimal surface thickness.

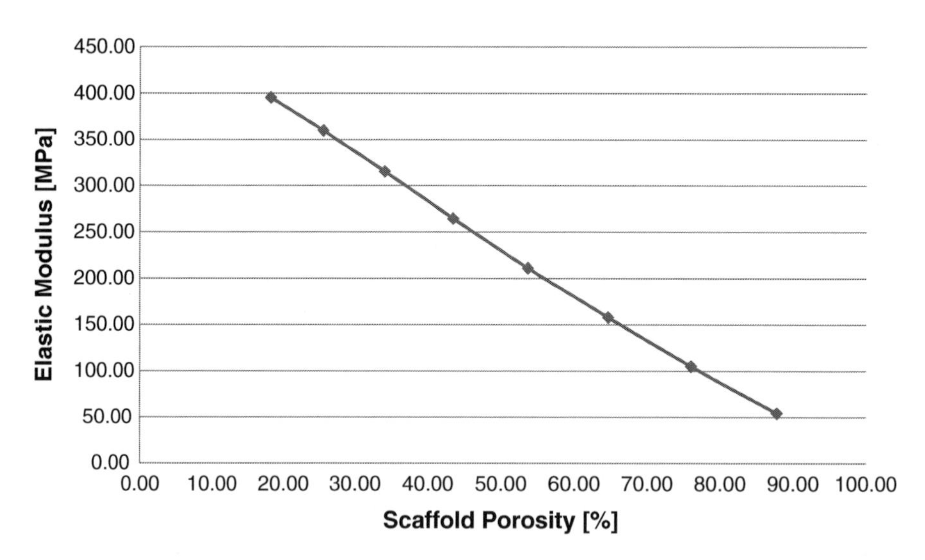

Fig. 19. Variation of the elastic modulus with porosity.

with linearised strains

$$\varepsilon_{ij} = \frac{1}{2}\left(\frac{\partial u_i}{\partial x_j} + \frac{\partial u_j}{\partial x_i}\right), \quad i, j = 1, 2, 3 \tag{20}$$

and the load linear form

$$l(u) = \int_{\Omega} fu \, dx + \int_{\Gamma_t} tu \, ds \tag{21}$$

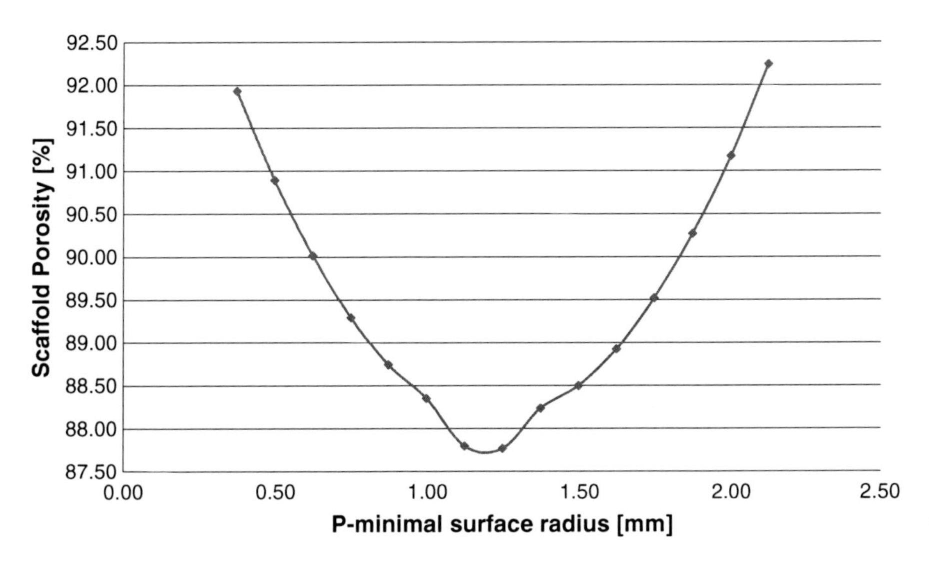

Fig. 20. Variation of the scaffold porosity with the P-minimal surface radius.

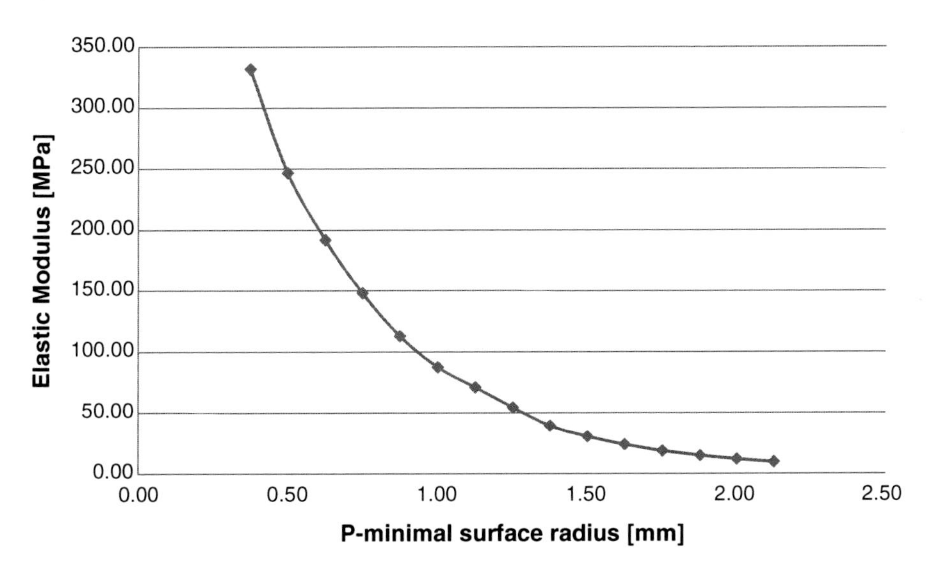

Fig. 21. Variation of the elastic modulus with the P-minimal surface radius.

The optimisation problem considered here is defined as follows:

$$\min_{\eta^*(x_e), i=1,\ldots,n} l(u) = \int_{\Omega} fu \, \mathrm{d}x + \int_{\Gamma_t} tu \, \mathrm{d}s$$

subjected to constraints:

$$(i) \quad \int_{\Omega} \sum_{i,j,k,l=1}^{3} \tilde{E}_{ijkl}(\eta(x))\varepsilon_{ij}(u(x))\varepsilon_{kl}(v(x))\mathrm{d}x = l(v), \ \forall v \in U \tag{22}$$

$$(ii) \quad 0 \leq \eta(x) \leq 1, \quad \forall x \in \Omega$$

$$(iii) \int_{\Omega} \eta(x)\mathrm{d}x \leq Vol.$$

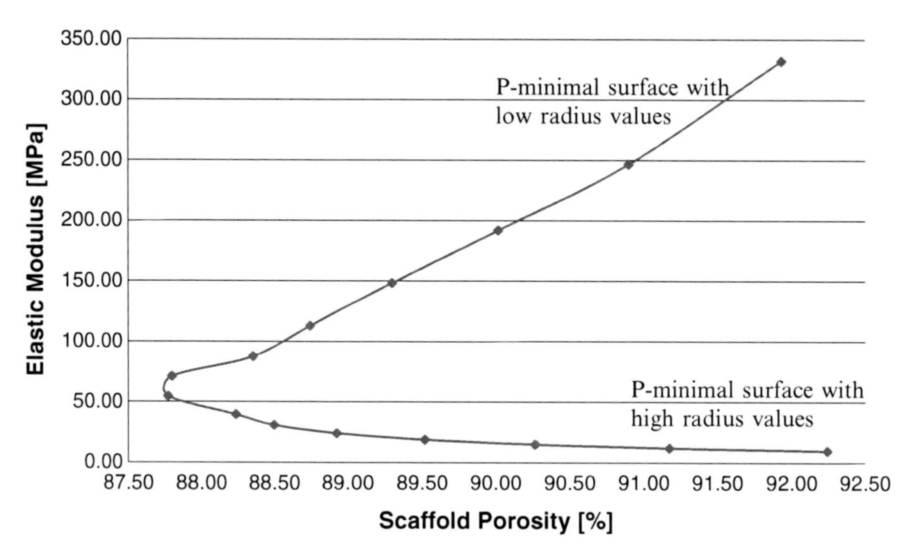

Fig. 22. Variation of the elastic modulus with the porosity.

4.1. Finite Element Discretisation for the Optimisation Problem

The domain Ω is represented as a collection of a finite number of subdomains. This is called discretisation of the domain. Each subdomain is called an element and the collection of elements is called the finite element mesh. In this case, $\eta(x)$ was discretised by assigning a constant value on each element of the finite element model, establishing a suitable piecewise constant function $\eta^*(x)$ to approximate $\eta(x)$.

4.2. Algorithm

The algorithm considered to obtain the solution of the minimum compliance problem is based on the following update strategy (27):

For $e = 1, \ldots, n$:

$$\eta_e^{k+1} = \begin{cases} \max\{(1-\zeta)\eta_e^k, eps\}, & \text{if } \eta_e^k(D_e^{k+1})^\alpha \leq \max\{(1-\zeta)\eta_e^k, eps\} \\ \eta_e^k(D_e^{k+1})^\alpha, & \text{if } \max\{(1-\zeta)\eta_e^k, eps\} \leq \eta_e^k(D_e^{k+1})^\alpha \leq \min\{(1+\zeta)\eta_e^k, 1\} \\ \min\{(1+\zeta)\eta_e^k, 1\}, & \text{if } \min\{(1-\zeta)\eta_e^k, 1\} \leq \eta_e^k(D_e^{k+1})^\alpha \end{cases} \quad (23)$$

with an appropriate weighing factor α, a move limit ζ, and an upper limit $eps>0$. To perform the update strategy in (23) for a given data η_e^k, $e = 1, \ldots, n$, eps, ζ, α, it is necessary first to compute D_e^{k+1}, $e = 1, \ldots, n$, which is given by the following equation:

$$D_e^{k+1} = (\Lambda^{k+1})^{-1} E \sum_{i,j,k,l=1}^{3} \frac{\partial \tilde{E}_{ijkl}(\eta_e^k)}{\partial \eta} \varepsilon_{ij}^k(u(x_e))\varepsilon_{kl}^k(u(x_e)). \quad (24)$$

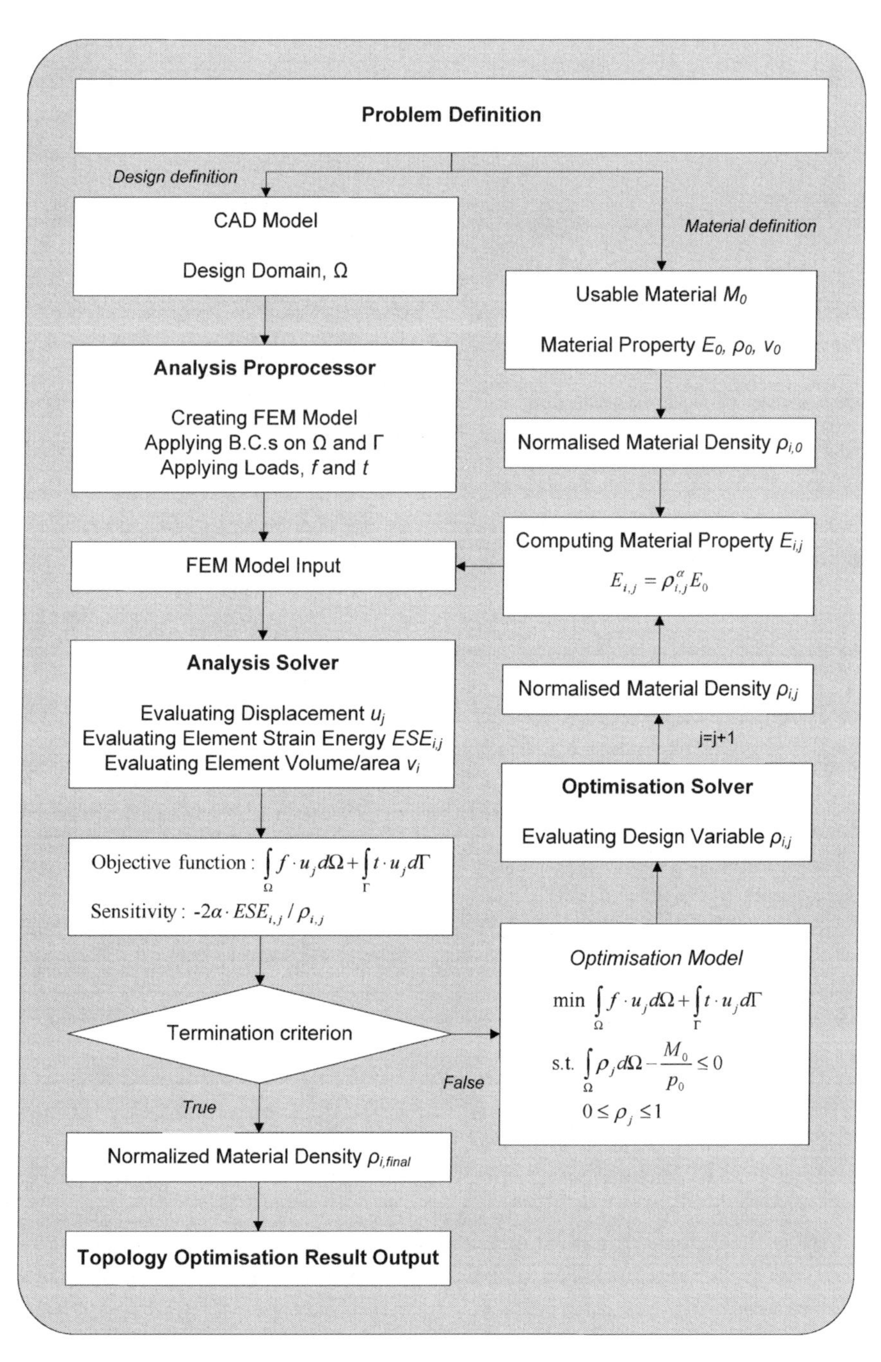

Fig. 23. A general topological optimisation process.

The Lagrange parameter Λ is updated by solving the equation:

$$V(\Lambda) := \sum_{e=1}^{n} \eta_e^{k+1}(\Lambda) - Vol = 0 \tag{25}$$

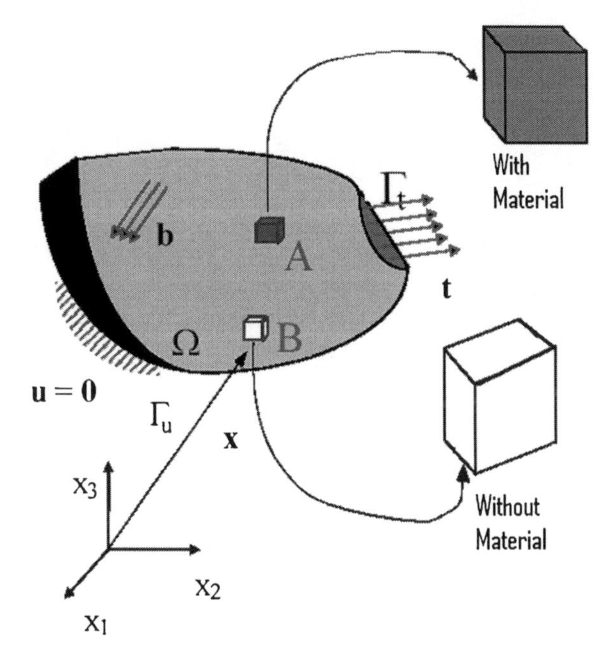

Fig. 24. Illustration of a topological optimisation.

in an inner interaction using an appropriate root finding algorithm (Bisection method, Newton's method, Secant method, False Position method). The selected method was the Bisection method which is less efficient than Newton's method and it is also much less prone to odd behaviour.

5. Results

The topological optimisation procedure starts by introducing the elastic modulus of the material, Poisson's ratio, and displacement submitted to the scaffold. Figure 25 illustrates the structural cubic mesh, constraints, and solicitations of the cubic scaffold unit without porosity. For the numerical computation of the elastic modulus, a uniform displacement in a single direction is considered (in this case, the *X* direction), which is equivalent to the strain on the same direction (ε_x) imposed to a face of the cubic unit (Face A). The opposite face (Face B) of the cubic unit is constrained and unable to have any displacement. By imposing different volume reductions, which corresponds to different levels of porosity, the topological algorithm evaluates the best distribution of material within the scaffold volume configuration in order to optimise the mechanical properties.

The material considered here is hydroxiapatite (elastic modulus of 2,000 MPa and Poisson's ratio of 0.25). For the search of the

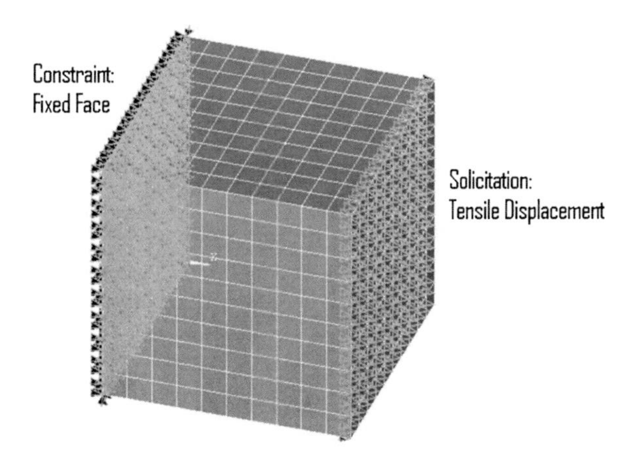

Fig. 25. Fully packed unit cube without the absence of material containing mesh details, constraints, and displacements definitions for the topological optimisation problem.

optimised structures, it is defined the number of 100 iterations for each calculation. The convergence tolerance parameter for each calculation is 1E-4. A mesh of 100 cubic solid elements was considered. The optimisation approach and the mechanical performance of scaffolds were implemented using the finite element method software ANSYS.

Topological optimisations for different levels of porosity (10%, 20%, 30%, 40%, 50%, 60%, 70%, and 80%) were performed considering the following conditions:

- Axial displacement of 0.05% of the unit's dimension, equivalent to a simple tensile solicitation effect (Fig. 26).
- Shear displacement of 0.05% of the unit's dimension, equivalent to a simple shear solicitation effect (Fig. 27).
- Combined axial and shear displacement of 0.05% of the unit's dimension, equivalent to a combined tensile and shear solicitation effect (Fig. 28).

In the previous Figs. 26–28, the yellow colour represents the absence of material and the light blue colour represents the existence of material in the scaffold unit. The remaining colours represent regions in which there is a transition of material, meaning, neither the mesh unit is completely filled with material nor completely absent of material. From the results of the topological optimisations, it was possible to obtain 3D CAD models. The 3D CAD models for the tensile effect are illustrated in Fig. 29 and for shear effect in Fig. 30. These models were obtained considering only the mesh that represented the existence of material (light blue colour elements). However, by neglecting the transition regions, the CAD models present a slightly different value of porosity regarding the initial imposed value.

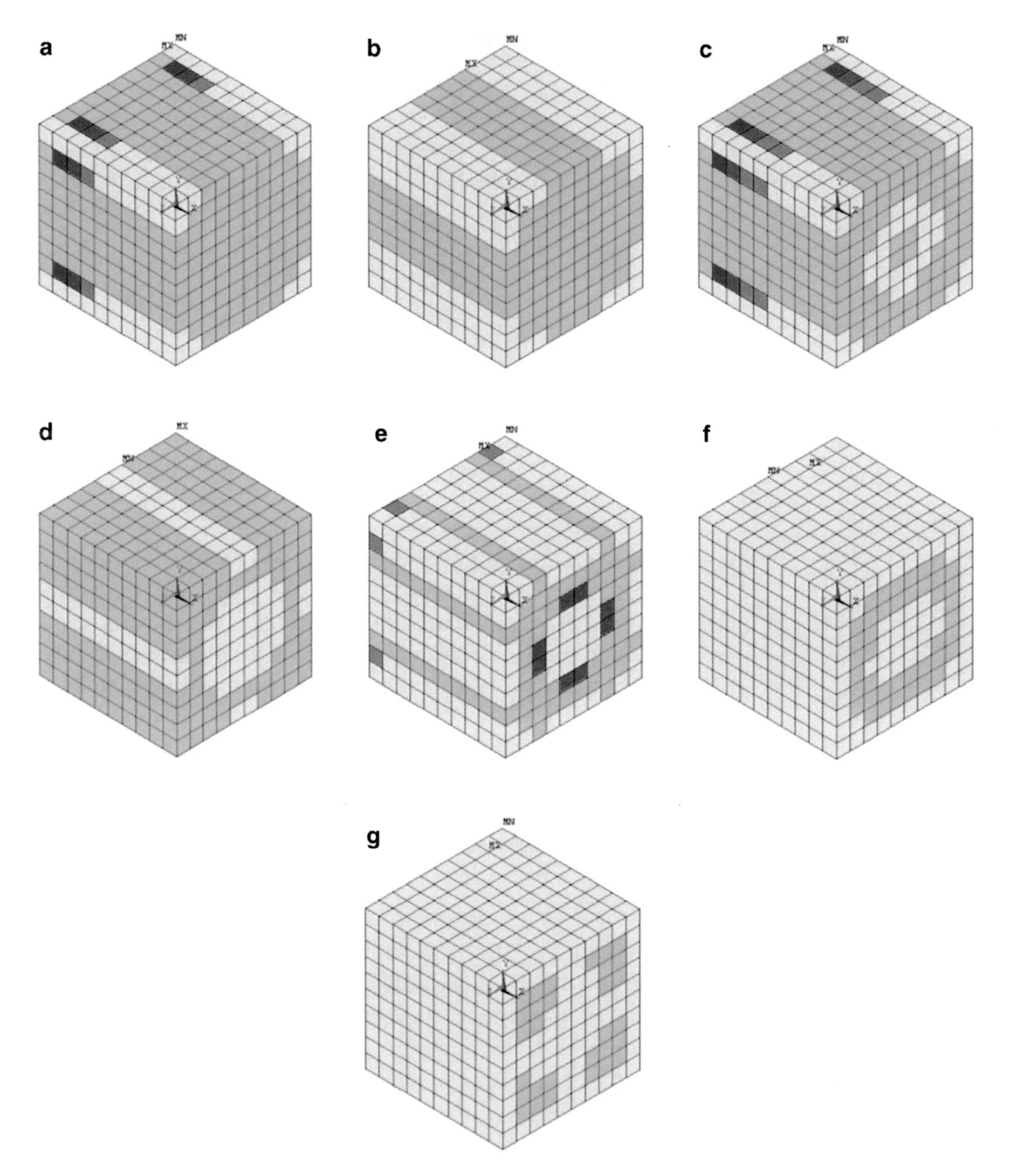

Fig. 26. Topological optimised scaffolds for a tensile load and a specified porosity. (**a**) 10% porosity, (**b**) 20% porosity, (**c**) 30% porosity, (**d**) 40% porosity, (**e**) 50% porosity, (**f**) 60% porosity, and (**g**) 70% porosity.

For porosity levels higher than ±70%, it was possible to observe a lack of integrity of the scaffold's main structure, in other words, the single scaffold unit divides itself into more than one component, as illustrated in Fig. 31.

From a 3D CAD model of the topologically optimised unit, it was possible to obtain a scaffold assembly unit combining several single

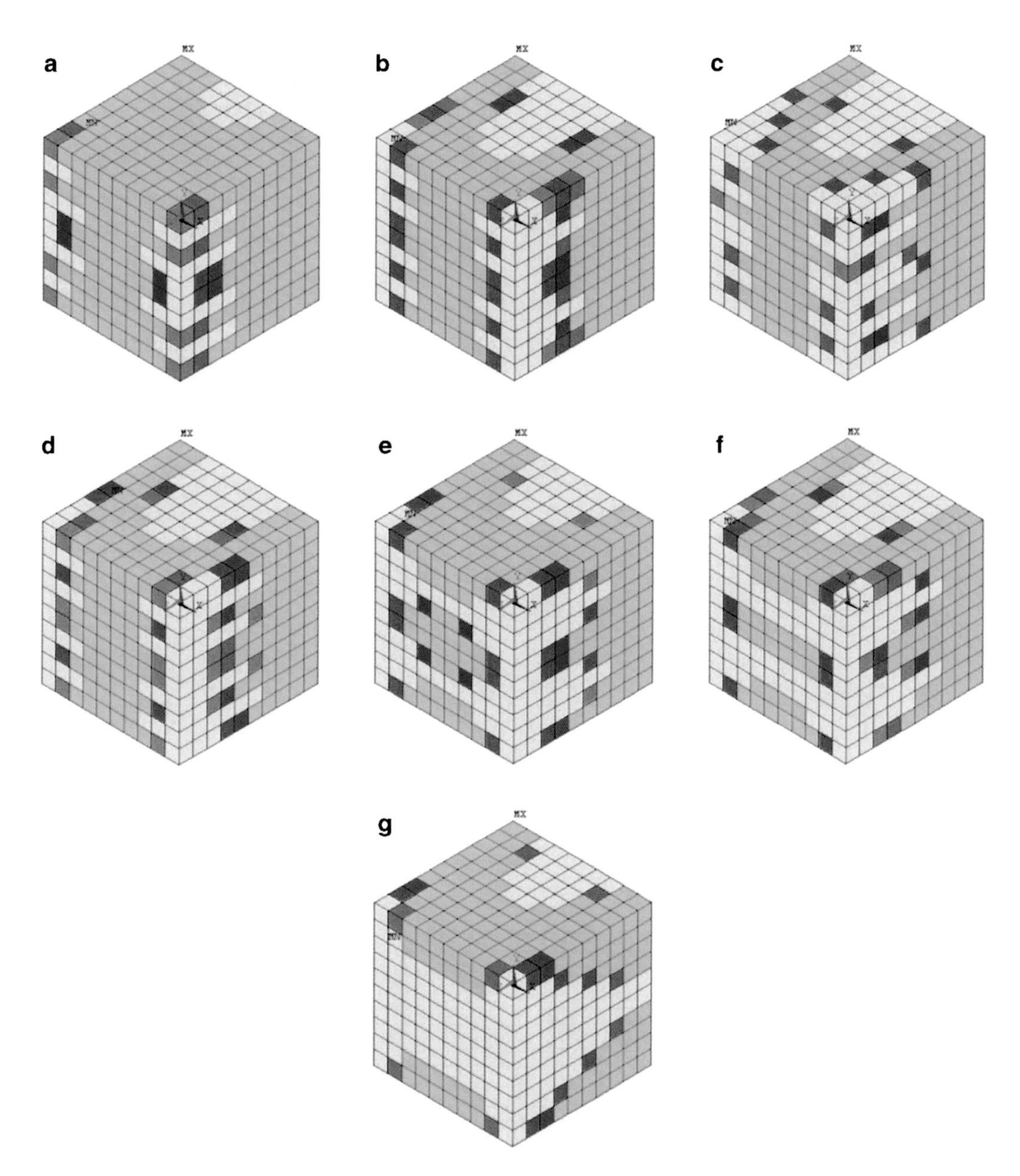

Fig. 27. Topological optimised scaffolds for a shear load and a specified porosity. (**a**) 10% porosity, (**b**) 20% porosity, (**c**) 30% porosity, (**d**) 40% porosity, (**e**) 50% porosity, (**f**) 60% porosity, and (**g**) 70% porosity.

scaffold units. Examples of assembly scaffold units are illustrated in Figs. 32 and 33 for tensile and shear effects.

Comparing the topological optimisation and the structural results obtained previously, it is possible to observe that as the scaffold porosity increases, the structural stiffness decreases, in spite of the pore topology: square (4 face) pores, 8 face pores,

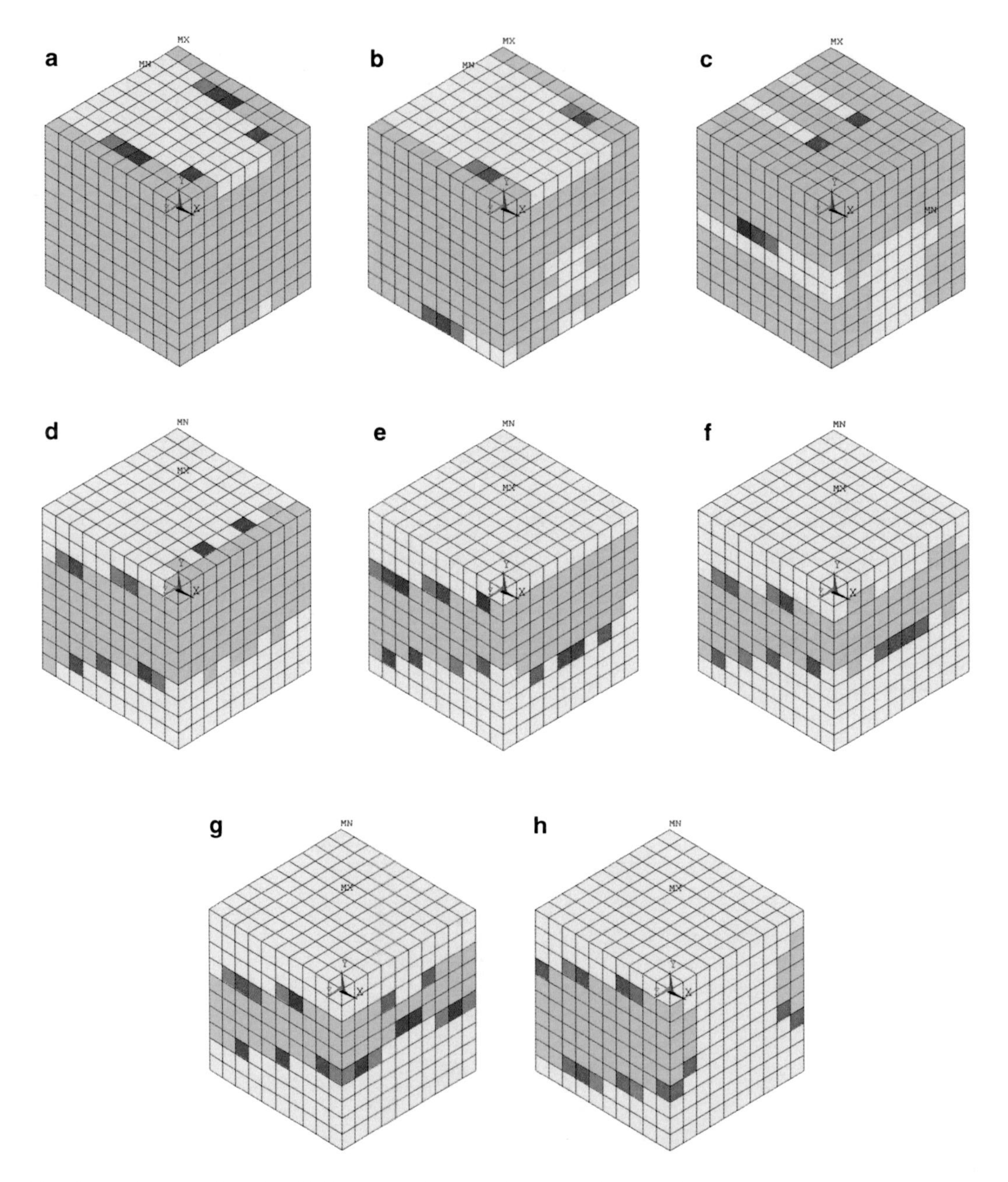

Fig. 28. Topological optimised scaffolds for a combination of tensile and shear effects and a specified porosity. (**a**) 10% porosity, (**b**) 20% porosity, (**c**) 30% porosity, (**d**) 40% porosity, (**e**) 50% porosity, (**f**) 60% porosity, (**g**) 70% porosity, and (**h**) 80% porosity.

12 face pores, and circular pores (Fig. 34). This is not the case of topological optimised scaffold units, where the structural stiffness maintains its value in spite of the porosity variation until a certain value of porosity. Higher values of porosity result in the structural loss of integrity as mentioned before.

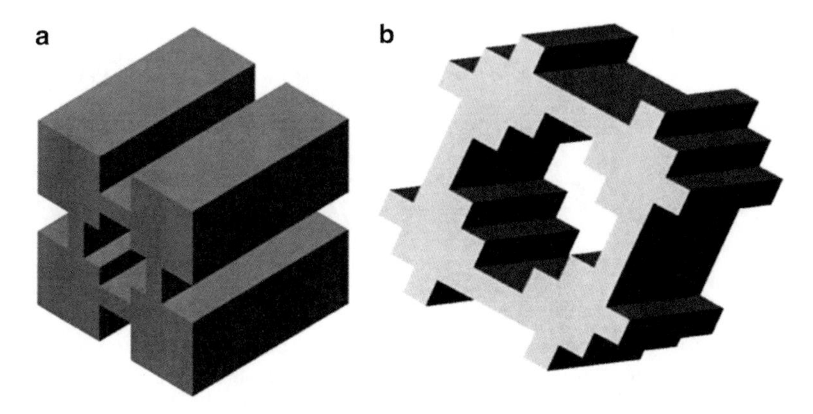

Fig. 29. Topological optimised CAD models of scaffolds for a tensile condition and a specified porosity. (**a**) 25% porosity and (**b**) 50% porosity.

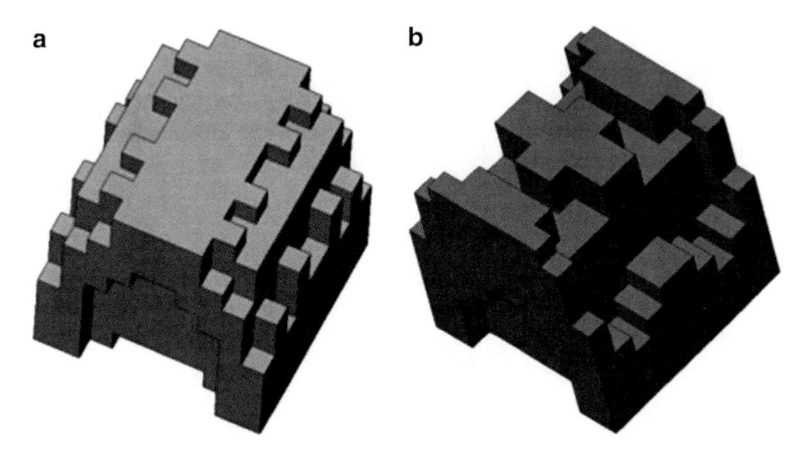

Fig. 30. Topological optimised CAD models of scaffolds for a shear solicitation and a given porosity. (**a**) 25% porosity and (**b**) 50% porosity.

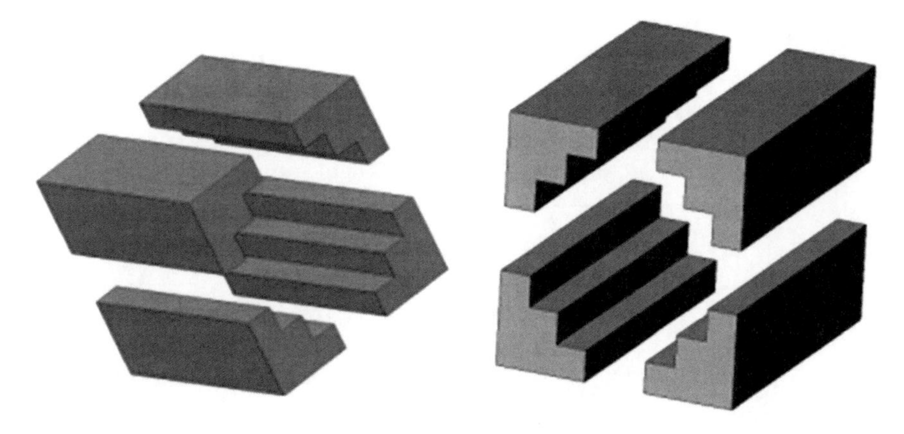

Fig. 31. CAD model of a topological optimised unit with a porosity of 70% illustrating the lack of integrity of the scaffold unit.

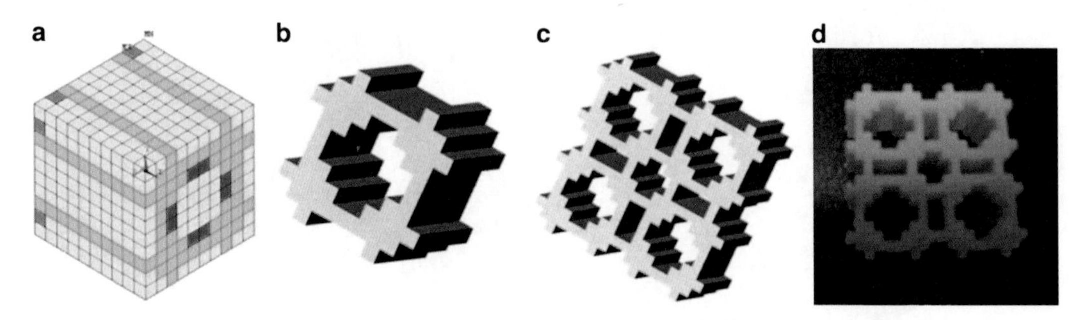

Fig. 32. Topological optimised scaffolds for a tensile solicitation and 50% of porosity. (**a**) Numerical evaluation, (**b**) optimised scaffold unit, (**c**) CAD scaffold assembly unit, and (**d**) physical assembly unit.

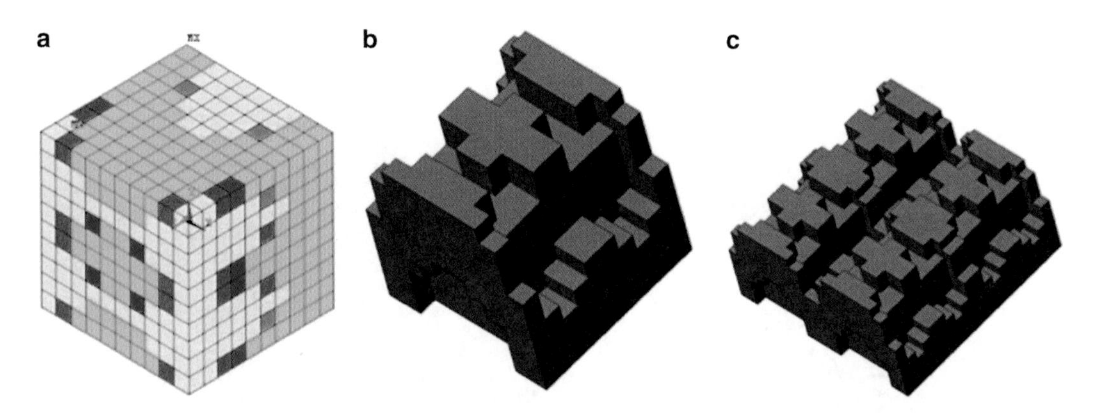

Fig. 33. Topological optimised scaffolds for a shear solicitation and 50% porosity. (**a**) numerical evaluation, (**b**) CAD scaffold unit, and (**c**) CAD scaffold assembly unit.

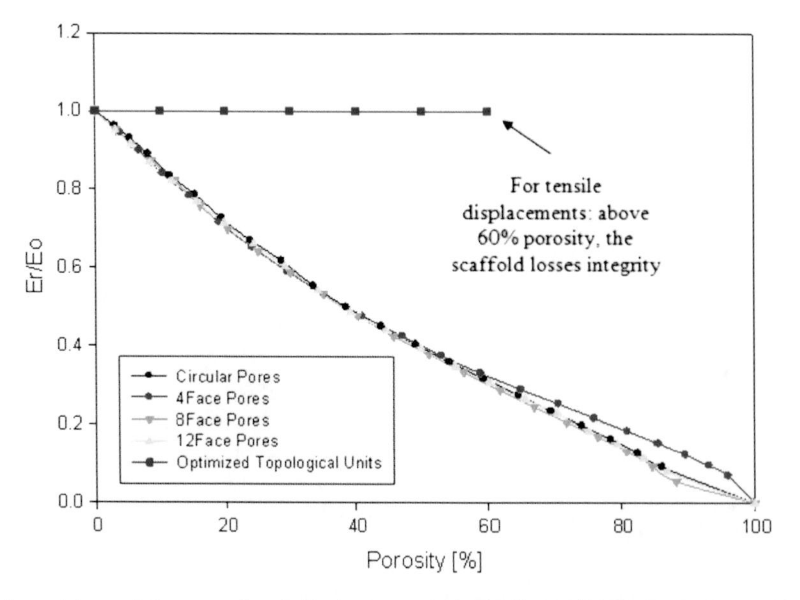

Fig. 34. Young's modulus variation according to the increase of porosity for scaffold units obtained through topological optimisation in comparison with scaffold units containing pores in function of the number of faces per pore.

6. Conclusions

Rapid prototyping technologies were recently introduced in the medical field, being particularly viable to produce porous scaffolds for tissue engineering. These scaffolds should be biocompatible, biodegradable, with appropriate porosity, pore structure, and pore distribution, on top of presenting both surface and structural compatibility. Surface compatibility means a chemical, biological, and physical suitability with the host tissue. Structural compatibility corresponds to an optimal adaptation to the mechanical behaviour of the host tissue. Thus, structural compatibility refers to the mechanical properties and deformation capabilities of the scaffold.

This chapter presents a computer tool to support the design of scaffolds to be produced by rapid prototyping technologies. The software enables to evaluate scaffold mechanical properties as a function of porosity and pore topology and distribution, for a wide range of materials, suitable for both hard and soft tissue engineering.

Acknowledgements

This research is supported by the Portuguese Foundation of Science and Technology through a Ph.D. grant of Henrique Almeida (SFRH/BD/37604/2007). The authors also wish to thank the sponsorship given by CYTED through a Biomanufacturing Network "Rede Iberoamericana de Biofabricação."

References

1. Almeida HA, Bártolo PJ, Ferreira J (2007) Design of scaffolds assisted by computer. In: Brebbia CA (ed) Modelling in medicine and biology VII. Wit, Southampton, pp 157–166
2. Almeida HA, Bártolo PJ, Ferreira J (2007) Mechanical behaviour and vascularisation analysis of tissue engineering scaffolds. In: Bártolo PJ et al (eds) Virtual and rapid manufacturing. Taylor & Francis, London, pp 73–80
3. Almeida HA, Bártolo PJ (2010) Virtual topological optimization of scaffolds for rapid prototyping. Int J Med Eng Phys 32 (7):775–782
4. Almeida HA, Bártolo PJ (2007) Topological optimization of rapid prototyping scaffolds, International conference on manufacturing automation (ICMA2007), National University of Singapore, Singapore.
5. Almeida HA, Bártolo PJ (2008) Computer simulation and optimisation of tissue engineering scaffolds: mechanical and vascular behaviour. In: Halevi Y, Fischer A (eds) Proceedings of the ninth biennial ASME conference on engineering systems design and analysis (ESDA2008). Haifa Isreal
6. Bártolo PJ, Almeida H, Laoui T (2009) Rapid prototyping & manufacturing for tissue engineering scaffolds. Int J Comp Appl Technol 36 (1):1–9
7. Chua CK, Yeong WY, Leong KF (2005) Development of scaffolds for tissue engineering using a 3D inkjet model maker. In: Bártolo PJ et al. (eds) Virtual modelling and rapid manufacturing. Taylor & Francis, London

8. Hutmacher DW, Sittinger M, Risbud MV (2004) Scaffold-based tissue engineering: rationale for computer-aided design and solid free-form fabrication systems. Trends Biotechnol 22:354–362
9. Almeida HA, Bártolo PJ, Mendes A (2009) Mechanical evaluation of alginate scaffolds considering the water loss/shrinkage process. In: Bártolo PJ et al. (eds) ICTE 2009 international conference on tissue engineering an ECCOMAS thematic conference. IST Press, Lisbon, pp 317–323
10. Andersson S (1983) On the description of complex inorganic crystal structures. Angew Chem Int Ed 22(2):69–81
11. Scriven LE (1976) Equilibrium bicontinuous structure. Nature 263(5573):123–125
12. Larsson M, Terasaki O, Larsson K (2003) A solid state transition in the tetragonal lipid bilayer structure at the lung alveolar surface. Solid State Sci 5(1):109–114
13. Lord EA, Mackay AL (2003) Periodic minimal surfaces of cubic symmetry. Curr Sci 85 (3):346–362
14. Wang Y (2007) Periodic surface modeling for computer aided nano design. Comp-Aided Des 39:179–189
15. Humphrey JD, Delange SL (2003) An introduction to biomechanics – solids and fluids, analysis and design. Springer, New York
16. Fung YC (1990) Biomechanics: motion, flow, stress and growth. Springer-Verlag, New York
17. Neches L, Cisilino A (2008) Topology optimization of 2D elastic structures using boundary elements. Eng Anal Bound Elem 32:533–544
18. Ansola R, Veguería E, Canales J, Tárrago J (2007) A simple evolutionary topology optimization procedure for compliant mechanism design. Finite Elem Anal Des 44: 53–62
19. Kruijf N, Zhou S, Li Q, Mai YW (2007) Topological design of structures and composite materials with multiobjectives. Solids Struct 44:7092–7109
20. Hsu MH, Hsu YL (2005) Generalization of two and three-dimensional structural topology optimization. Eng Optimizat 37:83–102
21. Bendsøe MP, Sigmund O (2003) Topology optimization: theory, methods and applications. Springer, Berlin
22. Rozvany GIN (2001) Aim, scope, methods, history and unified terminology of computer-aided topology optimization in structural mechanics. Struct Multidiscipl Optimizat 21:90–108
23. Bendsoe MP (1989) Optimal shape design as a material distribution problem. Struct Multidiscipl Optimizat 1:193–202
24. Bendsoe MP, Kikuchi N (1988) Generating optimal topologies in structural design using homosenization method. Comp Methods Appl Mech Eng 71:197–244
25. Mlejnek HP, Schirrmacher R (1993) An engineering approach to optimal material distribution and shape finding. Comput Methods Appl Mech Eng 106:1–26
26. Mlejnek HP (1992) Some aspects of the genesis of structures. Struct Multidiscipl Optimizat 5:64–69
27. Vogel F (1997) Topology optimisation of linear-elastic structures with ANSYS 5.4. NAFEMS conference on topology optimisation. Aalen, Germany

Chapter 14

Modeling of Bioreactor Hydrodynamic Environment and Its Effects on Tissue Growth

Bahar Bilgen and Gilda A. Barabino

Abstract

The design of optimal bioreactor systems for tissue engineering applications requires a sophisticated understanding of the complexities of the bioreactor environment and the role that it plays in the formation of engineered tissues. To this end, a tissue growth model is developed to characterize the tissue growth and extracellular matrix synthesis by chondrocytes seeded and cultivated on polyglycolic acid scaffolds in a wavy-walled bioreactor for a period of 4 weeks. This model consists of four components: (1) a computational fluid dynamics (CFD) model to characterize the complex hydrodynamic environment in the bioreactor, (2) a kinetic growth model to characterize the cell growth and extracellular matrix production dynamics, (3) an artificial neural network (ANN) that empirically correlates hydrodynamic parameters with kinetic constants, and (4) a second ANN that correlates the biochemical composition of constructs with their material properties. In tandem, these components enable the prediction of the dynamics of tissue growth, as well as the final compositional and mechanical properties of engineered cartilage. The growth model methodology developed in this study serves as a tool to predict optimal bioprocessing conditions required to achieve desired tissue properties.

Key words: Computational fluid dynamics, Artificial neural networks, Cartilage tissue engineering, Tissue growth kinetics, Hydrodynamic environment, Shear stress, Wavy-walled bioreactor

1. Introduction

For engineered tissues to serve as a clinically relevant substitute or replacement for damaged or diseased tissues, they must possess properties and mimic functionality of the native tissue. By providing an environment for biochemical, physical, and mechanical stimuli, bioreactors represent an important means for in vitro cultivation of suitable tissues. Among other benefits, hydrodynamic forces applied during culture of 3D tissue constructs in bioreactors has led to increased cell proliferation and matrix deposition and improved mechanical properties (1–4). The lack of understanding of the mechanisms and signals by which these forces affect tissue properties

Michael A.K. Liebschner (ed.), *Computer-Aided Tissue Engineering*, Methods in Molecular Biology, vol. 868,
DOI 10.1007/978-1-61779-764-4_14,

and of tissue growth models linked to bioprocessing conditions is a hindrance to the achievement of optimal tissues that are phenotypically appropriate. Using cartilage tissue engineering as a model tissue, a number of studies have addressed construct biochemical properties as a function of time (kinetics) (5–7) or nutrients and growth factors (8). These studies are limited, however, since they do not account for the effect of shear stress and other hydrodynamic parameters on cartilage construct formation. We have developed an integrated model that accounts for growth kinetics and hydrodynamic forces in the determination of tissue properties using a wavy-walled bioreactor as a tool for characterization of the fluid environment and cartilage tissue growth within bioreactors. This model combines computational fluid dynamics (CFD) and artificial neural network (ANN) models to enhance predictive capabilities relating culture conditions and tissue properties (9).

CFD has been employed to characterize fluid flow in tissue engineering bioreactors in an effort to better understand and extend insights gained from experimental observations (10). When backed by validation through experimental techniques, such as particle image velocimetry (PIV), CFD is a powerful tool to unravel and model the complex interrelationships between the hydrodynamic parameters associated with the flow environment in the immediate vicinity of the developing tissue constructs and the resultant tissue properties (11) (Table 1). The power of CFD is greatly enhanced when used in combination with ANN, a computational tool inspired by the network of biological neurons and capable of providing accurate solutions based on trained data (12). By way of example,

Table 1
Hydrodynamic parameters determined by the CFD simulations

Hydrodynamic parameters	Units	Case 1, C-50	Case 2, C-80	Case 3, L-50	Case 4, L-80
Mean-shear stress	Dynes/cm^2	0.24	0.49	0.18	0.41
Axial velocity	cm/s	0.73	1.24	0.88	1.7
Tangential velocity	cm/s	0.82	1.45	0.14	0.34
Radial velocity	cm/s	0.34	0.54	0.32	0.52
Uniformity of velocity distribution	Dimensionless	0.46	0.37	0.31	0.33

These parameters are used as ANN 1 training data inputs. The uniformity of velocity distribution is defined as the geometric mean of normalized velocity components (11)

this chapter describes methodology based on the effective use of CFD and ANN in the development of a hybrid tissue growth model designed to enable the prediction of the appropriate bioreactor conditions to support optimal tissue construct properties. In this case, a wavy-walled bioreactor is employed to provide distinct hydrodynamic environments for cultivating cartilage tissue constructs (11, 13) in the elucidation of bioreactor hydrodynamic effects on engineered tissue growth and the development of a growth model which characterizes the growth and function of chondrocytes cultivated on polyglycolic acid (PGA) scaffolds.

The tissue growth model consists of four components (Fig. 1):

1. A CFD model that models the hydrodynamic environment in the bioreactor with scaffolds and determines the key hydrodynamic properties.
2. A kinetic growth model that describes kinetic constants regarding cell and tissue growth over time based on experimental data.
3. An ANN that empirically correlates bioreactor hydrodynamic parameters with kinetic constants.
4. A second ANN that correlates biochemical composition of constructs with their material properties.

In tandem, these models enable the prediction and optimization of the dynamics of cell attachment and tissue growth in the PGA scaffold, as well as the final compositional and mechanical properties of constructs. The kinetic model is formulated to describe the attachment, growth, and metabolism of chondrocytes seeded on PGA scaffolds and cultivated in bioreactors. This is a deterministic model that describes cell growth as a set of differential equations, which can be solved to simulate the entire dynamics of the system. This model requires a number of kinetic constants, which need to be determined prior to solving the equations. Furthermore, these kinetic constants are functions of hydrodynamic parameters. For example, at increased shear stress around the constructs, the kinetic constant for cell proliferation in the outer ring of a construct is likely to decrease and the kinetic constant for cell proliferation in the inner circle increases. A possible explanation for this observation may be that the high shear stresses experienced by the cells on the surface of scaffolds suppress their proliferation in the outer ring (1, 14). As these types of dependencies require extremely detailed phenotype models—which are not available and are difficult to develop—ANN techniques are utilized to empirically correlate these kinetic constants with the hydrodynamic parameters using experimental data (see Subheading 2.3 for a description of ANNs). A similar ANN model is also employed to correlate the biochemical composition with the material properties. This hybrid modeling approach therefore enables the prediction of

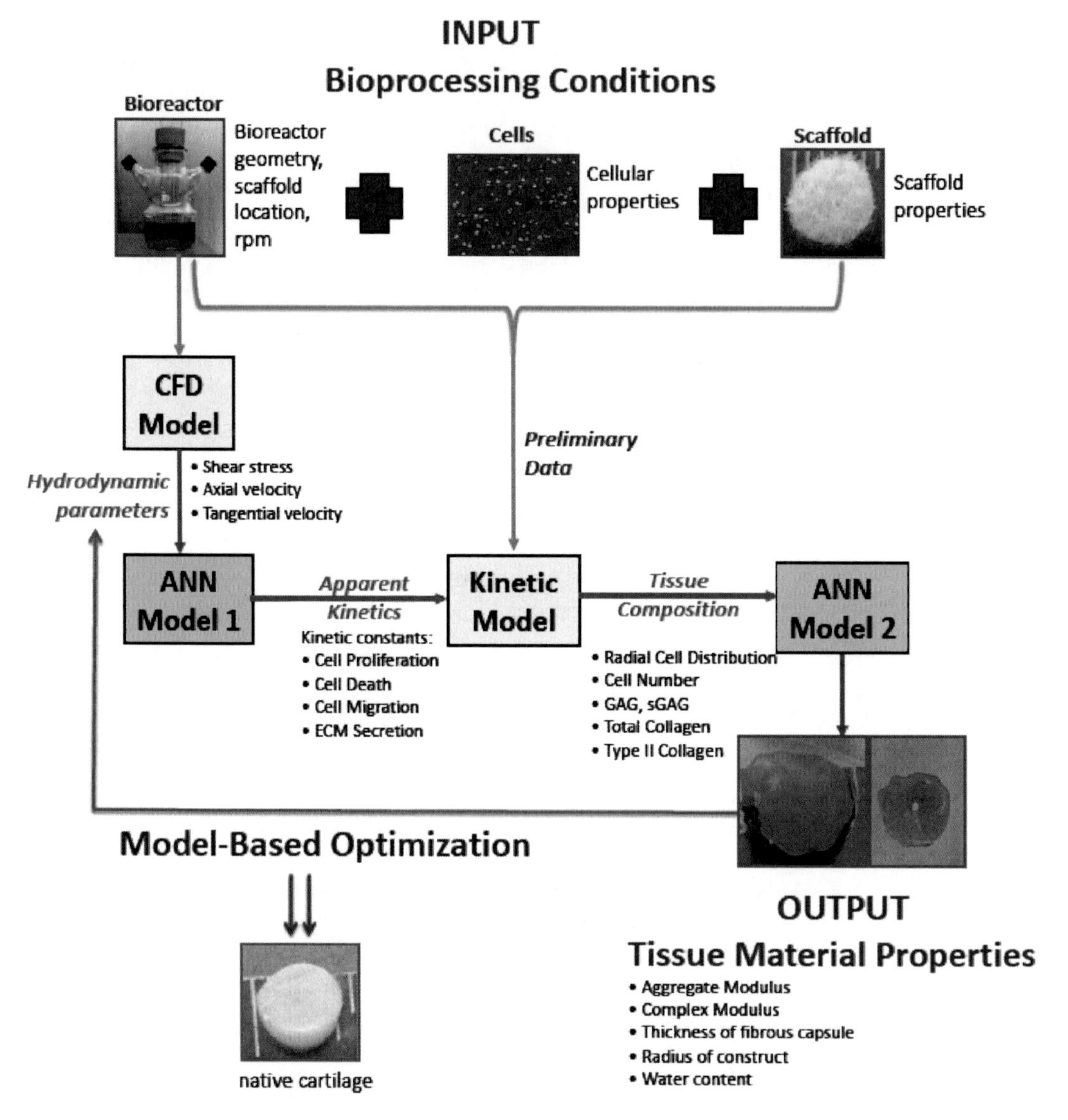

Fig. 1. Hybrid tissue growth modeling approach modified from Bilgen et al. (9).

cell attachment and growth dynamics in engineered constructs, and subsequently the material properties, when the operating conditions are given as input (see Note 1).

2. Methods

2.1. Computational Fluid Dynamics Modeling

2.1.1. Mesh Generation

The geometries for spinner flask and the two different configurations of the wavy-walled bioreactor (modified spinner flask), where the constructs are located in the lobe or in the center of the bioreactor

(WWB_{lobe} and $WWB_{central}$), are created in GAMBIT mesh generation software (Fluent Inc, Lebanon, NH) (Fig. 2) (15). The fluid volume is divided into two zones: one moving volume around the stir bar for sliding mesh simulations, and the bulk volume remaining in the vessel, surrounding the constructs, to perform sliding-mesh simulations. The mesh is generated using TGrid scheme, which consists of predominantly tetrahedral mesh elements (15).

2.1.2. CFD Simulations

The meshes created in GAMBIT are imported to FLUENT6.1 for the unsteady simulation of 3D turbulent flow. A turbulence model is necessary to modify the instantaneous transport equations to remove the small scales in order to produce a set of equations that are computationally less expensive to solve. The realizable k-ε model with standard wall functions (15, 16) is selected for modeling turbulence based on previous performance of the model in simulating flow in spinner flasks (17). The model constants and parameters used in this example are established by Fluent Inc. and were used successfully by Sucosky et al. (17, 18).

The surfaces of the vessel wall, stir bar, constructs, needles, and spacers are modeled as solid walls with no-slip boundary condition and as stationary with respect to the adjacent fluid. The surface of the fluid is modeled as a wall with zero shear stress, which yields similar results to modeling the surface as an atmospheric pressure inlet. The sliding zone around the stir bar is modeled as a moving mesh. The time steps are determined as 1/80th of the time required for a full revolution of the stir bar. After initialization with stationary conditions, 2,000 unsteady state time steps are run before transient effects disappeared, corresponding to 25 full revolutions of the stir bar. The simulations for 80 rpm agitation rates started with the converged solution for the corresponding case with the agitation rate of 50 rpm. The computer simulations for this study are performed at dual-processor compute nodes with 4 GB memory in an HP rx2600 Itanium2 research computing cluster (see Note 2).

2.1.3. Hydrodynamic Parameters

The tissue growth model requires key hydrodynamic parameters to characterize the hydrodynamic environment in a bioreactor. These parameters are determined using CFD modeling results (11) for four operating conditions: scaffolds positioned at the center (C) or lobe (L) of the wavy-walled bioreactor, agitated at 50 or 80 rpm, resulting in C-50, C-80, L-50, and L-80 configurations (Fig. 2). A principal component analysis needs to be performed to determine the key hydrodynamic parameters, and here such an analysis concludes that three hydrodynamic parameters, (1) the average shear stress, (2) axial, and (3) tangential fluid velocity on the control volume surface created around the constructs explain >99.9% of the variability of the hydrodynamic environment around the constructs, and the radial velocity and uniformity of

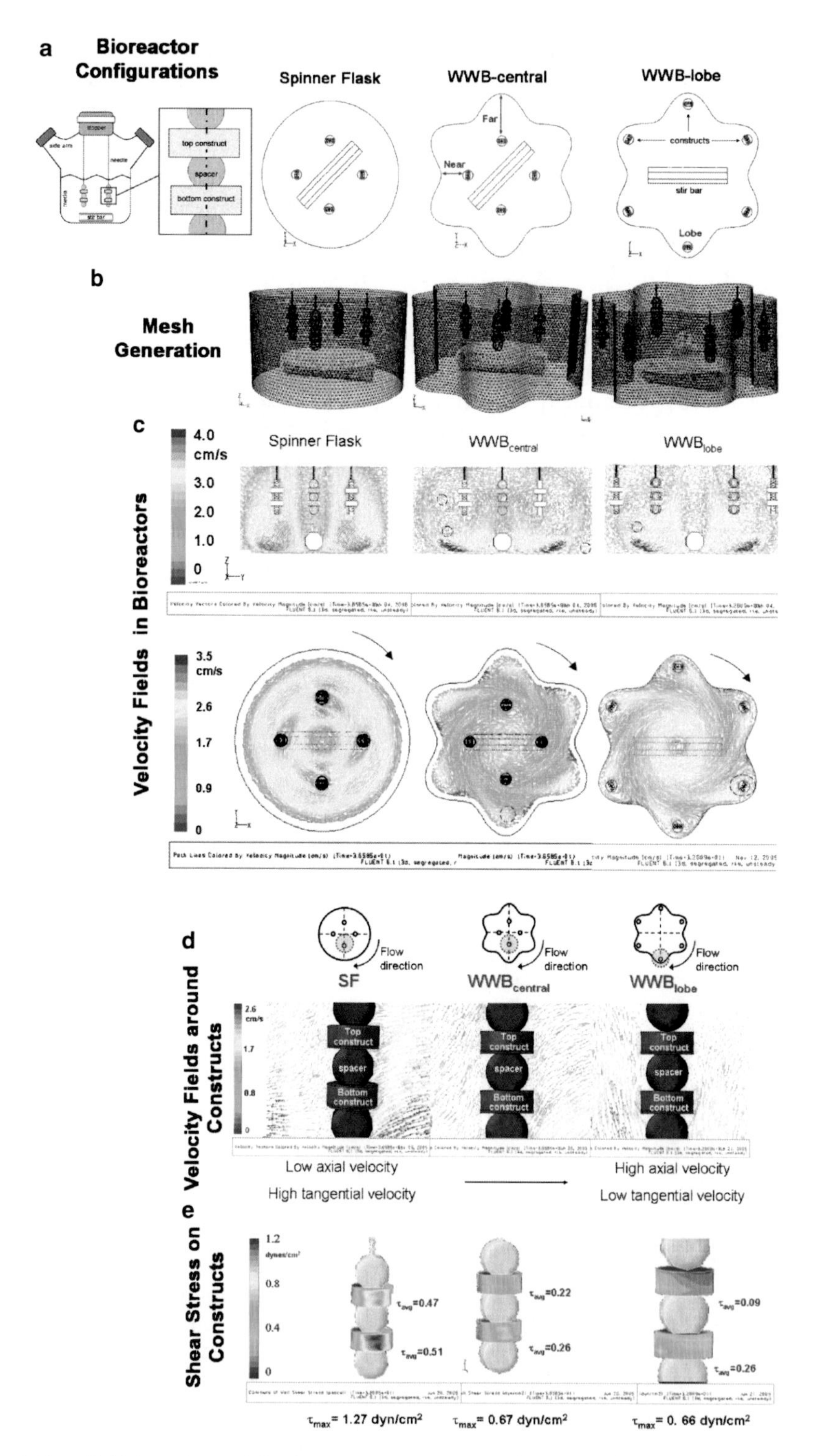

Fig. 2. Computational fluid dynamics modeling for different bioreactor configurations yield different flow fields and shear stress distribution on the constructs. This figure shows representative results from the CFD modeling for the spinner flask and two WWB configurations (*lobe* and *center*) at 50 rpm. (**a**) Bioreactor configurations. (**b**) Mesh generation. (**c**) Velocity fields in vertical (*top*) and horizontal (*bottom*) cross-sections of bioreactors. (**d**) Velocity field in the vicinity of constructs in bioreactors. (**e**) Shear stress distribution on the surface of constructs. Modified from Bilgen and Barabino (11).

velocity distribution are interdependent on the other variables. The combination of these hydrodynamic parameters describes how the actual operating conditions of the bioreactor determine the flow environment in the immediate vicinity of the constructs. The variations of local shear stresses applied to constructs and oxygen and nutrition distribution throughout the bioreactor and in the immediate vicinity of the construct are only indirectly considered through the fluid velocity components.

2.1.4. Particle Image Velocimetry for CFD Validation

This is a computer-aided experimental method to characterize the two-dimensional (2D) flow field within the bioreactors (Fig. 3). The main steps in the PIV technique are image acquisition and processing. The fluid flow in the bioreactor is observed with tracer particles that follow the flow direction. A cross-section in the bioreactor is illuminated with a thin laser sheet. A video camera is placed perpendicular to the illuminated plane to acquire an image pair synchronized to a pair of pulses from the laser. The image processing is then performed in a rectangular grid generated on the two consecutive frames of a pair and centered on a grid point to calculate the instantaneous average displacement (and thus, the instantaneous average velocity) of the particles contained within the window. The repetition of this procedure for all the points on the grid produces the instantaneous displacement (and thus velocity) field of the particles at the time the first frame is captured. Measurements of velocity fields can be used to estimate the mean-shear stress, Reynolds stress, and turbulent kinetic energy components in the vicinity of the constructs within the bioreactor (13). These results are used to compare with and validate the 3D CFD results. Before performing CFD, we based our results on these experimental measurements; therefore, these are essential.

2.2. A Kinetic Model for Tissue Growth Dynamics

Tissue growth experiments are necessary to develop and validate the tissue growth model. For the specific example covered in this chapter, the set of experiments were described elsewhere (1, 9). Briefly, engineered cartilage constructs were cultivated in four well-defined hydrodynamic environments for 4 weeks (Fig. 4). The tissue properties that are relevant to the uniformity and strength of the cartilage constructs were determined as (1) cell distribution ratio, i.e., the ratio of the number of cells in an inner circle to the number of cells in the remaining portion of the histological cross-section (outer ring), (2) ECM deposition [glycosaminoglycans (GAG), type I and type II collagen] in the construct, (3) mechanical properties under confined compression (aggregate and complex modulus), and (4) physical properties such as the radius and water content of the constructs, and the thickness of the fibrous outer capsule, caused by the flow-induced shear stress. The set of properties listed in (3) and (4) will be referred to as “material properties.”

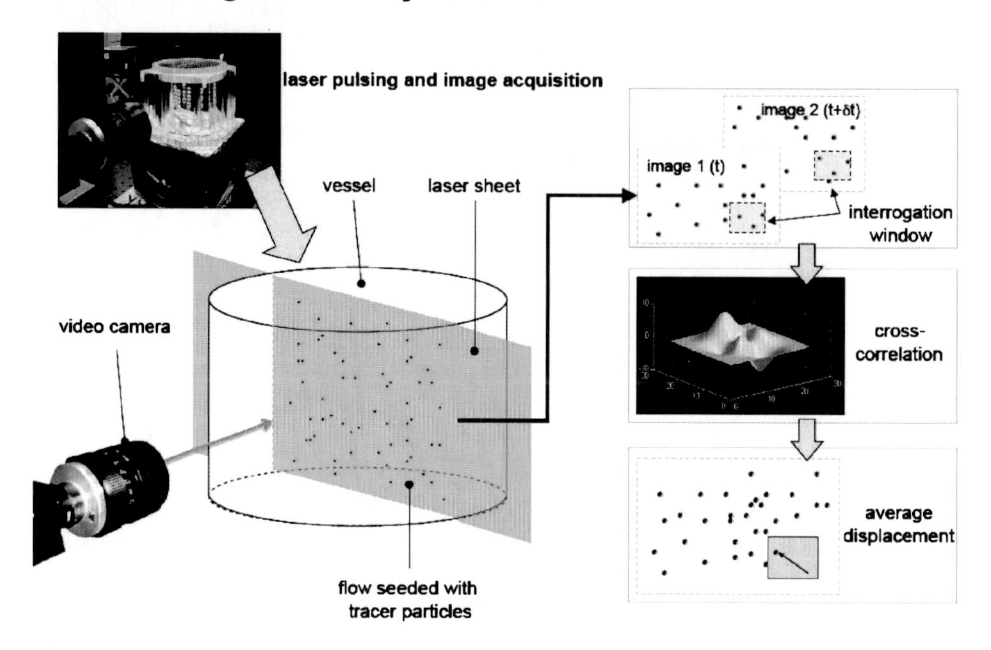

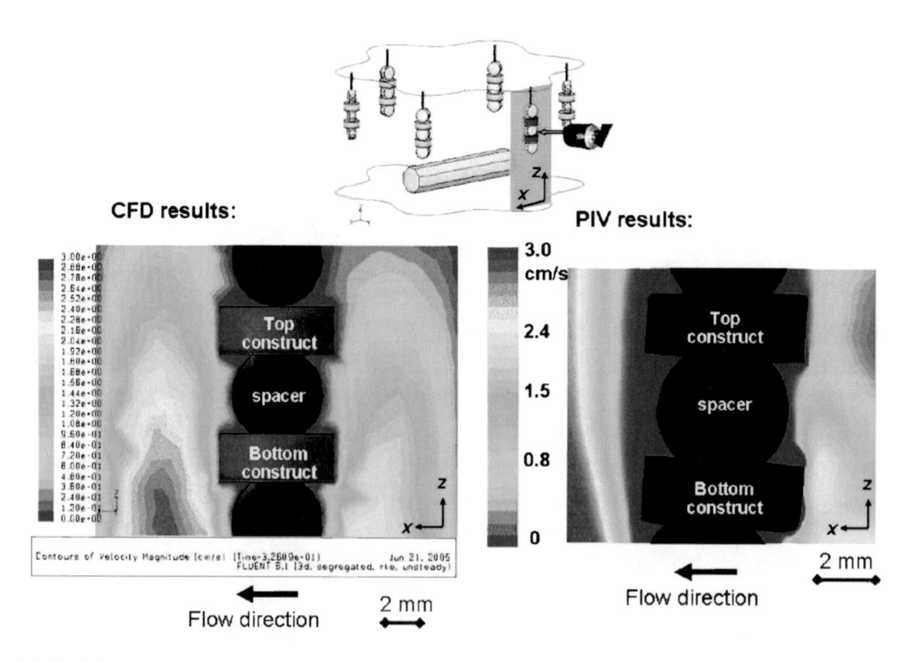

Fig. 3. (**a**) Particle image velocimetry method, image courtesy of Philippe Sucosky. (**b**) PIV results are used to validate the CFD results by comparison of velocity field results. Modified from Bilgen et al. (13) and Bilgen and Barabino (11).

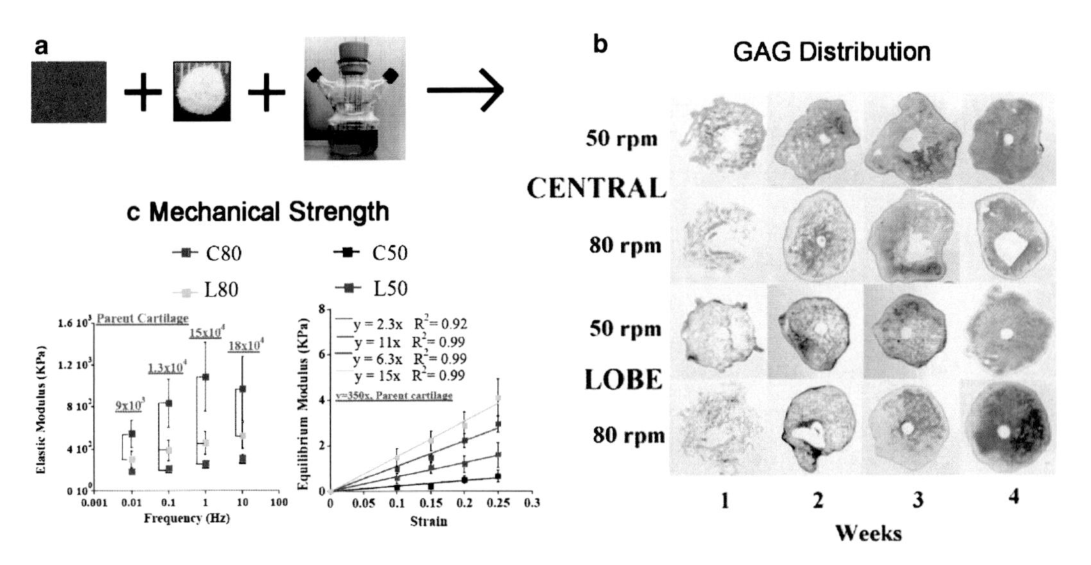

Fig. 4. Preliminary data used to train the model. (**a**) Chondrocytes were seeded on PGA scaffolds and cultivated in a wavy-walled bioreactor. The WWB setup for cultivation of engineered cartilage constructs included two scaffold positions: at the center (**c**) or the lobes (*L*) of the bioreactor and agitated at 50 or 80 rpm, resulting in four experimental cases: C-50, C-80, L-50, and L-80. The position of scaffolds and the agitation rate affected the biochemical composition and mechanical properties of the engineered tissue. (**b**) Histological cross-sections of constructs stained with Safranin-O/fast green depicting the distribution of glycosaminoglycans (GAG). (**c**) Mechanical testing. Adapted from Bueno et al. (1).

A kinetic model is developed to simulate the dynamic tissue growth in the bioreactor. This is a lumped parameter model that includes spatial variations that differentiated between a predefined outer ring and the inner circle of scaffold cross-sections. To reduce the model complexity, a number of primary assumptions are made.

2.2.1. Assumptions

1. Cells can either attach directly to sites within the scaffold or to already-attached cells. Previous cell aggregation studies support this assumption (19).
2. Since the cells can attach to each other as well as to free sites on the scaffold, cell attachment does not change the number of available binding sites.
3. Cells can migrate from the outer ring to the inner circle.
4. Cell death occurs only in the culture medium and cell growth occurs only within the scaffold. The rate of cell growth within the scaffold is considered as the net growth (growth minus death), as it is not possible to differentiate growth and death from the currently available data.
5. The effects of substrate distribution are not directly included in the model, but these effects result in the differences in the identified kinetic constants; hence, the model indirectly includes transport of substrates.

6. Attached cells proliferate, and secrete ECM. After a certain period of cultivation, the cell proliferation and matrix deposition slows over time. The cells that do not secrete ECM and do not proliferate are defined as mature cells, and the actively proliferating and ECM synthesizing cells are defined as immature cells.

Based on the above assumptions, the reactions that take place in the bioreactor can be formulated as:

Cell Attachment

1. $N_s \xrightarrow{k_o} N_o$.
2. $N_s \xrightarrow{k_i} N_i$.

Cell Migration

3. $N_s \xrightarrow{k_o} N_o$.

Cell Maturation

4. $N_o \xrightarrow{k_{Mo}} N_{Mo}$.
5. $N_i \xrightarrow{k_{Mi}} N_{Mi}$.

Cell Growth and Death

6. $N_o \xrightarrow{\mu_o} 2N_o$.
7. $N_i \xrightarrow{\mu_i} 2N_i$.
8. $2N_s \xrightarrow{k_d} N_s$.

Product Formation

9. $N_o + N_i \xrightarrow{k_{GAG}} N_o + N_i + C_{GAG}$.
10. $N_o + N_i \xrightarrow{k_{sGAG}} C_{sGAG} + N_o + N_i$.
11. $N_o + N_i \xrightarrow{k_{CI}} N_o + N_i + C_{CI}$.
12. $N_o + N_i \xrightarrow{k_{CII}} N_o + N_i + C_{CII}$.

Where:

N_s = number of cells in suspension, million cells/construct.

N_o = number of cells attached to the outer ring of the scaffold/construct, million cells/construct.

N_i = number of cells attached to the inner circle of the scaffold/construct, million cells/construct.

N_{Mo} = number of mature cells in the outer ring of construct, million cells/construct.

N_{Mi} = number of mature cells in the inner circle of construct, million cells/construct.

C_{GAG} = GAG concentration in the constructs, mg/construct.

C_{sGAG} = soluble GAG concentration in the media, mg/construct.

C_{CI} = type I collagen concentration, mg/construct.

C_{CII} = type II collagen concentration, mg/construct.

The differential equations corresponding to the postulated reactions above are formulated as:

$$\frac{\mathrm{d}N_{\mathrm{o}}}{\mathrm{d}t} = k_o N_{\mathrm{s}} - k_{\mathrm{m}} N_{\mathrm{o}} + \mu_{\mathrm{o}} N_{\mathrm{o}} - k_{\mathrm{Mo}} N_{\mathrm{o}}. \quad (1)$$

$$\frac{\mathrm{d}N_{\mathrm{i}}}{\mathrm{d}t} = k_{\mathrm{i}} N_{\mathrm{s}} + k_{\mathrm{m}} N_{\mathrm{o}} + \mu_{\mathrm{i}} N_{\mathrm{i}} - k_{\mathrm{Mi}} N_{\mathrm{i}}. \quad (2)$$

$$\frac{\mathrm{d}N_{\mathrm{Mo}}}{\mathrm{d}t} = k_{\mathrm{Mo}} N_{\mathrm{Mo}}. \quad (3)$$

$$\frac{\mathrm{d}N_{\mathrm{Mi}}}{\mathrm{d}t} = k_{\mathrm{Mi}} N_{\mathrm{Mi}}. \quad (4)$$

$$\frac{\mathrm{d}N_{\mathrm{s}}}{\mathrm{d}t} = -k_{\mathrm{i}} N_{\mathrm{s}} - k_{\mathrm{o}} N_{\mathrm{s}} - k_{\mathrm{d}} N_{\mathrm{s}}. \quad (5)$$

$$\frac{\mathrm{d}C_{\mathrm{GAG}}}{\mathrm{d}t} = k_{\mathrm{GAG}} N_{\mathrm{o}} + k_{\mathrm{GAG}} N_{\mathrm{i}}. \quad (6)$$

$$\frac{\mathrm{d}C_{\mathrm{sGAG}}}{\mathrm{d}t} = k_{\mathrm{sGAG}} N_{\mathrm{o}} + k_{\mathrm{sGAG}} N_{\mathrm{i}}. \quad (7)$$

$$\frac{\mathrm{d}C_{\mathrm{CI}}}{\mathrm{d}t} = k_{\mathrm{CI}} N_{\mathrm{o}} + k_{\mathrm{CI}} N_{\mathrm{i}}. \quad (8)$$

$$\frac{\mathrm{d}C_{\mathrm{CII}}}{\mathrm{d}t} = k_{\mathrm{CII}} N_{\mathrm{o}} + k_{\mathrm{CII}} N_{\mathrm{i}}. \quad (9)$$

Equations 1 and 2 describe the growth and attachment of cells in the outer ring and the inner circle. All freshly attached "immature" cells are considered as chondrocytes that can proliferate and function regularly. The rate of cell proliferation and matrix synthesis slows down with time. Cell maturation, as defined in Eqs. 3 and 4, creates a self-limiting growth model, which is more realistic for this case than a standard Monod model that suggests exponential growth and substrate limitation (20, 21) (see Note 3). Equation 5 describes the population balance in the suspension, including a cell death term for the decrease in viable cells in the medium. Equations 6–9 describe the GAG, total collagen and type II collagen synthesis by chondrocytes. These reactions are considered first-order with respect to GAG and collagen secreting cells; however, the cell maturation process gives the model the flexibility to fit more complex secretion data (such as saturation behavior). Equation 7 describes secretion/release of soluble GAG to the culture media: a high concentration of soluble GAG in the media is undesirable because it may decrease GAG deposition in the constructs, thereby reducing construct mechanical strength. GAG is released in the media likely via diffusion from the construct. There are a variety of GAG monomers with different diffusion constants. It is plausible that while some are released in the media, others remain within the construct. To avoid an unnecessarily complex model, this simpler formulation is used, which assumes production of soluble GAG that is released into the media is distinct from and independent of the GAG deposition within the construct.

The kinetic model is trained separately for each of the four cases using the Genetic optimization Algorithm (GA) in MATLAB. Overall, this method can be classified as a parameter-fitting problem for a differential equation. Briefly, the kinetic constants are defined as the decision variables and fed to the differential equation solver subroutine for simulation of the kinetic model, which produced predictions of cell density, GAG, and collagen synthesis. The GA is then run to minimize the differences between predictions and data, expressed as the sum of squared errors. The GA is run multiple times (20+) to increase the likelihood of achieving the global minimum, although with stochastic algorithms it is not possible to guarantee a global solution. As a result, a set of 11 kinetic constants is determined for each case.

2.3. Artificial Neural Network Models

ANNs are adaptive models that can learn from data and generalize what is learned (Fig. 5). In ANN models, simple nodes (neurons) are connected together to form a network of nodes, which are used as an arbitrary function approximation mechanism to "learn" from observed data (see Note 1). Due to the nonlinear nature of the ANNs, they are able to express much more complex phenomena than some linear modeling or regression techniques, and can characterize unknown mechanisms analogous to black box systems (12).

2.3.1. ANN Model 1

The first ANN model employed in this study correlates the hydrodynamic parameters (inputs, Table 1) with the kinetic constants (outputs) (Fig. 1). The kinetic constants are obtained by applying the kinetic model to the experimental data (Table 2). The CFD calculations used for the hydrodynamic parameters did not include

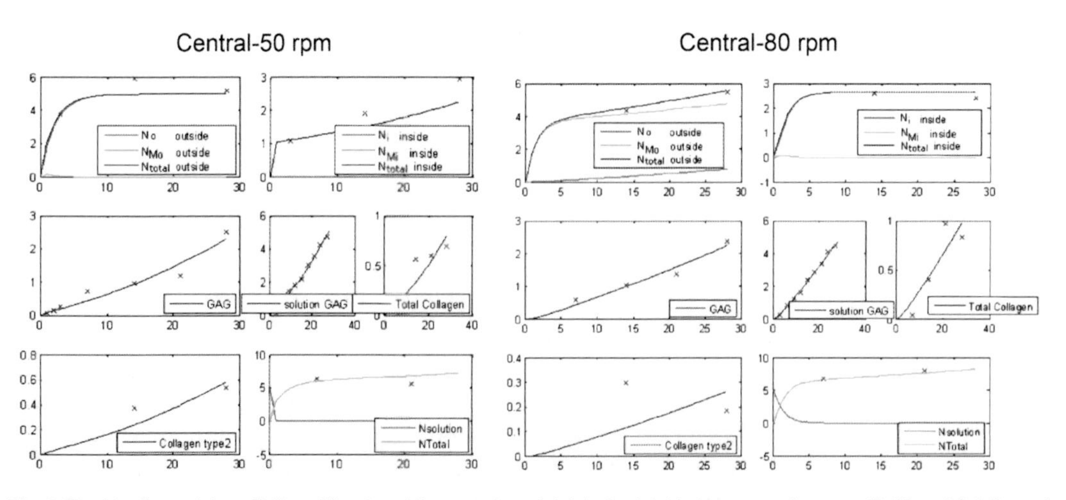

Fig. 5. The kinetic model predictions (*lines*) and the experimental data (*points*) in two example cases (C-50 and C-80) used for training the ANN models. The experimental data represents mean ± SEM. Adapted from Bilgen et al. (9).

Table 2
Kinetic constants obtained from dynamic data fitting of the preliminary data (the outputs for ANN 1)

Kinetic constant	Units	C-50	C-80	L-50	L-80
μ_o	per day	0.0000	0.0166	0.0095	0.0000
μ_i	per day	11.5143	9.6782	23.2496	7.2925
k_o	per day	0.4611	0.5217	2.2448	0.2570
k_i	per day	0.0114	0.0717	0.1143	0.0105
k_d	per day	0.0000	0.1099	0.0732	0.0000
k_{Mo}	per day	0.0000	0.0070	0.0000	0.0000
k_{Mi}	per day	12.3346	11.9942	26.3020	7.8136
k_{GAG}	per mg·cells/day	0.0175	0.0199	0.0110	0.0209
k_{sGAG}	per mg·cells/day	0.0383	0.0422	0.0340	0.0455
k_{CI}	per mg·cells/day	0.0061	0.0087	0.0039	0.0079
k_{CII}	per mg·cells/day	0.0045	0.0023	0.0025	0.0032

These constants are used to calculate the predictions depicted in Fig. 5. Adapted from Bilgen et al. (9)

the tissue growth and cell proliferation in the simulation, hence the change in flow dynamics due to change in cell concentration or construct size is not included in the model.

The network is setup and trained with MATLAB generalized regression network function, commonly used in parameter fitting studies (22). This is a two-layer network structure; the first layer contains radial basis neurons with biases and the second layer contains pure linear neurons. A tuning study is performed to identify the number of parameters necessary to yield the best cross-validation (see Subheading 4) accuracy (the spread parameter); hence, a good compromise between correlation and ability to generalize is displayed. Other ANN types are tested, but are found to be less accurate after cross-validation, or to produce abnormal kinetic constants (i.e., negative values) in certain cases (results not shown).

2.3.2. ANN Model 2

The second ANN model correlates the final composition of the tissue (number of cells in the inner circle and outer ring, total cells and cell distribution ratio and final GAG, total and type II collagen deposition) with the material properties of the construct (mechanical properties-complex and aggregate modulus-, construct radius, capsule thickness, and water content) (Fig. 1, see Note 4). A generalized regression network similar to ANN 1 is used for this model, with seven inputs and five outputs.

2.4. Validation of the Models

A widely used leave-one-out cross-validation (LOOCV) technique is used to test the capability of the proposed models (23–26). This cross-validation technique is designed to measure the expected accuracy of models in cases with limited data and involves setting one experiment aside for validation, and using the remaining experiments to train the model, and repeating the procedure until each experiment is used once for validation. This ensures that a lucky selection of a validation set does not produce overly optimistic accuracy results, or vice versa.

For cross-validation, the two ANNs are trained on three selected cases, and the kinetic constants of the fourth case are predicted using ANN1. Then, the kinetic model predicted the collagen, GAG, and cell distribution, followed by ANN2 that predicted the final construct properties. This procedure is repeated four times until each experiment is used once for validation. The errors are calculated by comparing the predicted construct properties to the experimental results. In this example, the average accuracy is 70.3% (Table 3), which is a common acceptable accuracy level in sophisticated predictive models of biological systems. Overall, this accuracy is satisfactory for a model of this complexity (see Note 5).

2.5. Optimization Studies

The tissue growth model developed in this study predicts the dynamics of tissue growth and cell attachment as well as the final material properties as a function of the hydrodynamic environment

Table 3
Leave-one-out-cross-validation (LOOCV) results: the prediction errors in each case are calculated by training the ANNs using the other three cases, as detailed in the cross-validation methods (data from Bilgen et al. (9))

Prediction errors (%)	Case 1, C-50	Case 2, C-80	Case 3, L-50	Case 4, L-80
Aggregate modulus	240.7	37.3	1.4	66.4
Complex modulus at 10 Hz	32.4	63.0	29.4	30.3
Capsule thickness	17.1	9.4	15.5	14.2
Construct radius	3.9	10.7	4.9	6.3
Water content	1.2	1.1	1.9	7.9
Average error (%)	29.7			

within the system. Therefore, the optimal operating conditions in the bioreactor can be identified based on the tissue growth model. An optimization problem is formulated to determine optimum hydrodynamic parameters where the objective is to maximize the complex and aggregate moduli while maintaining a nearly uniform cell distribution ratio. To avoid errors due to excessive extrapolation, these variables are limited to their experimentally observed ranges in the four cases (i.e., the experiments performed under four different operating conditions), leaving limited room for extrapolation. The full optimization problem is as follows:

$$\min_{\mathbf{u}} \; I = \left[-\frac{\text{Aggregate Modulus}}{15.4} - \frac{\text{Complex Modulus}}{968.0467} - \frac{N_{\text{I}}(t=28)}{N_{\text{O}}(t=28)} \right]$$

$$\begin{aligned}
\text{s.t.} \quad \frac{\mathbf{dx}}{dt} &= \mathbf{f}(\mathbf{x}, \mathbf{p}_1) && \text{kinetic model} \\
\mathbf{x}(0) &= \mathbf{x}_0 && \\
\mathbf{p}_1 &= \mathbf{h}_1(\mathbf{u}) && \text{ANN model 1} \\
\mathbf{p}_2 &= \mathbf{h}_2(x) && \text{ANN model 1} \\
\mathbf{u}^{\min} \leq \mathbf{u} &\leq \mathbf{u}^{\max} && \text{bounds on decision variables.}
\end{aligned}$$

$\mathbf{u}$ = mean-shear stress, axial velocity, tangential velocity

$$\mathbf{u}^{\min} = [0 \quad 0 \quad 0]^T.$$

$$\mathbf{u}^{\max} = [1 \quad 3 \quad 3]^T.$$

The values used to normalize the aggregate and complex moduli (i.e., weights) are the maximum values observed in the experimental data. Function vectors $\mathbf{h_1}$ and $\mathbf{h_2}$ are the ANN models in explicit form. The problem is solved by the genetic optimization package in MATLAB with a sequential algorithm similar to that employed for the parameter fit problem. In addition, the following alternative objective functions can be considered:

Maximize material strength, i.e., aggregate modulus and complex modulus (objective 2)

$$\min_{\mathbf{u}} \; I = \left[-\frac{\text{Aggregate modulus}}{15.4} - \frac{\text{Complex modulus}}{968.05} \right].$$

Maximize cellular uniformity ratio (objective 3)

$$\min_{\mathbf{u}} \; I = -\left(\frac{N_I(t=28)}{N_O(t=28)} - 1 \right)^2.$$

Maximize ECM Secretion (objective 4)

$$\min_{\mathbf{u}} \; I = \left[-\frac{\text{GAG}(t=28)}{2.15} - \frac{\text{CollagenII}(t=28)}{0.3575} \right].$$

Table 4
Optimization results with respect to different objective functions. Adapted from Bilgen et al. (9)

	Objective 1: max agg. and complex moduli and cell dist. ratio	Objective 2: max agg. and complex moduli	Objective 3: max cell dist. ratio	Objective 4: max ECM
Optimal hydrodynamic parameters				
Mean-shear stress (dynes/cm^2)	1.0	1.0	1.0	0.1
Axial velocity (cm/s)	3.0	3.0	1.47	0.1
Tangential velocity (cm/s)	0.84	0.80	3.0	1.56
Radial velocity (cm/s)	2.18	2.20	1.82	1.41
Uniformity of velocity distribution	0.38	0.38	0.43	0.74
Optimal material properties				
Aggregate modulus (kPa)	11.4	11.4	11.4	6.5
Complex modulus at 10 Hz (kPa)	705.6	705.6	747.0	376.5
Cell distribution ratio	0.665	0.666	0.903	0.425

Optimization with respect to the above objectives results in the optimal bioreactor conditions listed in Table 4. These optimal solutions support the use of the highest possible shear stresses and high velocities and relatively lower uniformity of velocity, corresponding to high agitation rates for maximizing the material strength and cellular uniformity (objectives 1–3). To maximize ECM synthesis (GAG and type II collagen), the model suggests the use of lowest possible shear stresses together with high uniformity of velocity (objective 4) (see Note 6).

3. Notes

1. The extrapolative capacity of this methodology relies on the extent of the preliminary data that are used for training the ANNs. The particular tissue growth model given as an

example in this study is based on experimental data that employed chondrocytes seeded on PGA scaffolds and cultivated in a specific serum-supplemented media in a wavy-walled bioreactor for cartilage tissue engineering. In order to predict the tissue growth in different in vitro culturing conditions (e.g., different type of bioreactor, scaffold material, media composition, cells), preliminary experiments need to be carried out to provide training data for the ANNs. This way, this modeling approach can be used for rational design of experiments and data analyses and therefore, can be used to improve the efficiency of experimental processes. The data fitting algorithms will be updated as biological and physical governing principles become more available, resulting in more generalized tissue growth models.

2. The CFD simulations are performed on a fixed volume mesh; however, a model that employs a mesh that can reflect the dynamics of size, shape, and permeability of the tissues would yield higher accuracy. A fixed mesh is used in the study described here because of its simplicity. In addition, computational cost would be a major issue for simulations that reflect weeks of tissue growth; the simulations used in this study span over one half revolution of a stir bar (0.6 s). Another important consideration for mesh generation is the determination of an optimum mesh that gives an accurate solution, but is not so highly refined to increase the computational cost (10). In this example, the mesh was refined around one needle and a set of scaffolds to generate more accurate shear stress and velocity field results around the vicinity of constructs.
3. The cell maturation concept that was introduced in the kinetic model can also be modified to include several steps involving cell differentiation and hypertrophy, depending on the cell and tissue type.
4. An alternative multistage model could be built to correlate tissue biochemical composition with some material properties (construct radius, capsule thickness, and water content), and then these could be correlated with mechanical properties (aggregate and complex modulus). While being physically more relevant, such an approach may complicate the model and reduce its ability to generalize.
5. The cross-validation algorithms employed produce a reasonable real-world estimate of the accuracy that can be expected from the model predictions. Additional data sets, which cover a wider range of operating conditions that result in a wider range of tissue properties, will certainly improve the model, with increased accuracy. For example, the mechanical properties are poorly predicted, indicating that these two parameters are

weakly correlated with characteristics depicted by the kinetic model. In addition, as a benchmark, an ANN-only model can be run, which directly correlates the hydrodynamic parameters with the construct parameters, without a kinetic model. Such a model is less accurate in the predictions of these parameters in this case, demonstrating that the hydrodynamic parameters alone are also not well correlated with the mechanical properties; therefore, a variable that is not captured within the available model parameters may need to be considered.

6. Optimum cultivation strategies are investigated for different functional requirements based on this model. To maximize material properties and cellular uniformity (objective 1), the tissue-growth-model-based optimization predicted that a hydrodynamic environment that combines the maximum shear stress with higher axial and radial fluid velocities than those employed in the experiments would be beneficial. The predicted aggregate and complex moduli are on the order of those obtained for the constructs at 80-rpm, still much lower than those of native cartilage. On the other hand, to optimize ECM secretion (objective 4), lowest shear, and higher tangential velocity are suggested by the model-based optimization. In general, these trends are consistent with the experimental data and literature and these results suggest that the functionality requirements, i.e., the objective function, need to be well defined. This model provides insights on the effect of the hydrodynamic environment on tissue growth and suggests that it may be viable to use different flow regimes and multiple strategies during the long-term growth phase.

References

1. Bueno EM, Bilgen B, Barabino GA (2009) Hydrodynamic parameters modulate biochemical, histological, and mechanical properties of engineered cartilage. Tissue Eng A 15 (4):773–785
2. Pazzano D, Mercier KA, Moran JA, Fong SS, DoBiasio DD, Rulfs JX, Kohles SS, Bonassar LJ (2000) Comparison of chondrogenesis in static and perfused bioreactor culture. Biotechnol Prog 16:893–896
3. Vunjak-Novakovic G, Freed LE, Biron RJ, Langer R (1996) Effects of mixing on the composition and morphology of tissue-engineered cartilage. AIChE J 42:850–860
4. Vunjak-Novakovic G, Obradovic B, Martin I, Freed LE (2002) Bioreactor studies of native and tissue engineered cartilage. Biorheology 39:259–268
5. Freed LE, Marquis JC, Vunjak-Novakovic G, Emmanual J, Langer R (1994) Composition of cell-polymer cartilage implants. Biotechnol Bioeng 43:605–614
6. Wilson CG, Bonassar LJ, Kohles SS (2002) Modeling the dynamic composition of engineered cartilage. Arch Biochem Biophys 208:246–254
7. Kino-Oka M, Maeda Y, Yamamoto T, Sugawara K, Taya M (2005) A kinetic modeling of chondrocyte culture for manufacture of tissue-engineered cartilage. J Biosci Bioeng 99 (3):197–207
8. Saha AK, Mazumdar J, Kohles SS (2004) Prediction of growth factor effects on engineered cartilage composition using deterministic and stochastic modeling. Ann Biomed Eng 32:871–879

9. Bilgen B, Uygun K, Bueno EM, Sucosky P, Barabino GA (2009) Tissue growth modeling in a wavy-walled bioreactor. Tissue Eng A 15 (4):761–771
10. Hutmacher DW, Singh H (2008) Computational fluid dynamics for improved bioreactor design and 3D culture. Trends Biotechnol 26:166–172
11. Bilgen B, Barabino GA (2007) Location of scaffolds in bioreactors modulates the hydrodynamic environment experienced by engineered tissues. Biotechnol Bioeng 98:282–294
12. Hagan MT, Demuth HB, Beale M (1995) Neural Network Design. PWS Publishing Company, Boston, MA
13. Bilgen B, Sucosky P, Neitzel GP, Barabino GA (2006) Flow characterization of a wavy-walled bioreactor for cartilage tissue engineering. Biotechnol Bioeng 95:1009–1022
14. Bilgen B (2006) Hydrodynamic characterization of a novel bioreactor for tissue engineering. Chemical Engineering, Northeastern University, Boston, MA
15. Fluent Inc. (2003) Fluent 6.1 User's Guide, Lebanon, NH
16. Shih T-H, Liou WW, Shabbir A, Yang Z, Zhu J (1995) A new k-e Eddy-viscosity model for high reynolds number turbulent flows – model development and validation. Comput Fluids 24:227–238
17. Sucosky P, Osorio DF, Brown JB, Neitzel GP (2004) Fluid mechanics of a spinner-flask bioreactor. Biotechnol Bioeng 85:34–46
18. Sucosky P (2005) Flow characterization and modeling of cartilage development in a spinner-flask bioreactor. Mechanical Engineering, Georgia Institute of Technology, Atlanta, GA
19. Bueno EM, Bilgen B, Carrier RL, Barabino GA (2004) Increased rate of chondrocyte aggregation in a wavy-walled bioreactor. Biotechnol Bioeng 88:767–777
20. Monod J (1949) The growth of bacterial cultures. Ann Rev Microbiol 3:371–394
21. Bailey JE, Ollis DF (1977) Biochemical engineering fundamentals. McGraw-Hill, New York
22. Wassermann PD (1993) Advanced methods in neural computing. Van Nostrand Reinhold, New York
23. Habib T, Zhang C, Yang JY, Yang MQ, Deng Y (2008) Supervised learning method for the prediction of subcellular localization of proteins using amino acid and amino acid pair composition. BMC Genomics 9(Suppl 1):S16
24. Witten IH, Frank E (1999) Data mining: practical machine learning tools and techniques with Java implementations, 1st edn. Morgan Kaufmann, San Francisco, CA
25. Zhang Y, Xuan J, de los Reyes BG, Clarke R, Ressom HW (2008) Network motif-based identification of transcription factor-target gene relationships by integrating multi-source biological data. BMC Bioinformatics 9:203
26. Lee MC, Nelson SJ (2008) Supervised pattern recognition for the prediction of contrast-enhancement appearance in brain tumors from multivariate magnetic resonance imaging and spectroscopy. Artif Intell Med 43:61–74

Chapter 15

Recent Advances in 3D Printing of Tissue Engineering Scaffolds

Min Lee and Benjamin M. Wu

Abstract

Computer-aided tissue engineering enables the fabrication of multifunctional scaffolds that meet the structural, mechanical, and nutritional requirements based on optimized models. In this chapter, three-dimensional printing technology is described, and several limitations in the current direct printing approach are discussed. This chapter also describes indirect three-dimensional printing protocol to overcome convergent demands with a traditional method, without sacrificing the key advantage of material versatility.

Key words: Multifunctional scaffolds, Optimization model, 3D printing, Osteochondral tissue engineering

1. Introduction

In tissue engineering, the success of functional tissue or organ regeneration relies on the development of suitable scaffolds to direct three-dimensional tissue ingrowth. For the regeneration of large and complex tissues, control over macro- and microstructural properties of scaffolds is required to optimize structural and nutrient transport conditions. Structural properties, such as macroscopic shape, pore size, porosity, pore interconnectivity, surface area, surface chemistry, and mechanical properties, are considered to be required in the design of scaffolds for tissue engineering (1, 2). Fabrication techniques have been developed and conventional fabrication methods include fiber bonding (3), melt molding (4), solvent casting and particulate leaching (5), membrane lamination (6), phase separation (7), gas foaming/high-pressure processing (8), as well as a combination of these techniques (9). Although scaffolds produced from conventional techniques address individual issue of structural properties, no single scaffold can yet meet all properties.

Michael A.K. Liebschner (ed.), *Computer-Aided Tissue Engineering*, Methods in Molecular Biology, vol. 868,
DOI 10.1007/978-1-61779-764-4_15, © Springer Science+Business Media, LLC 2012

By 2000, the field of computer-aided tissue engineering (CATE) emerged with advance in computer-aided technology (10, 11). CATE integrates advanced imaging technologies such as computed tomography (CT) or magnetic resonance imaging (MRI), computer-aided design (CAD) technology, and rapid prototyping (RP) and/or solid freeform fabrication (SFF) technology with tissue engineering applications. Over the past two decades, various RP techniques were introduced that provide unique ways to build objects with controlled macroarchitecture as well as microstructures. RP techniques also enable the production of customized shapes for individual patients using medical imaging and CAD (12, 13). For example, RP techniques can be integrated with CT or MRI data directly from the patient and they allow the production of a tissue engineering graft that matches precisely a patient's contours.

Thus, CATE enables the fabrication of multifunctional scaffolds with control over scaffold composition, porosity, macroarchitecture, and mechanical properties based on optimized models (14). Automated computer-aided fabrication produces accurate and consistent scaffolds with minimal manpower and cost requirements. In addition, controlled spatial distribution of pharmaceutical and biological agents can be obtained by the simultaneous addition of cells or growth factors during fabrication.

2. Solid Freeform Fabrication

SFF technologies create laminated 3D structures from numerical models (15, 16). Since the initial development of stereolithography (SLA) in the 1980s (17), numerous competing SFF technologies have been introduced, and technological and materials development have combined to fuel a paradigm shift away from making nonfunctional "models" to fabricating finished, functional devices. The first biomedical devices made directly by SFF were reported in the early 1990s using a custom-built three-dimensional printing (3DP) machine (18–20). Since then, numerous commercially available SFF technologies, such as fused deposition modeling (FDM) (21), SLA (22), and selective laser sintering (SLS) (23), have been utilized to fabricate scaffolds.

3. 3D Printing

3DP technology was introduced at the Massachusetts Institute of Technology. This technology is based on inkjet printing liquid binder to join loose powder to create 3D structures (18–20). Since most biomaterials exist in either a solid or liquid state, a wide range of biomaterials has been utilized directly in 3DP. A thin layer of powder is spread onto the building platform and a printer head selectively prints a liquid binder solution onto a powder bed to form the 2D pattern. This process is repeated layer by layer until the entire objects are fabricated. The completed objects are extracted from the powder bed by removing the unbound powder. Since objects are being supported by surrounding unbounded powders, fabrication of complex scaffolds, such as channels or hanging features, is available with this technique.

3DP is one of the most investigated SFF techniques in tissue engineering scaffolds. Kim et al. created highly porous scaffolds in combination with particulate leaching techniques by 3DP and demonstrated the ingrowth of hepatocytes into the scaffolds (24). Recent studies showed 3D porous ceramic scaffolds with fully interconnected channels for bone replacement using this technique (25). Also, room temperature processing conditions allow the incorporation of temperature-sensitive materials, such as pharmaceutical and biological agents, into scaffolds (19). Lam et al. fabricated starch-based scaffolds by printing distilled water with the feasibility of using biological agents and living cells during fabrication (26). Another favorable characteristic of this technology for tissue engineering is multi-“color” printing, where each color ink can be positioned on a precise location. Therefore, this technology has good potential to arrange multiple types of cells, extracellular matrix, and bioactive agents simultaneously for biological tissue manufacturing.

3.1. 3D Printing in Osteochondral Tissue Engineering

Recent 3DP technology allowed the design and creation of osteochondral composite scaffold to repair cartilage defect and prevent associated osteoarthritis (27). Osteoarthritis is degenerative joint disease and one of the most significant causes of disability in middle-aged and older people, affecting approximately 33 million Americans (28–30). Unfortunately, articular cartilage has a poor regenerative capacity and spontaneously healed cartilage sometimes undergoes degeneration a few months after repair (31, 32). Current therapies include chondral shaving, subchondrol drilling, microfracturing, mosaicplasty, and prosthetic replacement (33–35). However, these treatments are associated with problems, including inadequate number of donor organs, donor site morbidity, and limited durability of prosthetics (36).

Tissue engineering is an emerging interdisciplinary research field that has the potential to regenerate many tissues and organs (37–39). Cartilage tissue engineering is one of the most promising therapies that may provide an ultimate solution for osteoarthritis (40, 41). In particular, engineering of osteochondral tissue, simultaneous regeneration of articular cartilage, and underlining subchondral bone can facilitate the fixation and integration with host tissue. A variety of scaffolds for osteochondral tissue regeneration have been developed to meet the complex functional demands of cartilage and bone tissues that have distinctively different structural, chemical, and mechanical properties. One strategy is to engineer two individual cartilage and bone scaffold layer, and join two separately fabricated scaffolds together by suturing, glue, simple press fitting, or biodegradable external fixation (42–45). However, these methods are limited by the inferior integration between cartilage and bone tissues resulting with eventual separation of the two tissues.

Recent studies have reported biphasic scaffolds to engineer individual cartilage and bone tissue simultaneously on a single integrated construct. These biphasic structures promote complete integration of cartilage and bone tissues without a requirement of a joining process. Holland et al. utilized a bilayered scaffold based on the polymer oligo(poly(ethylene glycol) fumarate) (OPF) (46). A bilayered scaffold consisted of OPF for the bone layer and OPF/TGF-β-loaded gelatin microparticles to promote cartilage layer formation. Hyaline cartilage was observed in the chondral region with the underlying subchondral bone formation completely integrated with the surrounding bone at 14 weeks of implantation. In another study, Chen et al. developed a biphasic scaffold from biodegradable synthetic and naturally derived polymers (47). The upper layer of the scaffold was a collagen sponge for the cartilage layer and the lower layer was a collagen/PLGA composite for the bone layer. PLGA sponge provides a high mechanical strength at the bone layer. Collagen sponge formed in the PLGA sponge promotes cell attachment and proliferation while connecting two parts.

Recent novel technologies, such as SFF fabrication technique, allow the creation of a complex scaffold incorporating a mechanical and structural properties' gradient. Sherwood et al. have developed osteochondral composite scaffold for articular cartilage repair using the TheriForm 3DP process (27). The upper region of a construct was composed of D,L-PLGA/L-PLA with 90% porosity for cartilage regeneration while the lower cloverleaf-shaped region was composed of a L-PLGA/TCP composite to maximize bone ingrowth. The structural integrity between two regions was improved by using a gradient of materials and porosity in the transition region. These scaffolds supported formation of cartilage tissue during a 6-week in vitro culture and had a similar tensile strength of the bone region to native cancellous human bone.

While the use of a biphasic scaffold appears to facilitate osteochondral tissue regeneration, additional design strategy to mimic the biophysical properties of the native cartilage and promote optimized tissue regeneration may be required for successful osteochondral defect regeneration. It has been shown that introduction of chondrogenic growth factors, such as TGF-β, induced the intrinsic repair of articular cartilage without the need for cell or tissue transplantation (48, 49). However, there is a high potential of the upgrowth of bone tissue into the cartilage layer, which is undesirable for appropriate cartilage growth. Thus, interface between cartilage and bone layer of scaffolds must be able to prevent blood vessels and perivascular cells from penetrating into the cartilage layer. Blood vessel-and cell excluding membrane may be helpful to prevent any potential osseous upgrowth into the cartilage layer (50). In addition, articular cartilage is composed of phenotypically different zones, where resident chondrocytes and collagen have different morphology and organization (51). This organized 3D network of cells and ECM determines the mechanical properties of articular cartilage. For the success of osteochondral tissue engineering, additional design strategy to mimic the native zonal organization may be required to meet the mechanical and structural properties of osteochondral tissues.

3.2. Indirect 3D Printing

While 3DP technique offers innovative ways to produce biomimetic structures, the current direct 3DP approach incurs several limitations on fabrication of tissue engineering scaffolds that provide all the requirements necessary to optimize tissue growth. The first constraint involves the pore size. Highly porous microstructure with significantly interconnected pores is essential for cell attachment, growth, and migration in 3D structures. In addition, these allow mass transfer of nutrients and metabolic waste products to the cells. A highly porous scaffold was demonstrated through a combination of 3DP with a particulate leaching technique, and created scaffolds supported the ingrowth of hepatocytes (24). It has been suggested that a large pore size in a range of 200–400 μm is needed to facilitate mineralized bone ingrowth (52), but a typical 3DP incremental layer thickness of 150 μm or less is needed to minimize vertical z-steps and maximize interlayer connectivity (53). To reliably spread porogen particles of the appropriate sizes (~200–400 μm), the incremental layer thickness of each layer must be increased to the porogen size range. This increased layer thickness necessitates additional vertical penetration by the binder droplets, which is problematic since typical 3DP nozzles deposit small droplets (diameter < 70 μm) with low kinetic energy in order to maximize feature fidelity and resolution. Incomplete penetration reduces layer-to-layer connectivity and leads to lamination defects. Larger or faster droplets can penetrate thicker layers at the expense of print resolution. Furthermore, larger incremental layer thickness results in correspondingly large vertical z-steps along external curved surfaces.

A second limitation of the direct 3DP approach is shape complexity when the powder material (e.g., degradable polymer) demands the use of organic solvents as liquid binders because solvents also dissolve most commercially available drop-on-demand printhead subsystems. Although solvent-resistant printhead subsystems are available, none has been integrated with the necessary motion control and powder handling subsystems for widespread use. Meanwhile, nonoptimal solutions involve the use of low-resolution printheads which are solvent compatible or the use of high-resolution jets through stencils (24).

The former solution reduces part resolution and increases minimum feature size. The latter solution is inefficient for complicated structures. Alternatively, high-resolution aqueous printheads can be used to print porogens/polymer, and then the entire printed structure is infused with the appropriate solvent phase. Using a commercially available 3DP machine, Lam et al. fabricated starch-based scaffolds directly by printing distilled water (26). This approach, however, is subject to the first constraint: the porogen size. Finally, the direct 3DP approach for biodegradable polymers and solvents requires custom machines, proprietary control software, and extensive operator expertise.

To address this issue, a practical, indirect 3DP protocol for use with common biodegradable polymers was introduced (54, 55). An indirect approach, where molds are printed and the final materials are cast into the mold cavity, can overcome the limitations of the direct technique. In this technique, molds are printed using commercially available plaster powder and biodegradable polymers are cast into the printed mold. A combination of other scaffold fabrication methods, such as solvent casting and particulate leaching technique, with indirect 3DP technique is available to create porous microstructures. Material choice is highly flexible with this approach because many different materials can be cast under the similar printing process parameters, whereas conventional direct method is highly dependent on powder material properties.

Scaffolds with small villi architecture were constructed to evaluate the resolution that is possible with the indirect approach (Fig. 1). Porous scaffolds were produced by solvent casting Poly (D,L-lactic-co-glycolic acid) (PLGA) solution mixed with sucrose particles in the specified range into plaster molds, followed by particulate leaching in deionized water to remove plaster molds and porogen. Individual process parameters, such as particle size of plaster powder, viscosity of binder, and spreading layer thickness, were optimized to maximize build resolution.

This indirect technique can also create highly porous microstructure with high pore interconnectivity, which is essential for

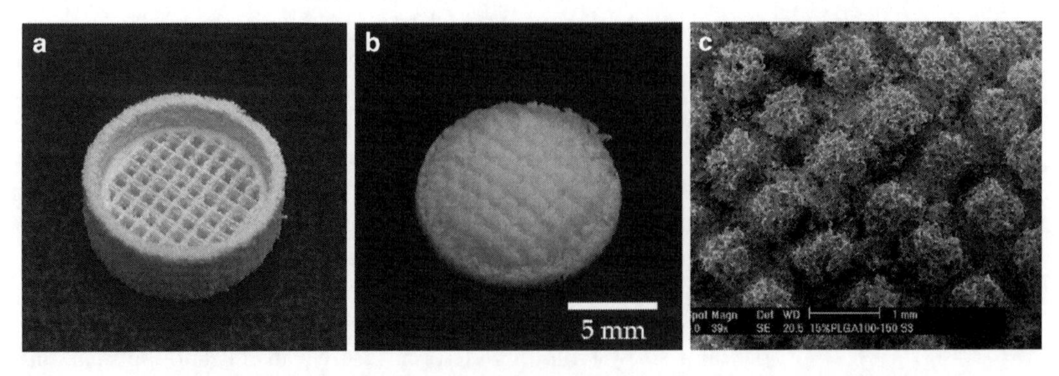

Fig. 1. (**a**) 3D printed plaster mold for villi-shaped scaffold. (**b**) Villi-shaped PLGA scaffold generated from plaster mold. (**c**) SEM image of scaffold. Reproduced with permission from ref. 54.

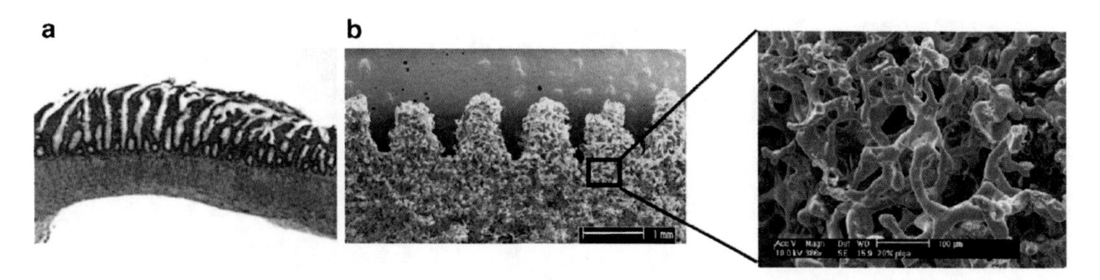

Fig. 2. (**a**) Histological structure of the small intestine. (**b**) SEM image of the cross section of villi-shaped PLGA scaffold. Reproduced with permission from ref. 54.

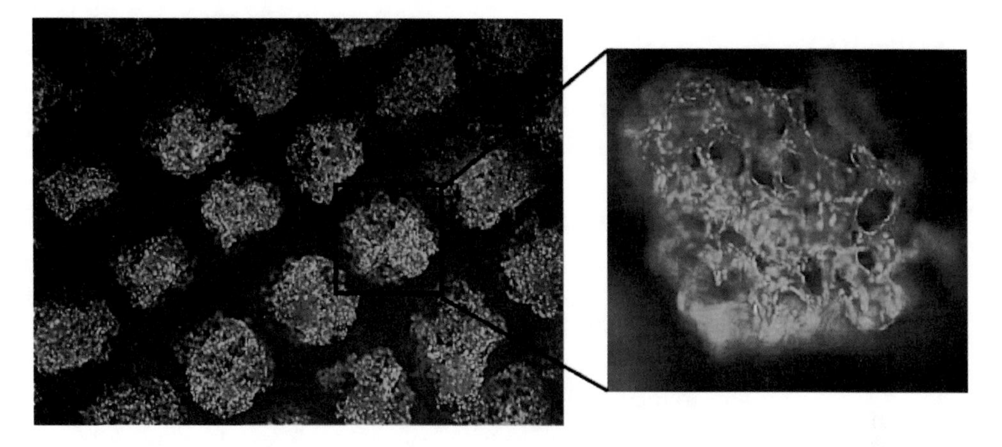

Fig. 3. Fluorescence image of villi-shaped scaffold after 7 days of IEC6 cell culture (scale bar = 1 mm).

uniform cell attachment and proliferation. In addition, these allow sufficient supply of oxygen and nutrient to the cells and removal of cellular and polymer by-product. Scanning electron microscope (SEM) showed highly open, well-interconnected, uniform pore architecture (Fig. 2). Cross-section image shows columnar-shaped villi features (1-mm tall, 500-μm diameter, Fig. 2b) mimicking native small intestine. Intestinal epithelial cells (IEC6) cultured in these constructs demonstrated the feasibility of the scaffolds to support cell attachment and proliferation (Fig. 3).

The freeform nature of this technique with large pore size was demonstrated using anatomically shaped zygoma scaffolds. Zygoma was reconstructed from computed tomography data and zygoma scaffold was created from the mold having a negative shape of reconstructed 3D model (Fig. 4). Internal pore morphology of scaffolds showed well-interconnected pores with the range of 300–500 μm (Fig. 4c). This range of pore size has not been reported with direct 3DP, in which the porogen particle size in powder bed is limited to typical incremental powder layer thickness of about 100–150 μm for optimum resolution and integrity of constructs.

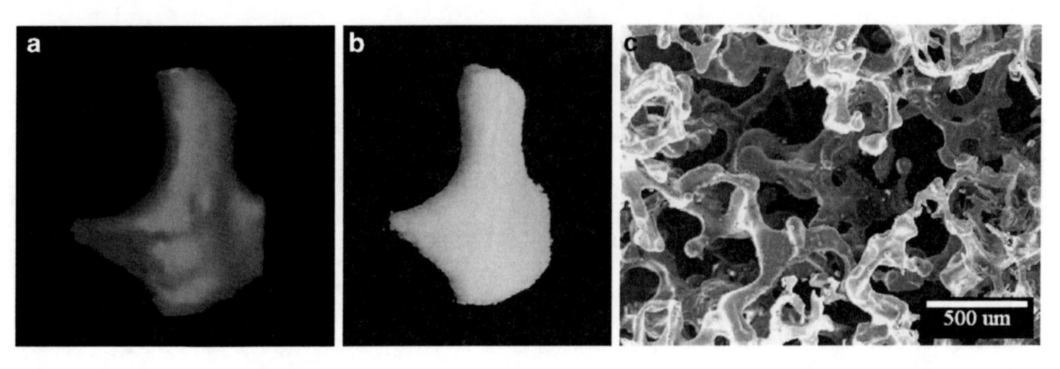

Fig. 4. (**a**) 3D reconstruction of zygoma. (**b**) Zygoma-shaped PLGA scaffold generated from indirect 3DP. (**c**) SEM image of the cross section of scaffold. Reproduced with permission from ref. 54.

4. Future Directions

The application of 3DP technology to a scaffold fabrication has shown a desirable method to design and develop the optimal scaffolds for multiple tissue regeneration. However, further studies on developing advanced fabrication methods must be conducted to create biomimetic constructs satisfying a number of necessary requirements in the growth of various tissues for the successful tissue regeneration.

Blood vessel ingrowth into the tissue-engineered structures is critical for the regeneration of thick complex organs. 3D biodegradable polymer scaffolds with an intrinsic network of interconnected channels have been produced to support hepatocyte growth using a 3DP technique (24). However, current 3DP techniques are not able to produce a structure containing microchannels (~10 μm) that can guide vessel ingrowth because of the limited resolution of this technology. Recent developments in the field of microelectromechanical systems (MEMSs) have enabled patterns of complex structure at the micron level. Kaihara et al. reported successful construction of branched architecture of a microvascular network using MEMS technique for liver fabrication (56). However, this approach is limited to very thin structures, and regenerated thin tissue needs to be folded into 3D configurations to form thick 3D tissues before implantation. Integration of microfabrication techniques to indirect 3DP technology would give great capability in scaffold design and fabrication to guide blood vessel ingrowth.

An ideal scaffold should provide a nanofibrous structure similar to native extracellular matrix (ECM), promoting proliferation and differentiation of cells. Electrospinning technology has been introduced to mimic the ECM-like environment (57–59). Electrospun nanofibrous scaffolds have high porosity and surface area for promoting cell adhesion and proliferation. However, this fabrication technique has some limitations on the creation of macropores or channels, which are important for nutrient supply and diffusion. The further development of this technology, combined with 3DP technology, will provide 3D scaffolds with well-defined macroshape and nanofibrous internal networks at the same time.

Hence, integration of different scaffold fabrication technologies to 3DP technique or other SFF technology can provide cells with mechanical and biochemical cues at the macro-, micro-, and nano scale. Such an integrating technology has the potential to create 3D scaffolds satisfying complex biological requirements of tissue engineering scaffolds.

References

1. Zhang R, Ma PX (2000) Synthetic nano-fibrillar extracellular matrices with predesigned macroporous architectures. J Biomed Mater Res 52 (2):430–438
2. Karande TS, Ong JL, Agrawal CM (2004) Diffusion in musculoskeletal tissue engineering scaffolds: design issues related to porosity, permeability, architecture, and nutrient mixing. Ann Biomed Eng 32(12):1728–1743
3. Mikos AG, Bao Y, Cima LG, Ingber DE, Vacanti JP, Langer R (1993) Preparation of poly(glycolic acid) bonded fiber structures for cell attachment and transplantation. J Biomed Mater Res 27(2):183–189
4. Thomson RC, Yaszemski MJ, Powers JM, Mikos AG (1995) Fabrication of biodegradable polymer scaffolds to engineer trabecular bone. J Biomater Sci Polym Ed 7(1):23–38
5. Mikos AG, Thorsen AJ, Czerwonka LA, Bao Y, Langer R, Winslow DN, Vacanti JP (1994) Preparation and characterization of poly(L-lactic acid) foams. Polymer 35(5):1068–1077
6. Mikos AG, Sarakinos G, Cima LG (1996) Biocompatible polymer membranes and methods of preparation of three-dimensional membrane structures. US Patent No.5,514,378
7. Nam YS, Park TG (1999) Biodegradable polymeric microcellular foams by modified thermally induced phase separation method. Biomaterials 20(19):1783–1790
8. Mooney DJ, Baldwin DF, Suh NP, Vacanti LP, Langer R (1996) Novel approach to fabricate porous sponges of poly(D, L-lactic-co-glycolic acid) without the use of organic solvents. Biomaterials 17(14):1417–1422
9. Harris LD, Kim BS, Mooney DJ (1998) Open pore biodegradable matrices formed with gas foaming. J Biomed Mater Res 42(3):396–402
10. Sun W, Lal P (2002) Recent development on computer aided tissue engineering – a review. Comput Methods Programs Biomed 67 (2):85–103
11. Hutmacher DW, Sittinger M, Risbud MV (2004) Scaffold-based tissue engineering: rationale for computer-aided design and solid free-form fabrication systems. Trends Biotechnol 22(7):354–362
12. Winder J, Cooke RS, Gray J, Fannin T, Fegan T (1999) Medical rapid prototyping and 3D CT in the manufacture of custom made cranial titanium plates. J Med Eng Technol 23 (1):26–28
13. Colin A, Boire JY (1997) A novel tool for rapid prototyping and development of simple 3D medical image processing applications on PCs. Comput Methods Programs Biomed 53(2):87–92
14. Hollister SJ, Maddox RD, Taboas JM (2002) Optimal design and fabrication of scaffolds to mimic tissue properties and satisfy biological constraints. Biomaterials 23(20):4095–4103
15. Leong KF, Cheah CM, Chua CK (2003) Solid freeform fabrication of three-dimensional scaffolds for engineering replacement tissues and organs. Biomaterials 24(13):2363–2378
16. Yang SF, Leong KF, Du ZH, Chua CK (2002) The design of scaffolds for use in tissue engineering. Part II. Rapid prototyping techniques. Tissue Eng 8(1):1–11
17. Dowler CA (1989) Automatic model building cuts design time, costs. Plastics Eng 45 (4):43–45
18. Cima LG, Sachs E, Cima LG, Yoo J, Khanuja S, Borland SW, Wu BM, Giordano RA (1994) Computer-derived microstructure by 3D printing: bio- and structural materials. In: Proceedings of the SFF symposium

19. Wu BM, Borland SW, Giordano RA, Cima LG, Sachs EM, Cima MJ (1996) Solid free-form fabrication of drug delivery devices. J Control Release 40(1–2):77–87
20. Griffith L, Wu BM, Cima MJ, Chaignaud B, Vacanti JP (1997) In vitro organogenesis of liver tissue. Ann N Y Acad Sci 831:382–397
21. Zein I, Hutmacher DW, Tan KC, Teoh SH (2002) Fused deposition modeling of novel scaffold architectures for tissue engineering applications. Biomaterials 23(4):1169–1185
22. Fisher JP, Vehof JWM, Dean D, van der Waerden JPCM, Holland TA, Mikos AG, Jansen JA (2002) Soft and hard tissue response to photocrosslinked poly(propylene fumarate) scaffolds in a rabbit model. J Biomed Mater Res 59 (3):547–556
23. Leong KF, Phua KKS, Chua CK, Du ZH, Teo KOM (2001) Fabrication of porous polymeric matrix drug delivery devices using the selective laser sintering technique. Proc Inst Mech Eng H 215(H2):191–201
24. Kim SS, Utsunomiya H, Koski JA, Wu BM, Cima MJ, Sohn J, Mukai K, Griffith LG, Vacanti JP (1998) Survival and function of hepatocytes on a novel three-dimensional synthetic biodegradable polymer scaffold with an intrinsic network of channels. Ann Surg 228 (1):8–13
25. Seitz H, Rieder W, Irsen S, Leukers B, Tille C (2005) Three-dimensional printing of porous ceramic scaffolds for bone tissue engineering. J Biomed Mater Res B Appl Biomater 74B (2):782–788
26. Lam CXF, Mo XM, Teoh SH, Hutmacher DW (2002) Scaffold development using 3D printing with a starch-based polymer. Mater Sci Eng C 20(1–2):49–56
27. Sherwood JK, Riley SL, Palazzolo R, Brown SC, Monkhouse DC, Coates M, Griffith LG, Landeen LK, Ratcliffe A (2002) A three-dimensional osteochondral composite scaffold for articular cartilage repair. Biomaterials 23 (24):4739–4751
28. Buckwalter J (2004) Cartilage degeneration and regeneration. Clin Orthop Relat Res 427: S26–S26
29. Buckwalter JA, Saltzman C, Brown T (2004) The impact of osteoarthritis – implications for research. Clin Orthop Relat Res 427:S6–S15
30. Buckwalter JA, Martin JA (2006) Osteoarthritis. Adv Drug Deliv Rev 58(2):150–167
31. Lapadula G, Iannone F, Zuccaro C, Grattagliano V, Covelli M, Patella V, Lo Bianco G, Pipitone V (1998) Chondrocyte phenotyping in human osteoarthritis. Clin Rheumatol 17 (2):99–104
32. Iannone F, Lapadula G (2008) Phenotype of chondrocytes in osteoarthritis. Biorheology 45 (3–4):411–413
33. Hangody L, Kish G, Modis L, Szerb I, Gaspar L, Dioszegi Z, Kendik Z (2001) Mosaicplasty for the treatment of osteochondritis dissecans of the talus: two to seven year results in 36 patients. Foot Ankle Int 22(7):552–558
34. Martinek V, Imhoff AB (2003) Treatment of cartilage defects. Deutsche Zeitschrift Fur Sportmedizin 54(3):70–76
35. Spahn G, Kahl E, Muckley T, Hofmann GO, Klinger HM (2008) Arthroscopic knee chondroplasty using a bipolar radiofrequency-based device compared to mechanical shaver: results of a prospective, randomized, controlled study. Knee Surg Sports Traumatol Arthrosc 16 (6):565–573
36. Jerosch J, Filler T, Peuker E (2000) Is there an option for harvesting autologous osteochondral grafts without damaging weight-bearing areas in the knee joint? Knee Surg Sports Traumatol Arthrosc 8(4):237–240
37. Langer R, Vacanti JP (1993) Tissue engineering. Science 260(5110):920–926
38. Vacanti JP, Langer R (1999) Tissue engineering: the design and fabrication of living replacement devices for surgical reconstruction and transplantation. Lancet 354:Si32–Si34
39. [Anon] (1999) The promise of tissue engineering. Sci Am 280(4):59–89
40. Temenoff JS, Mikos AG (2000) Review: tissue engineering for regeneration of articular cartilage. Biomaterials 21(5):431–440
41. Martin I, Miot S, Barbero A, Jakob M, Wendt D (2007) Osteochondral tissue engineering. J Biomech 40(4):750–765
42. Brittberg M, SjogrenJansson E, Lindahl A, Peterson L (1997) Influence of fibrin sealant (Tisseel(R)) on osteochondral defect repair in the rabbit knee. Biomaterials 18 (3):235–242
43. Shao XX, Hutmacher DW, Ho ST, Goh JCH, Lee EH (2006) Evaluation of a hybrid scaffold/cell construct in repair of high-load-bearing osteochondral defects in rabbits. Biomaterials 27(7):1071–1080
44. Schaefer D, Martin I, Jundt G, Seidel J, Heberer M, Grodzinsky A, Bergin I, Vunjak-Novakovic G, Freed LE (2002) Tissue-engineered composites for the repair of large osteochondral defects. Arthritis Rheum 46 (9):2524–2534

45. Kreklau B, Sittinger M, Mensing MB, Voigt C, Berger G, Burmester GR, Rahmanzadeh R, Gross U (1999) Tissue engineering of biphasic joint cartilage transplants. Biomaterials 20 (18):1743–1749
46. Holland TA, Bodde EWH, Baggett LS, Tabata Y, Mikos AG, Jansen JA (2005) Osteochondral repair in the rabbit model utilizing bilayered, degradable oligo(poly(ethylene glycol) fumarate) hydrogel scaffolds. J Biomed Mater Res A 75A(1):156–167
47. Chen GP, Sato T, Tanaka J, Tateishi T (2006) Preparation of a biphasic scaffold for osteochondral tissue engineering. Mater Sci Eng C 26(1):118–123
48. Pittenger MF, Mackay AM, Beck SC, Jaiswal RK, Douglas R, Mosca JD, Moorman MA, Simonetti DW, Craig S, Marshak DR (1999) Multilineage potential of adult human mesenchymal stem cells. Science 284 (5411):143–147
49. Davidson ENB, van der Kraan PM, van den Berg WB (2007) TGF-beta and osteoarthritis. Osteoarthritis Cartilage 15(6):597–604
50. Hunziker EB, Driesang IMK, Saager C (2001) Structural barrier principle for growth factor-based articular cartilage repair. Clin Orthop Relat Res 391:S182–S189
51. Bhosale AM, Richardson JB (2008) Articular cartilage: structure, injuries and review of management. Br Med Bull 87(1):77–95
52. Boyan BD, Hummert TW, Dean DD, Schwartz Z (1996) Role of material surfaces in regulating bone and cartilage cell response. Biomaterials 17(2):137–146
53. Wu BM, Cima MJ (1999) Effects of solvent-particle interaction kinetics on microstructure formation during three-dimensional printing. Polym Eng Sci 39(2):249–260
54. Lee M, Dunn JCY, Wu BM (2005) Scaffold fabrication by indirect three-dimensional printing. Biomaterials 26(20):4281–4289
55. Lee M, Wu BM, Dunn JCY (2008) Effect of scaffold architecture and pore size on smooth muscle cell growth. J Biomed Mater Res A 87A (4):1010–1016
56. Kaihara S, Borenstein J, Koka R, Lalan S, Ochoa ER, Ravens M, Pien H, Cunningham B, Vacanti JP (2000) Silicon micromachining to tissue engineer branched vascular channels for liver fabrication. Tissue Eng 6(2):105–117
57. Murugan R, Ramakrishna S (2006) Nano-featured scaffolds for tissue engineering: a review of spinning methodologies. Tissue Eng 12 (3):435–447
58. Teo WE, Ramakrishna S (2006) A review on electrospinning design and nanofibre assemblies. Nanotechnology 17(14):R89–R106
59. Sill TJ, von Recum HA (2008) Electro spinning: applications in drug delivery and tissue engineering. Biomaterials 29(13):1989–2006

Chapter 16

Graft–Artery Junctions: Design Optimization and CAD Development

Yos S. Morsi, Amal Ahmed Owida, Hung Do, Md. Shamsul Arefin, and Xungai Wang

Abstract

Designing and manufacturing of vascular prosthesis for arterial bypass grafts is a very complex problem. The process involves the selection of suitable geometry, materials of appropriate characteristics, and manufacturing technique capable of constructing prosthesis in a cost-effective manner. In this chapter, all engineering aspects related to the design and optimization of an artificial graft are presented and discussed. These aspects include CAD design of the graft, in vitro hemodynamic analysis to ensure good mechanical integrity and functionality, and optimization of the manufacturing techniques. Brief discussion is also given on the endothelization and vascularization of the artificial vessels and the future directions of the development of synthetic vessels for human implementation.

Key words: Design optimization, Rapid prototyping, Arterial grafts, Computer-aided design, Vascular graft, Biomaterials

1. Introduction

Arterial bypass or replacement is a common treatment for vascular diseases and several thousand coronary bypass grafts performed annually using autologous grafts that include the saphenous vein, internal mammary artery, and radial artery (1, 2). Still, due to various conditions such as age, diseases, and prior usage, the appropriate autologous arteries for certain patients are difficult to find (3, 4). Therefore, synthetic materials, such as expanded polytetrafluoroethylene (ePTFE) and Dacron, are used extensively. However, limited success has been reported particularly for small-diameter arteries (<6 mm) due to the acute thrombogenicity of the graft, anastomotic intimal hyperplasia, aneurysm formation, infection, and progression of atherosclerotic disease (5, 6). Arteries of larger diameter, >6 mm, can stimulate a different adverse response.

Michael A.K. Liebschner (ed.), *Computer-Aided Tissue Engineering*, Methods in Molecular Biology, vol. 868,
DOI 10.1007/978-1-61779-764-4_16, © Springer Science+Business Media, LLC 2012

However, regardless of the size of the graft, analysis of hemodynamic force, particularly shear stresses, is essential in the general assessment of the graft functionality and integrity. Carrying out the fluid structure interaction (FSI) is equally important in determining the degree of deformation and compliance (7, 8).

Moreover, endothelization of grafts has been recognized as a very important factor in sustaining the compatibility of bypass graft. This is because the layer of endothelial cells (ECs) on the wall of the artery has the ability to act as a wall-shear biosensor which maintains a more uniform shear stress level in the range 10–20 dyne/cm^2. Conversely, low shear stress over time can cause the arteries to remodel by intimal thickening (9–11).

Still, despite over 50 years of research to develop an optimum graft, from hemodynamic and stress analysis point of view made from a biomimics material with good degree of compliance matching and can be manufactured on a cost-effective way, is still progressing. Recently, however, various artery bypass anastomoses have been proposed and clinically implemented with various degree of success. In this chapter, a brief discussion of the important elements of designing and optimizing the synthetic graft, including endothelization of the inner surface of the vessel for in vivo implementation, is given.

2. Design Aspects of the Artery Graft

In literature, there are huge amounts of research that have been devoted to the development of an alternative artificial vessel of biocompatible polymer that could harmonize with the natural host artery. The general consensus among researchers is that in order to achieve an efficient vessel, certain research steps have to be carried out in a holistic and systematic way. The selection of biomimic materials, synthetics, or natural and manufacturing techniques is one of the main challenges facing the bioengineers and graft designers. Moreover, hemodynamic and structure analyses are other important elements of the process which should be considered.

Figure 1 illustrates a typical research plan for the fluid structure analysis and in vitro validations that is used in the initial design and optimization of CABG.

As shown in Fig. 1, the first part in the optimization process is to provide the required geometry based on CAD/or scan of specific patient and engineering parameters, based on hemodynamic analysis and FSI and used these data to design and construct the optimized scaffold of a CABG from different materials. Various researchers have used this concept; for example, Mahnken et al. (12), who with the aid of computer-aided engineering design techniques, have constructed a geometry of carotid artery

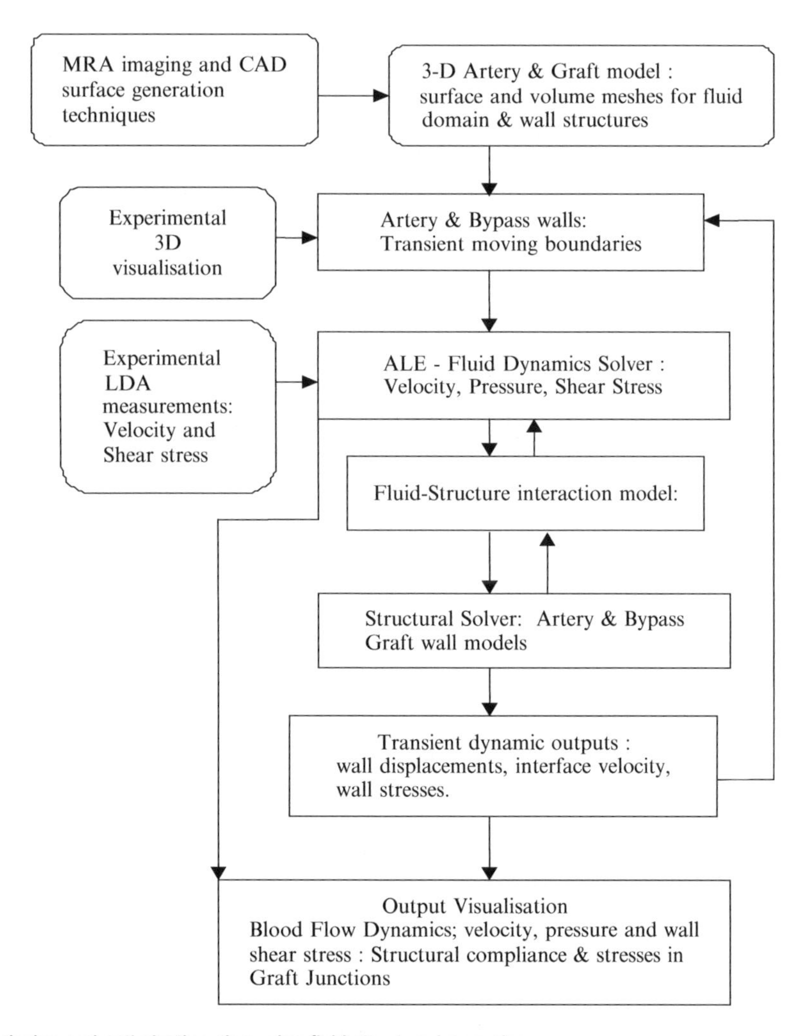

Fig. 1. Artery design and optimization plan using fluid structure interaction.

bifurcation with the abdominal aorta using computed tomography (CT) and magnetic resonance imaging (MRI) data. The authors then used the developed 3D geometrically accurate models for computational fluid dynamics (CFD) simulation studies. Other widely acceptable technique is the use of CAD modeling to construct an appropriate design of the vessels, based on prior knowledge or experimental data or both.

2.1. Physical and Numerical Modeling

It should be noted that the walls of arteries are composite materials with a nonlinear and viscoelastic stress and strain relationship, and significant changes in wall shear rate can occur when the movement due to the elasticity of the artery wall is taken into account. Hence, the wall motions of arteries are important in their influence on the blood flow, and as a result the determination of the degree of deformation and compliance, as well as the interaction between the blood flow and the artery wall structure, has received a lot of attention from researchers (7, 8, 13). In these types of investigations, the following steps are conducted in a systematic way.

- A complete analysis of the 3D fluid flow forces within and around a coronary graft and junction is necessary to obtain a clear picture of the complex flow and the interrelationships of the various parameters.
- Knowledge of the effect of the material properties of the artery, graft, and anastomoses in normal and stenotic conditions needs to be determined.
- The significant 3D movement of the coronary arteries, due to the pulsatile motion of the heart walls, has to be included in any geometrically realistic in vitro simulation of flows in coronary graft junctions.
- The in vitro simulation of hydrodynamic forces inside a bioreactor for tissue growth and endothelization for bypass artery has to be fully examined.

2.1.1. Computational Fluid Dynamics Studies

In recent years, CFD have been effectively used to investigate physical and geometrical parameters exemplifying the hemodynamics of various configurations of CABG. With this method, the pressure and fluid flow characteristics as well as the wall shear stress (WSS) during a cardiac cycle could be accurately simulated. In addition, it is widely accepted that using CFD is a better alternative to invasive or noninvasive flow measurements of blood flow by in vitro and/or ex vivo experimental flow setups, which can be very expensive and time consuming (10). The mathematical description of the flows to be solved for this type of problem is based on 3D Navier–Stokes equations. These equations and their solutions have been described and discussed widely in literature (14, 15). The most common variables that researchers currently focus on are:

- Acuteness of the angles of attachment
- Graft-to-artery diameter ratio
- Out-of-plane asymmetry of the graft geometry and geometry changes due to simulated restenosis

Moreover, information obtained from the FSI has the ability to satisfactorily predict coupling information between structure deformations and fluid forces. However, the accurate geometry and behavior of the natural material of the arteries and bypass graft are challenging as they are tethered to and supported by surrounding tissues and organs. To avoid such complexity, most of the researchers in their analysis selected to assume a rigid wall of uniform thickness. This assumption is based on the clinical observation that, under normal conditions, wall deformability does not greatly alter the hemodynamics within the arteries. Still, in order to determine the mechanical integrity of the native or artificial bypass grafts under consideration, a complete simulation of the correct physiological fluid and structure forces must be determined. Moreover, coupling medical planning with FSI numerical techniques to verify and compare in a clinically relevant time frame is in high demand (16).

Figure 2 shows the results of a total mesh displacement and mesh velocity distributions together with the WSS and velocity vectors of a typical CABG of 20° anastomosis angle and graft-to-artery ratio of 1.6 at the peak systole of the heart (17). These results show for the first time the direction of movement of the artery as well as the maximum area of deformation during one cycle of the heart. Such data may be useful for surgeons and graft designers to optimize the current and future CABG configurations and selection of materials.

2.1.2. *In Vitro* Experimental Studies

In the last few decades, various experimental modeling techniques have been used to gain initial understanding of the complex nature of the graft–artery junction, initially qualitatively and later quantitatively. In these types of in vitro studies, the complex flow in and around the graft junctions is visualized, and WSS and WSS gradients are normally determined by using a variety of experimental diagnostics techniques, such as Hot film, Laser Doppler Anemometry, and Particle Image Velocimetry.

For example, the laser illumination of particles technique was used to gain overall visualization characteristics inside the flow of end-side vascular graft anastomoses (11). Conversely, White et al. (18) used hydrogen bubble technique to determine the effect of angles on the flow patterns of a distal vascular graft. Keynton et al. (19) used magnetic resonance velocimetry to determine the general fluid dynamic characteristics of the unsteady flows typical in artery flows. In addition, magnetic resonance velocity mapping was used to determine the general fluid dynamic characteristics and to investigate the hypothesis of a vortex motion within the left ventricle interacting with mitral valve motion and inflow velocity (20).

Recently, dual-camera stereophotogrammetry (DCSP) has been adopted as a noninvasive in vitro technique to directly visualize the motion of the flexible model by using two identical cameras. In this technique, the surface of the model of heart valve is marked with a matrix of "ink dots" and the frame-by-frame mapping of the surface is then used to monitor the motion (21). Subsequently, the knowledge and the data acquired of these experiments can be used effectively to validate the moving mesh boundaries in the CFD simulations if required.

Equally experimental data obtained from pulsatile pressure and mass flow rates obtained by in vivo or in vitro studies using relatively noninvasive techniques, such as pulsed Doppler ultrasound and applanation tonometry, particle image velocimetry, and laser

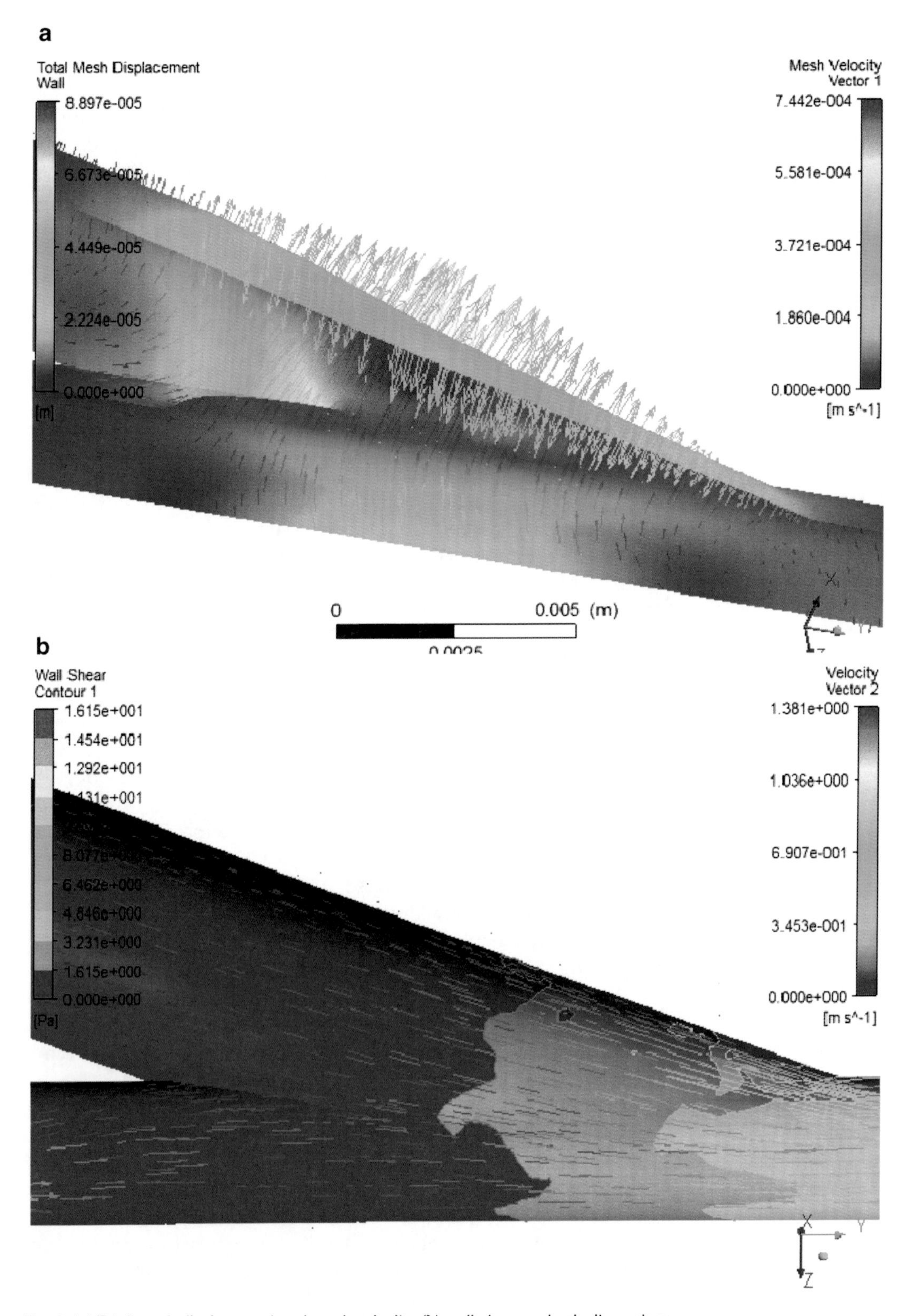

Fig. 2. (**a**) Total mesh displacement and mesh velocity; (**b**) wall shear and velocity vectors.

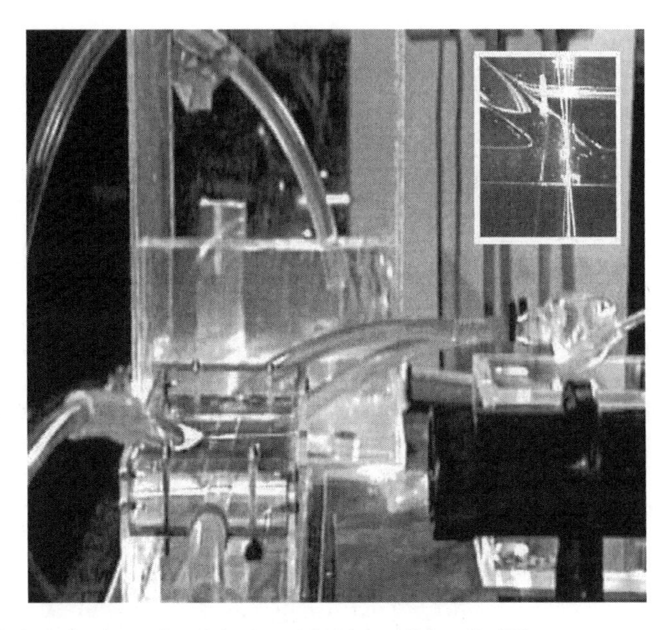

Fig. 3. A photo of experimental setup at Swinburne University (22).

anomemetry, were used to obtain data related to hemodynamic forces and shear stresses (11, 22). Moreover, the pathological data reported in literature show that the zones susceptible to disease and atherosclerotic lesions are associated with low or oscillatory shear stress and higher shear stress regions are free of deposits; hence, determination of the shear stress distributions in and around the graft/artery junction (experimentally or numerically) is deemed to be important in the overall analysis of the graft–artery junction.

However, it is indeterminate at this point which modeling technique (numerically or experimentally) is the most appropriate for progressive investigations of blood flows in CABG. However, it is always recommended to use both modeling techniques as one method can complement the other (11, 18). Typical experimental set of Laser Doppler in vitro testing of artery fluid flow is shown in Fig. 3.

3. Selection of Biomaterials for Artificial Blood Vessels

3.1. Materials Selection

The best artificial material is one that totally or partially mimics the natural artery. It should be biocompatible with minimum mismatching of the physical properties and good mechanical integrity and functionality. The extracellular matrix (ECM) of an artery has three dominant layers and these are elastin, collagen, and smooth muscle (23). Within these layers, elastin is purely elastic, whereas collagen and in particular smooth muscle have viscoelastic properties which provide creep, stress relaxation, and hysteresis characteristics and affect the dynamic properties of artery walls considerably.

Therefore, in the selection of the materials for synthetic blood vessel, parameters such as biocompatible, nonthrombogenic, and nonimmunogenic, mechanical properties, integrity, good degree of compliance, and easiness to fabricate must be considered carefully (24).

The key concept of the biomaterial is that the scaffold of vessel containing the structural or chemical information should mimic the same characteristics of native cell, like cell-to-cell communication, tissue formations, growth factors, and so on (25). Ease of manufacturing is another consideration.

3.2. Current Development of Biomaterials for Vascular Grafts

To date, limited success of synthetic vascular grafts has been reported, and currently various researchers worldwide are exploring the possibilities of different combinations of biological and polymer materials for an optimum graft performance. Recently, collagen, due to its properties such as biodegradability, low antigenicity, and resistance to calcification and its ability to be cross-linked with other polymers to increase the mechanical integrity of the vessels, has been searched widely. However, collagen has a major disadvantage as it is thrombogenic which is the main initiator of the clotting inside the graft (26). Nevertheless, in principle, cross-linking and incorporation of heparin can reduce collagen thrombogenicity (26).

Moreover, various researchers have investigated the effect of the inner surface characteristics and the degree of porosity on the rate of tissue ingrowth and attachment. Furthermore, it has been suggested that transmural capillary ingrowth can provide the cell source for the surface endothelialization. Conversely, to reduce thrombus formation, carbon coating is used and attachment of anticoagulant or antithrombotic agents to the synthetic graft has also been proposed. Moreover, recent literature suggests that heparin binding is the most promising material which is pretreated with a cationic agent, tridodecil-methyl-ammonium chloride (TMAC), and bound through Van der Waals forces to the polyester fiber (27). Consequently, heparin-bound ePTFE grafts showed improved patency rates and reduced thrombogenicity compared with the standard infrarenal aortic graft in rat (28). Moreover, according to Xue and Greisler (27), a heparin-bonded Dacron graft by InterVascular (La Ciotat, France) is currently available in the European market.

Moreover, in keeping with this, various bioactive substances have been incorporated within synthetic grafts using different delivery methods to adjust the graft healing process. Still, fibrin glue (FG) delivery of growth factors onto the ePTFE grafts has been used, where the growth factors can be slowly released from the FG which retained their bioactivities in vivo. Likewise, as indicated by Bos et al. (29), the growth factors in combination with heparin could be added to the sealant, which is known to bind and protect them from denaturation and enzymatic degradation. In addition, it has been long established that treated ePTFE grafts saturated with FG containing fibroblast growth factor (FGF)-1 and heparin bring out good endothelialization and tissue incorporation than untreated or FG/heparin without FGF-treated grafts (30, 31). Nevertheless, although all the findings are positive, so far a long-term in vivo human trial is the only viable indicator to determine the actual functionality and superiority of this type of graft.

However, bacterial cellulose (BC), developed from *Acetobacter xylinum*, is a polysaccharide containing suitable properties for arterial grafting and vascular tissue engineering (TE). It also contains high burst pressure (800 mmHg), high water content, high crystallinity, and highly porous fiber structure similar to that of collagen. Moreover, it was tested in vivo and it has been shown that the surface of BC possesses lowest thrombogenicity than Dacron and ePTFE (32). Furthermore, the most widely used vascular grafts in clinical applications, Dacron and ePTFE, the material-induced blood coagulation was tested and compared with BC, and it was found that BC showed the slowest and least activation of the coagulation (25, 32).

3.2.1. The Use of Elastic Polymer and Compliance Mismatch

The current compliance mismatch of ePTFE and Dacron grafts can potentially contribute to the development of intimal hyperplasia at the anastomotic regions of the bypass. This situation can introduce unfavorable hemodynamic conditions and unsteady shear stress. With this in mind, elastic polymers have been searched extensively to develop compliant vascular grafts that partially or totally mimic the natural graft. Elastic polymer, polyurethane (PU), has been introduced where the segmented PUs are copolymers and comprise three different monomers: a hard domain derived from a diisocyanate, a chain extender, and a soft domain, most commonly polyol. With this structure, the soft domain is incorporated for elasticity and flexibility of the graft and the hard domain incorporated to ensure the mechanical integrity and functionality of the graft. Still, there is always the option of selecting three different monomers to optimize different mechanical and physical properties which make PU a good candidate for blood vessels application and other medical devices. It should be noted that Dupon in 1962 commercialized Lycra which is the trade name of a segmented polyether PU (27).

3.2.2. Biodegradable and Bioresorbable Polymers

The selection and development of good biodegradable polymers for arterial applications and TE have attracted a tremendous amount of research. In general, as stated above, the materials selected for blood vessels or scaffolds for TE need to be robust and flexible, and should possess good mechanical properties similar to those of the host artery and mimic and synchronize with the nature artery. The biocompatibility of the scaffold of the vessel would also need to be considered as toxic by-products from the polymer may leach into the cellular environment. Additionally, the materials used should be conducive to endothelialization as this overcomes the thrombogenic nature of the polymers.

A short history of the use of bioresorbable polymers showed that in 1979 a fully bioresorbable vascular graft made from Vicryl sheets was introduced. However, due to the fact that this type of graft was prone to aneurysmal dilation and rupture (27) the attempt has been abandoned. Later, to address this drawback, grafts composed of woven PGA have been implemented in vivo (rabbit model) and the findings were promising as only 10% of the grafts showed aneurysmal dilation (33, 34).

However, the literature tends to suggest that the two most promising and investigated bioresorbable polymers are polyglycolic acid (PGA) and polylactic acid (PLA). However, PGA, although highly crystalline and hydrophilic, lacks mechanical strength. For example, PGA has been used to make the first synthetic absorbable suture. Still, it lost its mechanical strength in approximately 2–4 weeks after implantation. Such characteristic is mainly attributed to its in vivo hydrolytic degradation of PGA (35). Still, it is possible to fabricate porous scaffold and foams from PGA, but it might affect the properties and degradation characteristics depending on the type of processing techniques. PGA-based implants can also be fabricated from solvent casting, particular leaching method, and compression molding (36). PLA on the other hand due to the presence of methyl group in the lactide molecule, that limits water absorption and results in a lower hydrolysis rate which is more hydrophobic than PGA. Moreover, PLA is semicrystalline with excellent mechanical integrity and strength, and its hydrolytic product is naturally occurring L-lactic acid. Furthermore, due to its hydrophobic character, PLA degrades slower than PGA and as a result it slows down the backbone hydrolysis rate. Also, it is more resistant to hydrolytic attack than PGA (36).

Recent research focuses on the development of copolymer grafts manufactured either from two or more bioresorbable polymers with different rates of resorption or from combined bioresorbable polymers with a nonresorbable material to enhance mechanical integrity and functionality of a given application (27). For example, various attempts have been made to use polydioxanone (PDS), a material used clinically in bone pins and suture clips, with a slow resorbed compound for arterial graft applications. The findings from these studies showed that PDS grafts developed similar endothelialization of the regenerated luminal surface after implantation with PDS component remained for around 6 months (37). However, these studies still need further verifications and in vivo validations.

4. Manufacturing of the Graft

Various possible techniques are found in literature on the manufacturing methods of artificial grafts and substrates made from different types of polymers. Naturally, this practice, depending on the materials or the combination of materials used for the construction of the scaffolds, requires different processing requirements that include the condition of the processing and its accuracy as well as the degree of consistency and repeatability (38). Moreover, in designing and manufacturing scaffolds for human implementations, there are a number of essential requirements that scaffolds or the substrates need to satisfy, such as the mechanical integrity and functionality, surface morphology to assist in the cell adhesions and growth, and propagation and reorganization of cells. These requirements of the morphology of the substrate/scaffolds require other essential properties of the surface, such as macrostructure, degree of porosity and interconnectivity, pore size, as well as surface area and surface chemistry.

The manufacturing techniques are generally classified as conventional and solid freeform fabrication (SFF) systems [rapid prototyping (RP) techniques]. The conventional fabrication techniques cover methods, such as fiber bonding, gas foaming, emulsion freeze drying, melt molding, membrane lamination, and solvent casting. Consequently, each method has different degree of success and limitations. However, in general, the conventional techniques are known to suffer from the following: inadequacy in producing a precise and exact pore size and pore geometry, as well as poor spatial distribution of pores. More importantly, the scaffold produced by conventional techniques lacks the necessary interconnectivity of internal channels within the substrate and normally has inadequate mechanical integrity and functionality to stand the hemodynamic forces or the stress and strain in certain applications (39 and 38).

Currently, there are tremendous amounts of research aiming to address some of the above-mentioned drawbacks and improve these types of scaffold manufacturing. However, it should be noted that, generally with conventional techniques, there is always the challenge of reducing the harmful organic solvents and other chemicals required by the conventional manufacturing methods that may produce toxic by-products which can adversely affect the biocompatibility and the implementation of the scaffold in living beings (38).

Other method is known as SFF techniques which include fused deposition modeling, 3D printing, selective laser sintering, stereolithography (STL), and wax printing (40). SFF offers a precise control of scaffold pores with an excellent interconnectivity of pores. PU-based scaffold is shown in Fig. 4b.

The geometry of the scaffold is produced using either CAD packages, such as Pro engineer, or CT of the shape of the particular object in hand. This is then converted into stereolithographic file (STL), and with the aid of special software the scaffold is manufactured very accurately. However, SFF is not without disadvantages, as only thermoplastic materials can be used and there are a lot of limitations on the pore sizes; minimum size of rod is around 200 μm.

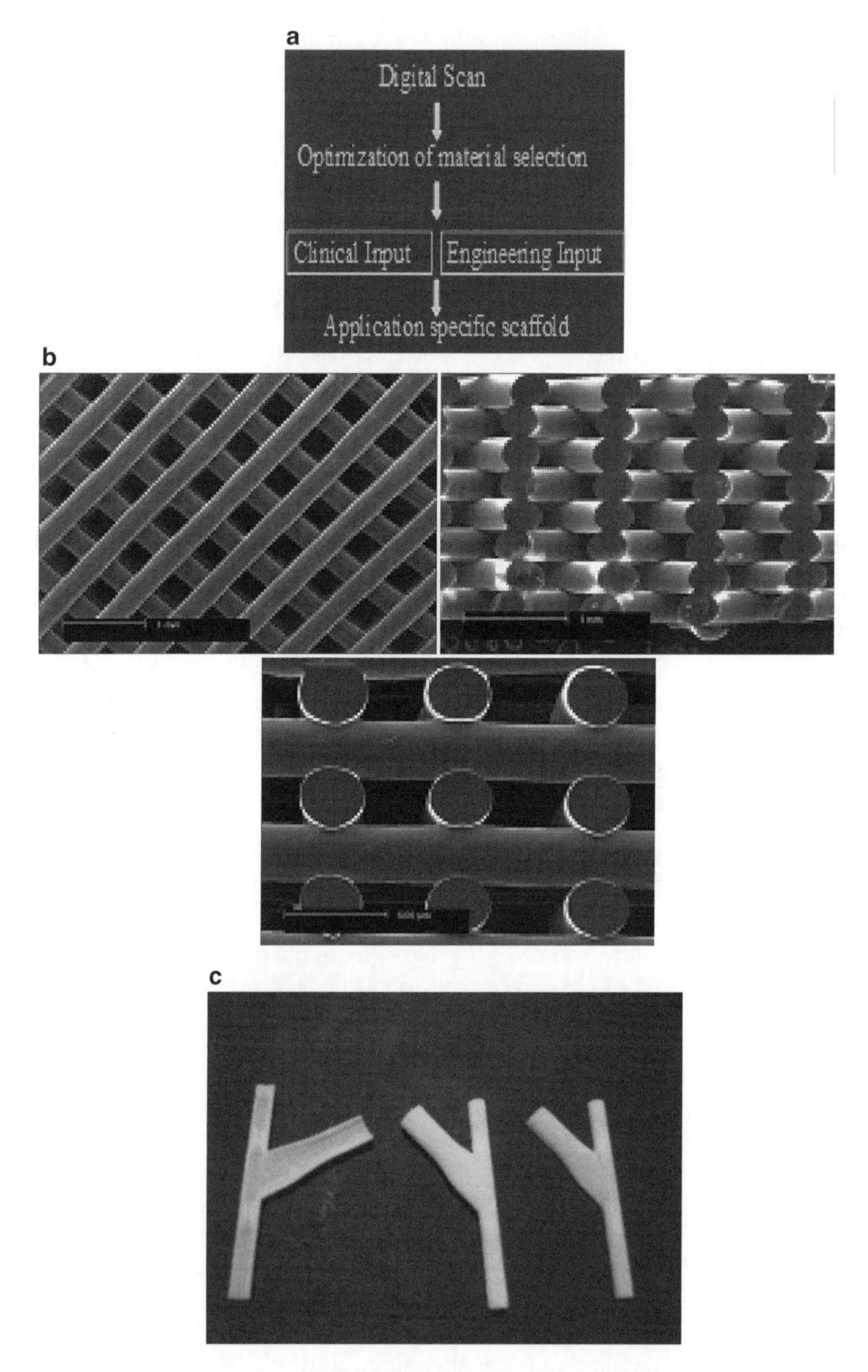

Fig. 4. (**a**) Steps involved for manufacturing application-specific scaffold. (**b**) Polyurethane-based scaffold using FDM at Swinburne University. (**c**) Prototypes of CABG configurations produced by FDM at Swinburne University.

The FDM process offers a safe, nontoxic, highly accurate, and fast method of constructing uniform 3D objects of any shape and complexity under computer control in selected thermoplastic materials by a layer-by-layer building technique (41). However, the manufacturing of a stent graft using FDM offers a challenging task because of the severe conditions required for the FDM processing of material through its liquefier head. Some of these conditions as stated above include that the stent graft must (1) be flexible enough to fit in the aortic wall; (2) be made of materials that are biocompatible and processable through the FDM head; and (3) have mechanical and hemodynamic properties to match those of the aorta tissues at the site of implantation. Moreover, FDM suffers from a number of limitations, such as pore opening which is not consistent in the entire dimension, and support materials may possess toxicity which may adversely affect the biocompatibility and cell proliferation and adhesion.

However, at Swinburne University, manufacturing of arterial graft was performed using thermal PU and typical models of CABG are shown in Fig. 4c.

Recently, electrospinning became one of the most widely used manufacturing methods in producing fibrous scaffolds, with fiber size in the range of 5–1,000 nm (42). Numerous studies suggest that TE scaffold, combined with both electrospun scaffolds and vascular cells, may become an alternative to prosthetic vascular grafts for vascular reconstruction (43). Note that the electrospinning process is essentially in the submicrometer regime and the system is capable of accurately generating scaffolds with optimized functionality and mechanical properties. Moreover, the structure of the substrate can be controlled to achieve the right degree of porosity and connectivity, which are important aspects for a scaffold construction (42). Figure 5a shows a typical electrospinning process.

Consequently, electrospinning has been successfully used to produce nanofibrous scaffolds and substrates made from many synthetic and natural polymers with fibers' diameters ranging from a few nanometers to micrometers (44–46). Vessels created from nanofibers have larger surface areas which may encourage cell growth. However, for an ideal tissue-engineered vascular graft, it should demonstrate comparable biomechanical properties, avoid restenosis after implanting in vivo, and also have antithrombotic properties.

Moreover, electrospinning has been utilized to produce nanofibers from PGA, PLA, and polycaprolactone for the application of engineering skin, blood vessel, bone, and cartilage (47). Polylactide-caprolactone nanofibrous scaffold was shown to be conducive to smooth muscle and ECs, possibly due to its biomimetic ECM (48). Furthermore, the technology has the flexibility to support modification of materials, for example incorporation of bioactive

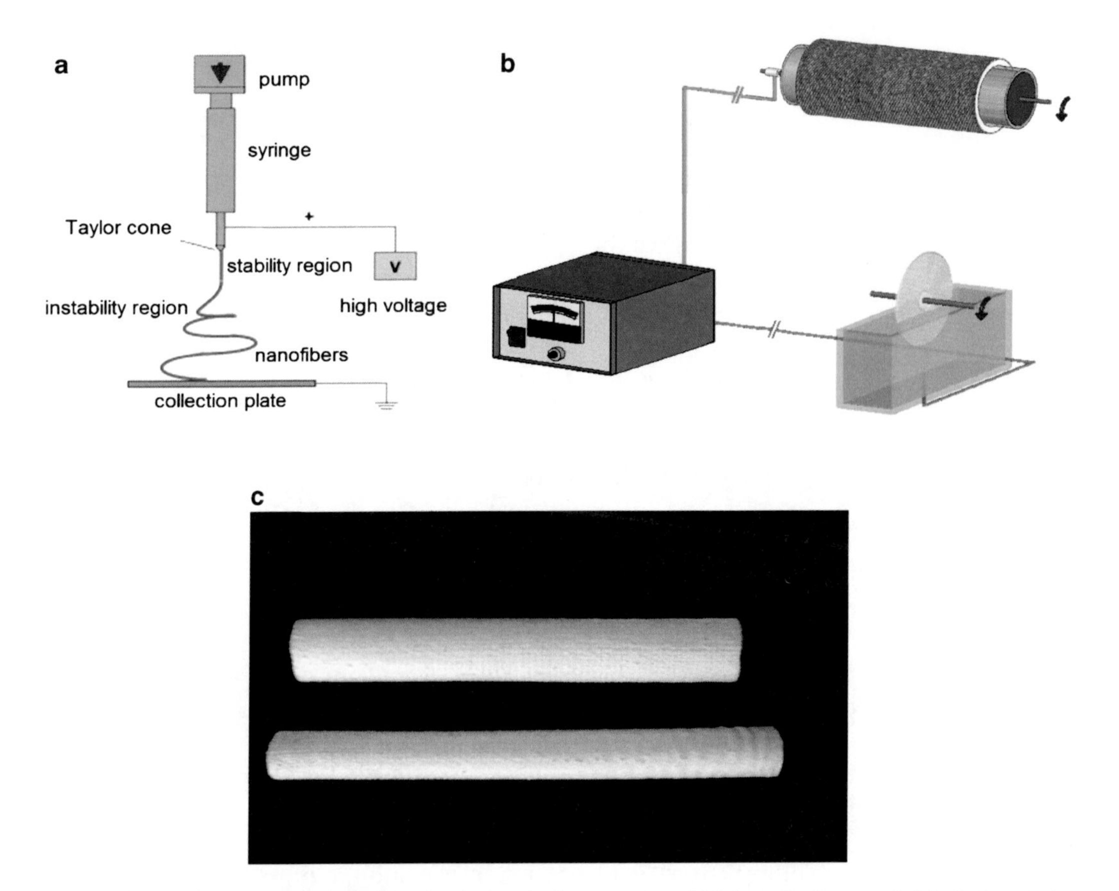

Fig. 5. (**a**) Schematic diagram of a typical needle electrospinning procedure. (**b**) Schematic diagram of disk electrospinning developed at Deakin University. (**c**) Knitted structure with surface coating (10 and 15 mm, respectively) developed at Deakin University.

components such as drugs to facilitate drug delivery and adhesion ligands to enhance cell attachment (49). Moreover, cells that are banded to scaffolds with microscale architectures flatten and spread as if cultured on flat surfaces, but substrates with nanoscale architectures have larger surface areas to adsorb proteins, presenting many more binding sites to cell membrane receptors.

On the other hand, regarding the tissue-engineered vascular grafts, poor mechanical properties, cell overgrowth, and thrombogenicity need to be overcome for clinical applications. Centola et al. (50) and Morsi coworker (51) reported that by combining the electrospinning and FDM technique a biopolymeric scaffold for vascular TE could be fabricated and could possess accurate configuration (Owida et al. (51)). Recently, for this experiment, PLLA/PCL scaffold, with releasing heparin using the combination of electrospinning and FDM, was used (50, 51). Moreover, efforts that produced clinically relevant biocompatible substrate can benefit from strategies that objectively measure and assign a value to fiber anisotropy; which can provide a source of guidance signals that can be exploited to regulate cell phenotype, cell distribution, and macroscopic properties of generated tissue (52).

The use of hydrogels which form a good scaffold material as the structure of many tissues is similar to the extracellular matrix of this material has been proposed. These scaffolds are generally processed under mild conditions relatively and they are delivered in a minimally invasive manner (53). Hydrogels have characteristics, such as hydrophilic nature and the mechanical properties, which are much similar to those of soft tissues. By encapsulating cells into this scaffold, we can overcome the problem of uniform infiltration and density which appear during seeding of cells into the scaffold (54). Hydrogels have a cross-linked polymer networks. Due this cross-linked polymer networks can have good mechanical strength, high capacity of water uptake, and tunable functionality. The hydrogels have a very slow degradation which plays a very important role in development of tissues (55, 56). Scaffolds are used to support the structure during the cell growth and tissue regeneration (56).

Researchers at Deakin University developed a needleless disk electrospinning (Fig. 5b) to coat the tubular spacer fabrics with biocompatible nanofibers. The group is currently experimenting with electrospinning collagen and chitosan-based composite nanofibers onto the surface of tubular fabrics to enhance their functionality. Moreover, synthetic biodegradable polymers, such as P (LLA-CL), are also being experimented at Deakin University.

Moreover, the uses of a knitted or woven fabric to construct a vascular graft may improve the biomechanical properties of the graft. This also enhances the design options for the vascular graft materials (Fig. 5c).

4.1. Endothelial Cell-Seeded Synthetic Vascular Grafts

In the last few decades, many researchers have been directed toward facilitating development of antithrombogenic endothelial linings in vascular prostheses. However, despite many publications, the most important progress in this regard has been the seeding of grafts at the time of implantation with endothelium. Rotmans et al. (57) have successfully used this method to seed both mechanically and enzymatically derived ECs into synthetic blood vessels. Unfortunately, however, many initial trials on human were unsuccessful after the first human study reported by Herring et al. (58). The literature suggests that the reason for the failure of the majority of these human implementations using vein-derived EC-seeded grafts could be attributed to the small quantity of seeded cells used and also the growth potential of human ECs could be lower than that of canine ECs (59).

Hashi et al. (60) reported that vascular graft containing small diameter, with high patency rate using chemical modification of microfibers, was developed. Moreover, they also mentioned that endothelialization and sufficient mechanical strength are two core aspects for the long-term stability for these grafts.

The above findings led to the ECs' culturing studies and experiments with cultured adult ECs in late 1980s. In addition to culturing ECs before seeding, in vitro culturing of cells already seeded on the prosthetic graft was introduced. After 1989, many randomized trials showed the significant higher patency rates for the seeded grafts compared to that of nonseeded control group (61). However, it should be noted that although there is a strong evidence to suggest that in vitro EC lining improves patency of small and large synthetic vessels the main problems associated with this are cost and time used to carry out the process. Furthermore, seeded cells have been cultured under all sorts of growth factors, inducing the risk of unwanted growth after implantation. In addressing this problem, various investigations have been carried out on other EC sources, including mesothelial cells (MCs), microvascular endothelial cells (MVECs), and endothelial progenitor cells (EPCs) (61, 62).

5. Final Remarks and Future Directions

Coronary bypass grafting is widely used for the treatment of coronary heart diseases (CHDs). However, as outlined above, in some cases, due to limitations and the availability of the use of autogenous vessels, prosthetic or artificial grafts are needed. These grafts require special design and requirements to ensure compliance matching, revascularlization, and antithrombogenicity. Numerous in vivo and in vitro experimentations have been carried out on the design and optimization of the grafts from hemodynamics point of view and are still progressing. The other challenge is in the selection of biomaterial and manufacturing into the exact 3D architecture and matching of the resultant mechanical properties with those of the natural arteries. The RP techniques are widely used to construct uncharacteristic models based on CT scan or CAD model with complicated features easily and quickly with high degree of accuracy. Moreover, the electrospinning can be effectively used to optimize the mechanical and physical properties of the vessel along with RP techniques. However, the ultimate solution would be to produce an artificial graft so that it partially or totally mimics the native vessel along with its ECM. This is currently being investigated via TE research.

There have already been case reports regarding first human pediatric applications of tissue-engineered large-diameter vascular grafts (63). To date, more than a few tens of children have been successfully treated with Shinoka's engineered tissues without any complications (62). First studies have also been published about completely engineered small-diameter blood vessels. Three models have been investigated; the first one is collagen-based and consists of an adventitia-like layer of fibroblasts, a media-like layer of smooth muscle cells, as well as a monolayer of ECs in Dacron sleeves to stand the required hemodynamic stress (64). With the second model, namely, self-assembly model, the cells are cultured to structure a continuous sheet of cells and ECM and then seeded with ECs (65). The third model, however, is made from polymeric biodegradable material that seeded with cells and in vitro tested dynamically which allows the cells to proliferate and produce an organized ECM while the scaffold degrades. In this model, ECs are seeded after the process has been completed (66).

Finally, as it has been widely reported in literature, the advantage of a TE vessel over an EC-seeded prosthetic graft is its ability to dilate and constrict as remodel. The main disadvantage of seeding a clean scaffold is that the EC layer must be implemented in vivo, so temporary closure of the vessel may be necessary. Besides, at this moment, we do not know whether intimal hyperplasia develops in TE vessels (61). In a nutshell, although novel approaches for producing small-diameter arterial grafts have been developed and have begun to show viability, demonstrating the potential clinical efficiency requires much more data from long-term implant studies.

References

1. Kim BS, Cho SW, Kim IK, Kang JM, Song KW, Kim HS, Park CH, Yoo KJ (2009) Evidence for in vivo growth potential and vascular remodeling of tissue-engineered artery. Tissue Eng A 15(4):901–912
2. Zheng JW, Zhong LP, Ye WM, Zhang ZY (2009) Autologous external jugular vein: a potential vascular graft for carotid artery reconstruction. Med Hypotheses 72(5):551–552
3. McKee JA, Banik SSR, Boyer MJ, Hamad NM, Lawson JH, Niklason LE, Counter CM (2003) Human arteries engineered in vitro. EMBO Rep 4(6):633–638
4. Campbell GR, Campbell JH (2007) Development of tissue engineered vascular grafts. Curr Pharm Biotechnol 8(1):43–50
5. Isenberg BC, Williams C, Tranquillo RT (2006) Small-diameter artificial arteries engineered in vitro. Circ Res 98(1):25–35
6. McClure MJ, Sell SA, Barnes CP, Bowen WC, Bowlin GL (2008) Cross-linking electrospun polydioxanone-soluble elastin blends: material characterization. J Eng Fiber Fabric 3(1):1–10
7. Kouhi E, Morsi YS, Masood SH (2008) Haemodynamic analysis of coronary artery bypass grafting in a non-linear deformable artery and Newtonian pulsatile blood flow. Proceedings of the Institution of Mechanical Engineers, part H. J Eng Med 222(8):1273–1287
8. Kouhi E, Morsi Y, Masood SH (2009) The effect of arterial wall deformability on hemodynamics of CABG. In: ASME international mechanical engineering congress and exposition proceedings, 2009
9. Giddens DP, Zarins CK, Glagov S (1993) The role of fluid mechanics in the localization and detection of atherosclerosis. J Biomech Eng 115(4B):588–594
10. Do H, Owida AA, Yang W, Morsi YS (2010) Numerical simulation of the haemodynamics in end-to-side anastomoses. Int J Num Meth Fluids
11. Owida A, Do H, Yang W, Morsi YS (2010) PIV measurements and numerical validation of end-to-side anastomosis. J Mech Med Biol 10 (1):123–138
12. Mahnken AH, Spuentrup E, Niethammer M, Buecker A, Boese J, Wildberger JE, Flohr T, Sinha AM, Krombach GA, Gunther RW (2003) Quantitative and qualitative assessment of left ventricular volume with ECG-gated multislice spiral CT: value of different image reconstruction algorithms in comparison to MRI. Acta Radiol 44(6):604–611
13. Tezduyar TE, Sathe S, Cragin T, Nanna B, Conklin BS, Pausewang J, Schwaab M (2007) Modelling of fluid–structure interactions with the space-time finite elements: arterial fluid mechanics. Int J Num Meth Fluid 54(6–8):901–922

14. Freshwater IJ, Morsi YS, Lai T (2006) The effect of angle on wall shear stresses in a LIMA to LAD anastomosis: numerical modelling of pulsatile flow. Proc Inst Mech Eng H J Eng Med 220(7):743–757
15. Morsi YS, Yang WW, Wong CS, Das S (2007) Transient fluid–structure coupling for simulation of a trileaflet heart valve using weak coupling. J Artif Organ 10(2):96–103
16. Figueroa CA, Vignon-Clementel IE, Jansen KE, Hughes TJR, Taylor CA (2006) A coupled momentum method for modeling blood flow in three-dimensional deformable arteries. Comput Meth Appl Mech Eng 195 (41–43):5685–5706
17. Do H (2011) Fluid structure interaction numerical simulation of coronary artery bypass graft, end-to-side configurations. PhD Thesis, Swinburne University of Technology, Swinburne
18. White SS, Zarins CK, Giddens DP, Bassiouny H, Loth F, Jones SA, Glagov S (1993) Hemodynamic patterns in two models of end-to-side vascular graft anastomoses: effects of pulsatility, flow division, Reynolds number, and hood length. J Biomech Eng 115(1):104–111
19. Keynton RS, Rittgers SE, Shu MC (1991) The effect of angle and flow rate upon hemodynamics in distal vascular graft anastomoses: an in vitro model study. J Biomech Eng 113(4):458–463
20. Kim WY, Walker PG, Pedersen EM, Poulsen JK, Oyre S, Houlind K, Yoganathan AP (1995) Left ventricular blood flow patterns in normal subjects: a quantitative analysis by three-dimensional magnetic resonance velocity mapping. J Am Coll Cardiol 26(1):224–238
21. Gao ZB, Pandya S, Hosein N, Sacks MS, Hwang NHC (2000) Bioprosthetic heart valve leaflet motion monitored by dual camera stereo photogrammetry. J Biomech 33 (2):199–207
22. Owida AM (2009) In vitro study of haemodynamic stresses and endothelialisation of artificial coronary arteries. PhD Thesis, Swinburne University of Technology, Swinburne
23. Black J, Hastings G (1998) Handbook of biomaterial properties. Chapman & Hall, London
24. Bordenave L, Fernandez P, Remy-Zolghadri M, Villars S, Daculsi R, Midy D (2005) In vitro endothelialized ePTFE prostheses: clinical update 20 years after the first realization. Clin Hemorheol Microcirc 33(3):227–234
25. Fink H (2009) Artificial blood vessels, studies on endothelial cell and blood interactions with bacterial cellulose. Doctor of Philosophy (Medicine), University of Gothenburg, Sahlgrenska Academy, Gothenburg
26. Gourd C, Wootton J, Uber B, Loening K, Weiss S (2008) Coronary angiogenesis: can it mend broken hearts? http://biomed.brown.edu/Courses/BI108/BI108_2003_Groups/Coronary_Angiogenesis/default.html
27. Xue L, Greisler H (2003) Biomaterials in the development and future of vascular grafts. J Vasc Surg 37(2):472–480
28. Walpoth BH, Rogulenko R, Tikhvinskaia E, Gogolewski S, Schaffner T, Hess OM, Althaus U (1998) Improvement of patency rate in heparin-coated small synthetic vascular grafts. Circulation 98(19 Suppl):II319–II323; discussion II324
29. Bos GW, Poot AA, Beugeling T, van Aken WG, Feijen J (1998) Small-diameter vascular graft prostheses: current status. Arch Physiol Biochem 106(2):100–115
30. Greisler HP, Cziperle DJ, Kim DU, Garfield JD, Petsikas D, Murchan PM, Applegren EO, Drohan W, Burgess WH (1992) Enhanced endothelialization of expanded polytetrafluoroethylene grafts by fibroblast growth factor type 1 pretreatment. Surgery 112(2):244–254; discussion 254–255
31. Gray JL, Kang SS, Zenni GC, Kim DU, Kim PI, Burgess WH, Drohan W, Winkles JA, Haudenschild CC, Greisler HP (1994) FGF-1 affixation stimulates ePTFE endothelialization without intimal hyperplasia. J Surg Res 57(5):596–612
32. Zahedmanesh H, MacKle JN, Sellborn A, Drotz K, Bodin A, Gatenholm P, Lally C (2011) Bacterial cellulose as a potential vascular graft: Mechanical characterization and constitutive model development. J Biomed Mater Res B Appl Biomater 97B(1):105–113
33. Greisler HP (1982) Arterial regeneration over absorbable prostheses. Arch Surg 117(11): 1425–1431
34. Greisler HP, Kim DU, Price JB, Voorhees AB Jr (1985) Arterial regenerative activity after prosthetic implantation. Arch Surg 120(3): 315–323
35. Pachence J, Kohn J (2000) Biodegradable polymers. In: Lanza R, Vacanti J (ed) Principles of tissue engineering. Academic Press, San Diego, CA
36. Morsi Y (2011) Tissue engineering of heart valve current and future directions. Nova Science Publishers (in press)
37. Greisler HP, Ellinger J, Schwarcz TH, Golan J, Raymond RM, Kim DU (1987) Arterial regeneration over polydioxanone prostheses in the rabbit. Arch Surg 122(6):715–721
38. Leong KF, Cheah CM, Chua CK (2003) Solid freeform fabrication of three-dimensional scaffolds for engineering replacement tissues and organs. Biomaterials 24(13):2363–2378
39. Morsi YS, Wong CS, Patel SS (2008) Virtual prototyping of biomanufacturing in medical

applications conventional manufacturing processes for three-dimensional scaffolds. In: Bidanda B, Bartolo P (eds) Virtual prototyping and bio manufacturing in medical applications. Springer, New York, 129–148

40. Yeong W-Y, Chua C-K, Leong K-F, Chandrasekaran M (2004) Rapid prototyping in tissue engineering: challenges and potential. Trends Biotechnol 22(12):643–652
41. Masood SH, Singh JP, Morsi Y (2005) The design and manufacturing of porous scaffolds for tissue engineering using rapid prototyping. Int J Adv Manuf Technol 27(3–4):415–420
42. Ramakrishna RMS (2006) Nano-featured scaffolds for tissue engineering: a review of spinning methodologies. Tissue Eng 12(3): 435–447
43. Cui W, Zhou Y, Chang J (2010) Electrospun nanofibrous materials for tissue engineering and drug delivery. Sci Technol Adv Mater 11 (1):014108
44. He W, Ma Z, Wee ET, Yi XD, Robless PA, Thiam CL, Ramakrishna S (2009) Tubular nanofiber scaffolds for tissue engineered small-diameter vascular grafts. J Biomed Mater Res A 90(1):205–216
45. Chen Z, Mo X, Qing F (2007) Electrospinning of collagen-chitosan complex. Mater Lett 61 (16):3490–3494
46. Mo X, Chen Z, Weber HJ (2007) Electrospun nanofibers of collagen-chitosan and P(LLA-CL) for tissue engineering. Front Mater Sci China 1(1):20–23
47. Venugopal J, Ramakrishna S (2005) Applications of polymer nanofibers in biomedicine and biotechnology. Appl Biochem Biotechnol 125 (3):147–158
48. Mo XM, Xu CY, Kotaki M, Ramakrishna S (2004) Electrospun P(LLA-CL) nanofiber: a biomimetic extracellular matrix for smooth muscle cell and endothelial cell proliferation. Biomaterials 25(10):1883–1890
49. Stevens MM, George JH (2005) Exploring and engineering the cell surface interface. Science 310(5751):1135–1138
50. Centola M, Rainer A, Spadaccio C, De Porcellinis S, Genovese JA, Trombetta M (2010) Combining electrospinning and fused deposition modeling for the fabrication of a hybrid vascular graft. Biofabrication 2(1):014102
51. Owida A, Sheetal Patel S, Morsi Y, Mo X (2011) Artery vessel fabrication using the combined fused deposition modeling and electrospinning techniques. Rapid Prototyping J 17 (1):37–44
52. Ayres C, Bowlin GL, Henderson SC, Taylor L, Shultz J, Alexander J, Telemeco TA, Simpson DG (2006) Modulation of anisotropy in electrospun tissue-engineering scaffolds: analysis of fiber alignment by the fast Fourier transform. Biomaterials 27(32):5524–5534
53. Drury JL, David DJ (2003) Hydrogels for tissue engineering: scaffold design variables and applications. Biomaterials 24:4337
54. Dhariwala B, Hunt E, Boland T (2004) Rapid prototyping of tissue-engineering constructs, using photopolymerizable hydrogels and stereolithography. Tissue Eng 10(9–10): 1316–1322
55. Xiao H, Yue Z, Henry JD, Tao LL (2007) Porous thermoresponsive-co-biodegradable hydrogels as tissue-engineering scaffolds for 3-dimensional in vitroculture of chondrocytes. Tissue Eng 13:2645–2652
56. Liao E, Yaszemski M, Krebsbach P, Hollister S (2007) Tissue-engineered cartilage constructs using composite hyaluronic acidcollagen i hydrogels and designed poly(propylene fumarate) scaffolds. Tissue Eng 13(3):537–550
57. Rotmans JI, Heyligers JMM, Stroes ESG, Pasterkamp G (2006) Endothelial progenitor cell-seeded grafts: rash and risky. Can J Cardiol 22 (13):1113–1116
58. Herring M, Gardner A, Glover J (1984) Seeding human arterial prostheses with mechanically derived endothelium. The detrimental effect of smoking. J Vasc Surg 1(2):279–289
59. Rosenman JE, Kempczinski RF, Pearce WH, Silberstein EB (1985) Kinetics of endothelial cell seeding. J Vasc Surg 2(6):778–784
60. Hashi CK, Derugin N, Janairo RRR, Lee R, Schultz D, Lotz J, Li S (2010) Antithrombogenic modification of small-diameter microfibrous vascular grafts. Arterioscler Thromb Vasc Biol 30(8):1621–1627
61. Heyligers JM, Arts CH, Verhagen HJ, de Groot PG, Moll FL (2005) Improving small-diameter vascular grafts: from the application of an endothelial cell lining to the construction of a tissue-engineered blood vessel. Ann Vasc Surg 19(3):448–456
62. Matsuda T (2004) Recent progress of vascular graft engineering in Japan. Int Soc Artif Organ 28(1):8
63. Shin'oka T, Imai Y, Ikada Y (2001) Transplantation of a tissue-engineered pulmonary artery. N Engl J Med 344(7):532–533
64. Weinberg CB, Bell E (1986) A blood vessel model constructed from collagen and cultured vascular cells. Science 231(4736):397–400
65. L'Heureux N, Paquet S, Labbe R, Germain L, Auger FA (1998) A completely biological tissue-engineered human blood vessel. FASEB J 12(1):47–56
66. Niklason LE, Gao J, Abbott WM, Hirschi KK, Houser S, Marini R, Langer R (1999) Functional arteries grown in vitro. Science 284 (5413):489–493

Chapter 17

Projection Printing of 3-Dimensional Tissue Scaffolds

Yi Lu and Shaochen Chen

Abstract

Our ability to create precise, predesigned, spatially patterned biochemical and physical microenvironments inside polymer scaffolds could provide a powerful tool in studying progenitor cell behavior and differentiation under biomimetic, three-dimensional (3D) culture conditions. The development of freeform fabrication technology has become a promising tool for the manufacturing of biological scaffolds for tissue regeneration and stem cell engineering. Freeform fabrication is a very promising technology due to the efficient and simple process for creating bona fide 3D microstructures, such as closed channels and cavities. It is also capable of encapsulating biomolecules and even living cells.

This chapter describes direct projection printing of 3D tissue engineering scaffolds by using a digital micromirror-array device (DMD) in a layer-by-layer process. This simple and fast microstereolithography system consists of an ultraviolet (UV) light source, a digital micromirror masking device, imaging optics, and controlling devices. Images of UV light are projected onto the photocurable resin by creating the "dynamic photomask" design with graphic software. Multilayered scaffolds are microfabricated through a photopolymerization process.

Key words: Projection printing, Scaffolds, DMD, Microstereolithography, Dynamic photomask, Microfabrication

1. Introduction

3D constructs that incorporate complex spatial patterning of extracellular matrix (ECM) components and growth factors could provide biomimetic complex microenvironments for studying cell behavior and differentiation (1–3). Most 3D scaffolding systems are only capable of differentiating a single progenitor cell population into one particular cell lineage due to either (a) bulk incorporation of biofactors within the scaffolding matrix (4) or (b) exogenous delivery of hormones, chemicals, or growth factors in culture medium (5–7). From a tissue engineering perspective, a significant advancement could be attained by creating precise, spatially distributed microenvironments within a single scaffold that

Michael A.K. Liebschner (ed.), *Computer-Aided Tissue Engineering*, Methods in Molecular Biology, vol. 868,
DOI 10.1007/978-1-61779-764-4_17, © Springer Science+Business Media, LLC 2012

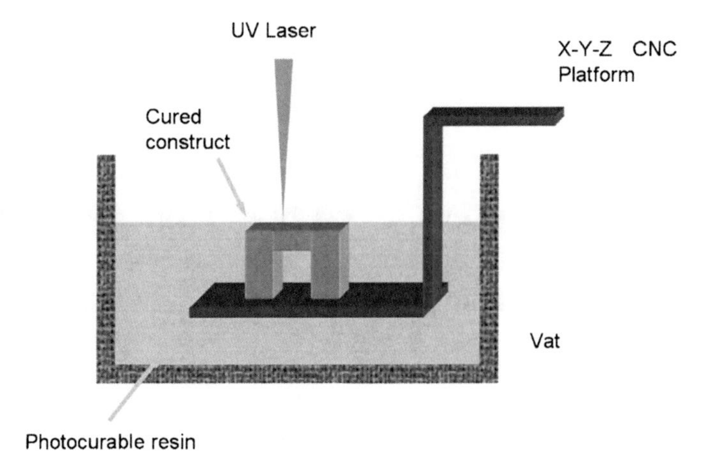

Fig. 1. Schematic of a laser stereolithography system.

would allow us to study simultaneous, patterned differentiation of stem and progenitor cells into multiple lineages and develop concepts to ultimately engineer complex, hybrid organ structures. A key step toward achieving such patterned 3D structures is the development of novel scaffold-manufacturing techniques by which distributed environments can be incorporated in a simple, yet precise, reproducible fashion.

Laser microstereolithography (μSL) has become an accepted rapid prototyping method that allows the 3D microfabrication of solid models from images created by computer-aided design (CAD) programs (Fig. 1) (8–11). In this method, a motorized *x–y–z* platform immersed in a liquid photopolymer is selectively exposed to a focused ultraviolet (UV) laser light. The polymer cures and becomes solid only at the focal point, whereas nonirradiated areas remain liquid. After the first layer is formed, the platform moves downward and a new layer of polymer is solidified according to the design. This layer-by-layer micromanufacturing method enables complex internal features, such as patterning of growth factors and ECM proteins.

Based on this principle, various systems were designed for different applications. For instance, one can use galvo mirrors, which scan a laser beam much faster than the motorized platform can move, to improve throughput. Yet this process is usually slow because of the nature of point-by-point laser scanning. This prevents the incorporation of cells within the scaffold walls during the fabrication process and could also lead to denaturation and inactivation of biological molecules during the prolonged fabrication period.

Rather than writing the 3D microstructure point-by-point, the DMD-based system allows for simultaneous photopolymerization of partial and entire layers of a scaffold (12). This novel scaffold fabrication technique enables precise predesigned patterning of multiple molecules and allows generation of complex architectures in a high-throughput layer-by-layer fashion. We describe that, by changing biofactors or controlled-released particles within the curable resin, each layer or even partial layers can be made up of a variety of controlled-release microparticles, thereby creating spatially distributed environments with micron-size resolution. In addition, precise and complex internal architectures, for example pore size and shape, were created using DMD. We also demonstrate the technique's capability of efficiently incorporating cells inside the scaffold walls during fabrication or seeding cells on the scaffolds following covalent modification of the surface with fibronectin. Murine marrow-derived stromal cells seeded in these DMD μSL-fabricated patterned scaffolds efficiently differentiated into osteoblasts and produced scaffold mineralization, thereby demonstrating the ability of such structures to support cell proliferation and differentiation.

2. Materials

Macromer solutions used to fabricate hydrogel scaffolds were formulated using 100% (w/v) poly(ethylene glycol) diacrylate (PEGDA, Mw 3400, Nektar Therapeutics, AL) dissolved in phosphate-buffered saline (PBS). Since PEGDA is already a macromer or polymer, it may seem confusing as "polymerization" occurs in the context, by which we mean a further polymerization or sometimes cross-linking process. "Polymer solution" and "resin" may be used interchangeably. To induce chain polymerization through the generation of free radicals, a UV photoinitiator, 2-hydroxy-1-[4 (hydroxyethoxy)phenyl]-2-methyl-1-propanone (Irgacure 2959, Ciba Geigy, USA), was used at a concentration of 0.1 wt%. The photoinitiator was first dissolved in PBS at a concentration of 0.7 wt% to ensure complete solubility before adding to the PEGDA solution. Prepared mixture was kept in a dark environment to prevent pre-cross-linking of the polymer by incidental exposure to ambient light. Fluorescently labeled polystyrene microparticles (1.0 μm Cy5 labeled and 1.0 μm FITC labeled) were obtained from Molecular Probes, Eugene, OR.

For higher spatial resolution scaffolds, we used perfluorohexane (99.5%) and PEGDA (Mn 258) from Sigma-Aldrich as received. Methacrylic acid (MAA) from Sigma-Aldrich was distilled before being used to remove inhibitors. UV absorber, Tinuvin 234, was provided by Ciba Chemistry and was used without further purification. To prepare the curable solution, 10 wt% of Irgacure 2959 and 0.2 wt% of Tinuvin 234 were added into a 4:1 (volume ratio) mixture of PEGDA and MAA. The monomer solution was sonicated for 30 min and degassed for 15 min.

3. Methods

3.1. DMD μSL System

The first-generation μSL system was developed based on a commercial projector (PB2120, BenQ, Taiwan). Shown in Fig. 2, the system consists of five major components: a DMD chip embedded in the projector as a dynamic mask, a light source, a projection lens assembly, a translation stage with a micrometer, and a vat containing macromer solution. All the components cooperate to ensure correct exposure, resolution, and layer thickness. To ensure cell viability through the use of a biocompatible UV photoinitiator, the original high-intensity white light source was replaced with a UV light source (Green Spot, UV Source, CA). The light was guided through a 1/4 in. (6.35 mm) liquid-filled fiber optics. Two biconvex lenses (18 mm diameter, 40 mm focal length) with 5-mm spacing were used to converge the light emanating from the fiber optics. The projection lens assembly with adjustable aperture and

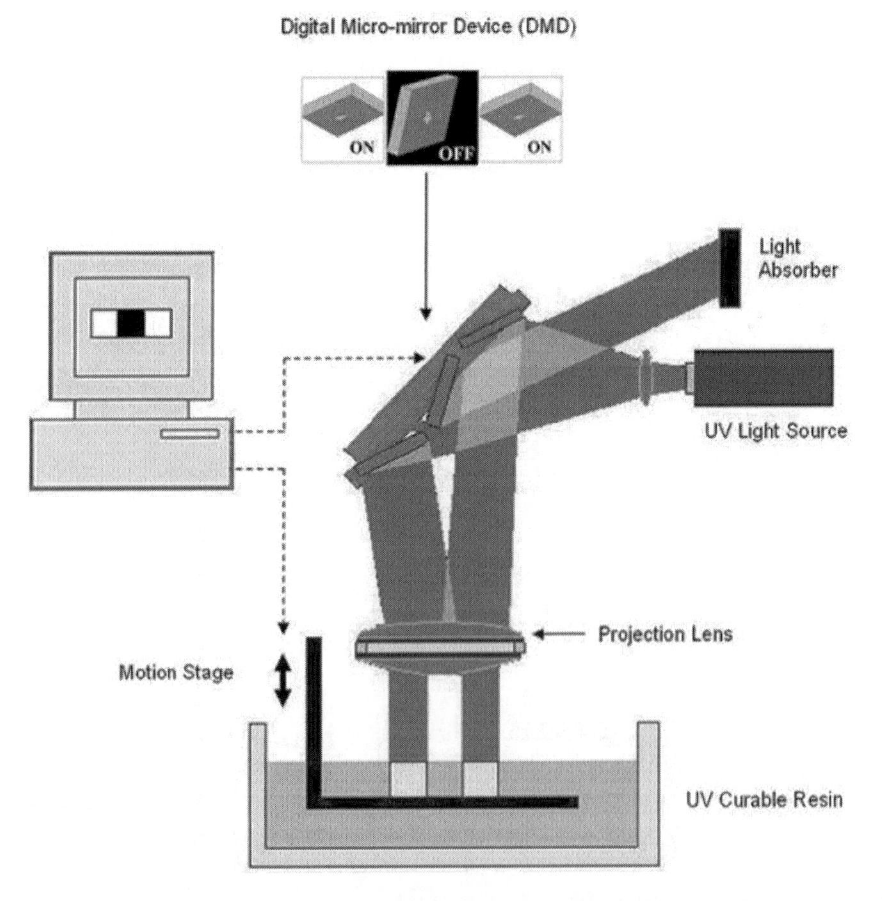

Fig. 2. Schematic of the digital micromirror device microstereolithography (DMD μSL) setup (this figure is reproduced with permission of John Wiley & Sons, Inc.).

focus consisted of two equal plano-convex lenses (25 mm diameter, 25 mm focal length). Each lens was oriented with the convex surface toward the longer conjugate distance. The aperture was placed in-between two lenses. All lenses were made of UV grade fused silica (Edmund Optics, Barrington, NJ). The average exposure intensity was determined to be 2 mW/cm^2.

Throughout the experiments, the magnification (size of the scaffold/size of the pattern) was fixed at 1/90. The working principle of the DMD chip is also detailed in Fig. 2. The DMD chip serves as an array of reflective aluminum micromirrors, which can be tilted with two bias electrodes to form angles of either +10° or −10° with respect to the surface. Illumination from the light source reflects into the projection lens only when the micromirror is in its +10° state. In the −10° state, the pixel appears dark because the illuminated light is not reflected into the projection lens. The reflected light from the +10° micromirror is collected by a light absorber. When the micromirror is in +10° state, it is classified as "tilt on" or ON. Conversely, when the micromirror is in −10° state, it is classified as "tilt off" or OFF. Though the schematic depicts only three micromirrors for illustrative purposes, the actual DMD chip contains more than 442,000 switchable mirrors on a 5/8-in. (15.875 mm)-wide surface. For instance, there are three pixels of white, black, white color, respectively, on the screen. The DMD chip is signaled by the computer to tilt the first micromirror to +10° state, the second micromirror to −10° state, and the third micromirror to +10° state. The first and third micromirror, which is in ON condition, reflects the illuminated light to the projection lens. Subsequently, the light reflected by the mirror into the projection lens is directed to the resin. Conversely, the second mirror, which is in OFF condition, reflects the illuminated light to the light absorber. Bertsch and colleagues previously reported a μSL process employing a liquid crystal display (LCD) as a dynamic mask and demonstrated fabrication of small mechanical parts, such as a turbine and spring (13, 14). Itoga and colleagues have also explored LCD projectors to study two-dimensional (2D) cellular behavior through the micropatterning of noncytoadhesive polymers onto plasma-treated glass surfaces (15). However, LCD as a dynamic mask has limited optical efficiency (16). The DMD offers better performance in terms of optical fill factor (85% with DMD vs. 64% with LCD) and light transmission (71% with DMD vs. 21% with LCD).

Similar to a conventional stereolithography process, the shapes of the constructed layers were determined by slicing the desired 3D scaffold design into a series of evenly spaced planes. Patterns of each layer were drawn in a series of PowerPoint slides, which were then executed on the DMD chip to generate a dynamic mask. The illuminated light is modulated according to the defined mask on the DMD chip and then goes through a reduction-projection lens assembly to form an image on the surface of the resin. The illuminated area was solidified simultaneously under one exposure while the dark regions remained in the liquid phase. After one layer was patterned, the substrate was lowered and the as-patterned layer was then covered by fresh resin. Microstructures with complex geometries were created by sequentially polymerizing the layers.

3.2. Scaffold Fabrication

The DMD μSL method was used to create polymer scaffolds with pores and channels having a wide variety of shapes and dimensions. The configuration of the scaffold pores was dictated simply by altering the "mask" drawn on a PowerPoint slide, thus illustrating the powerful capability of this system to design features of any shape or form. As shown in the scanning electron microscopic (SEM) micrographs of Fig. 3a–d, different pore geometries (hexagons, triangles, honeycombs with triangles, and squares) can be included within a single scaffold (pore size dimensions range from ~165 to 650 μm, scale bars shown). Precise internal features of the scaffolds were fabricated with one single 90-s exposure to the UV light of the DMD μSL. Additionally, controlled internal architectures can be generated in parallel. Figure 3d shows scaffolds fabricated in a multilayered fashion. The pore dimensions used in this scaffold are 250 μm by 250 μm with a measured wall thickness of

Fig. 3. Scanning electron microscopy (SEM) illustrates that DMD-μSL can create scaffolds with intricate pore geometries. *Hexagons* (honeycomb), *triangles*, *triangles inside hexagons*, and *square-shaped pores* were created by directly drawing in PowerPoint files and using the DMD as a dynamic "mask." Geometrical side dimensions of the pores range from approximately 165 to 650 μm (scale bars indicate 500 μm). Scaffolds depicted in (**d**) specifically show a two-layered scaffold. All scaffolds were irradiated for 90 s per layer and formulated using 100% (w/v) PEGDA in PBS and 0.1 wt% Irgacure 2959 (this figure is reproduced with permission of John Wiley & Sons, Inc.).

100 μm. Construct edges appeared to be slightly rounded due to swelling of the hydrogel structure. Keeping the light intensity (2 mW/cm^2) and exposure time constant, the smallest feature size attainable with DMD μSL was measured to be approximately 20 μm. This system is primarily limited by optical diffraction and diffusion of free radicals in the polymer solutions (see Notes).

The DMD μSL system can also produce multilayered scaffolds with predesigned spatially patterned molecules and particles. The feasibility of such precise spatial patterning was demonstrated using PEGDA solutions containing either Cy5- or FITC-labeled polystyrene particles that were encapsulated in a predesigned pattern during the polymerization process. As shown in Fig. 4a, resin containing different particles can be patterned in a quadrant-specific geometry, in which the resin with Cy5 particles was polymerized in the upper left and lower right regions and the resin with FITC particles was polymerized in the upper right and lower left regions. This figure demonstrates the ability of the DMD μSL system to pattern multiple agents within a single layer through sequential steps of polymerization and rinsing of unpolymerized resin.

We also demonstrated spatial patterning in multilayered scaffolds, as shown in Fig. 4b, c, by creating constructs that specifically consisted of two layers, each containing either Cy5- or FITC-labeled particles. The bottom layer was pattern polymerized with a single 90-s exposure using Cy5 particle–polymer solution, and then rinsed extensively to remove unpolymerized polymer and particles. The second layer, containing FITC particle–polymer solution, was then polymerized in the same method on top of the first layer using the same patterning mask.

Since PEG polymers have hydrophilic and nonionic properties, the scaffold surface must be modified to mediate efficient cell seeding. Covalent conjugation of fibronectin, an ECM component that signals for cell anchorage and spreading, to scaffolds can be

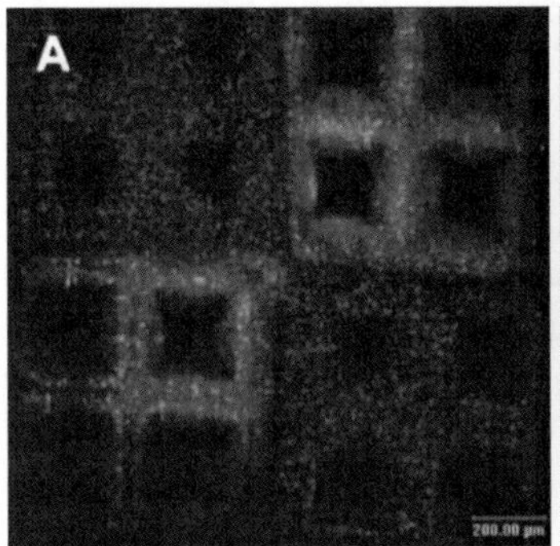

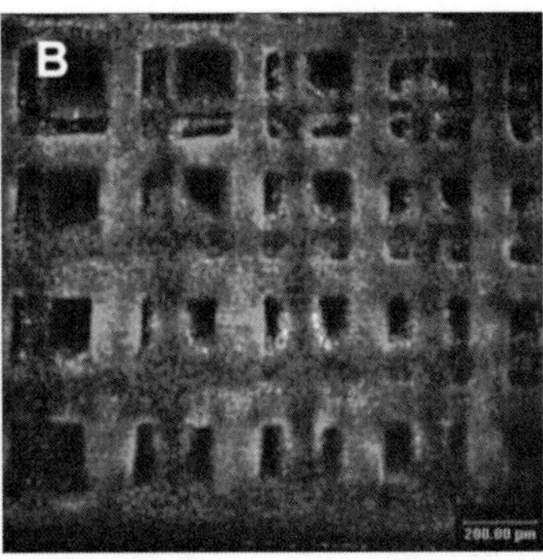

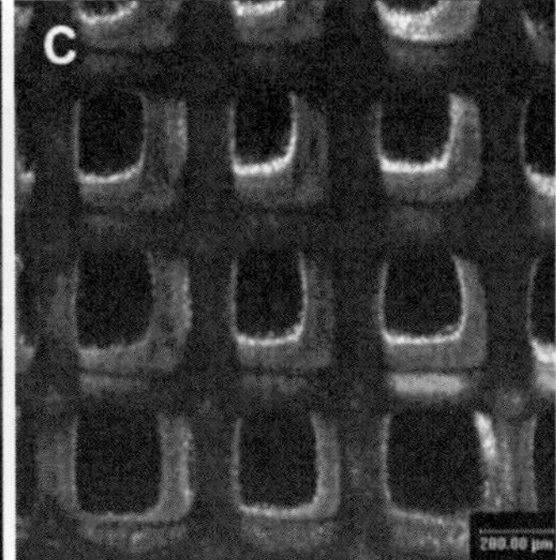

Fig. 4. DMD-μSL can create predesigned spatially patterning inside scaffold structures. Fluorescence confocal microscopy of scaffolds formulated with 100% (w/v) PEGDA in PBS, 0.1 wt% Irgacure 2959, and 0.03 wt% carrying either FITC- or Cy5-labeled polystyrene particles. (**a**) Spatial patterning of a single layer in a "quadrant" specific pattern. (**b**, **c**) Spatial patterning in multilayered scaffolds (this figure is reproduced with permission of John Wiley & Sons, Inc.).

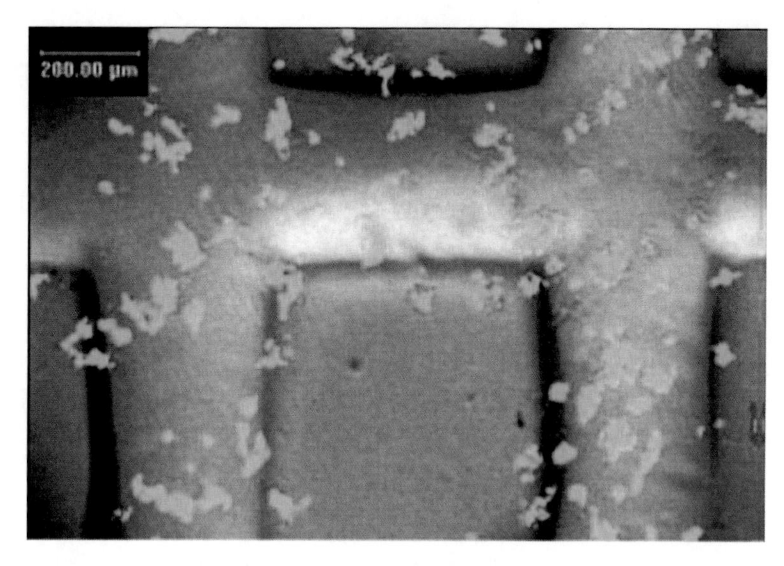

Fig. 5. Marrow-derived stromal cells remain viable following encapsulation in DMD-µSL-fabricated scaffolds (this figure is reproduced with permission of John Wiley & Sons, Inc.).

obtained via EDC/Sulfo-NHS chemistries. Methods of preparing bioreagents and cells are not elucidated in this chapter but can be readily found in refs 12, 17.

Patterned encapsulation of cells within the scaffold walls was achieved by the addition of OP-9 cells to the resin prior to DMD µSL UV irradiation. Figure 5 shows a fluorescence micrograph of cells overlaid onto a transmitted micrograph of the scaffold, and demonstrates the viability of encapsulated cells within a single-layered scaffold (~150-µm-thick layer). Encapsulation efficiency is dependent on scaffold channel or pore dimensions, the volume of cell resin used, and the presence of a container to hold the cell resin during the photopolymerization process to prevent solution spreading.

These constructs were analyzed using either a confocal microscope for fluorescence imaging (Leica SP2 AOBS) or by SEM (Phillips 515 SEM) for multilayered scaffolds.

The second-generation system was built later based on the same principle but with enhanced resolution and automation (18). The commercial projector was replaced by a Discovery 1100 kit from Texas Instruments. It is essentially a DMD chip mounted on its control circuit board, which freed real estate formally occupied by parts not particularly helpful to the application. Another important upgrade was the use of perfluorohexane (C_6F_{14}), an inert liquid with high density (1.685 g/cm^3) and a molecular polarity significantly lower than that of the resin. Because perfluorohexane is heavier and immiscible with the resin, the resin floats on top of perfluorohexane. It brings a few key advantages to the fabrication. Firstly, instead of filling the entire vat with resin, which often contains expensive bioreagents, it is sufficient to maintain a thin layer of resin throughout the fabrication as the perfluorohexane acts as a "supporter." In the meanwhile, it makes easier to control thickness of each layer. Secondly, low viscosity of perfluorohexane allows it to infiltrate cured layers and push out uncured resin. It is particularly useful in reducing cross contamination when switching between two or more types of resin.

Let us look at how it works. Resin and perfluorohexane are initially stored in multiple syringes, which are mounted on individual automated syringe pumps. Perfluorohexane is injected to fill the vat all but to the very top, leaving space for the resin. The first layer is cured, the computer-controlled motorized stage submerges into perfluorohexane, and thus the cured layer is cleaned. The stage then elevates to a position such that it leaves a space equal to the thickness of the second layer. Resin and perfluorohexane are refilled or withdrawn if needed. This cycle is repeated till finished.

Typical UV fluence was approximately 5 mW/cm^2. Exposure time was set for 60 s. As shown in the inset of Fig. 6, we constructed a scaffold by alternating three different cross-sectional images (1–3). The images were used in a sequence: 1-1-1-2-1-1-1-3-1-1-1-2-1-1-1-3-1. The layered structure of the scaffold includes four honeycomb

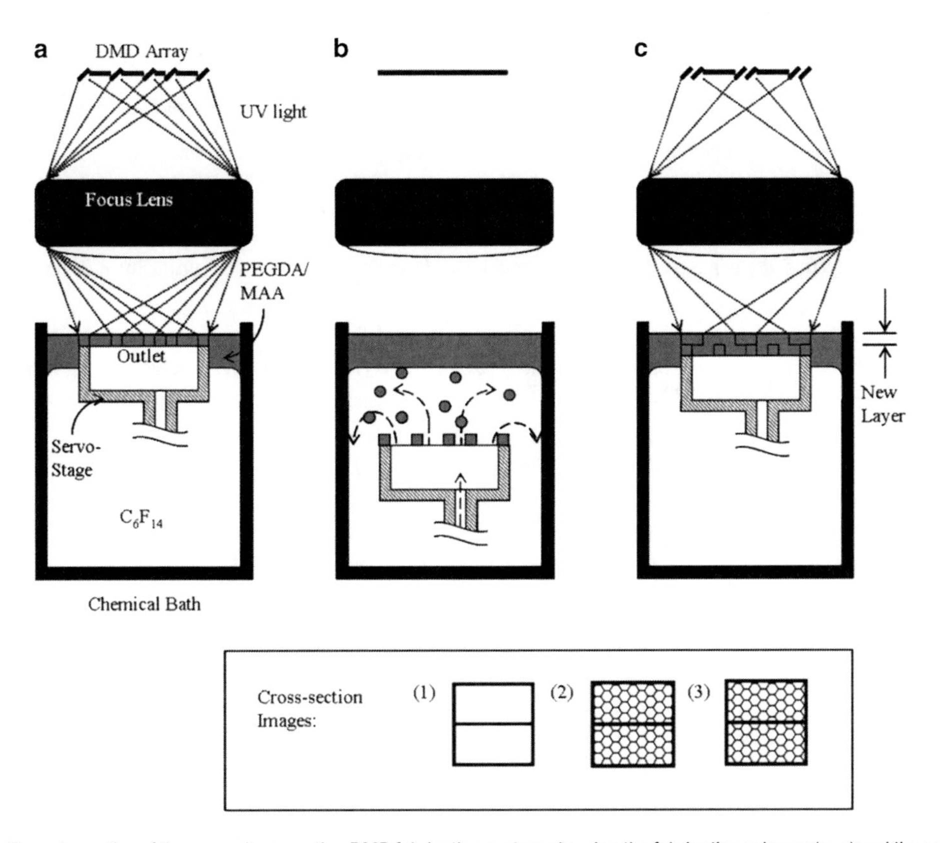

Fig. 6. The schematics of the second-generation DMD fabrication system showing the fabrication scheme (**a**–**c**) and the patterns of the scaffold cross section (*inset*) (this figure is reproduced with permission of American Society of Mechanical Engineering).

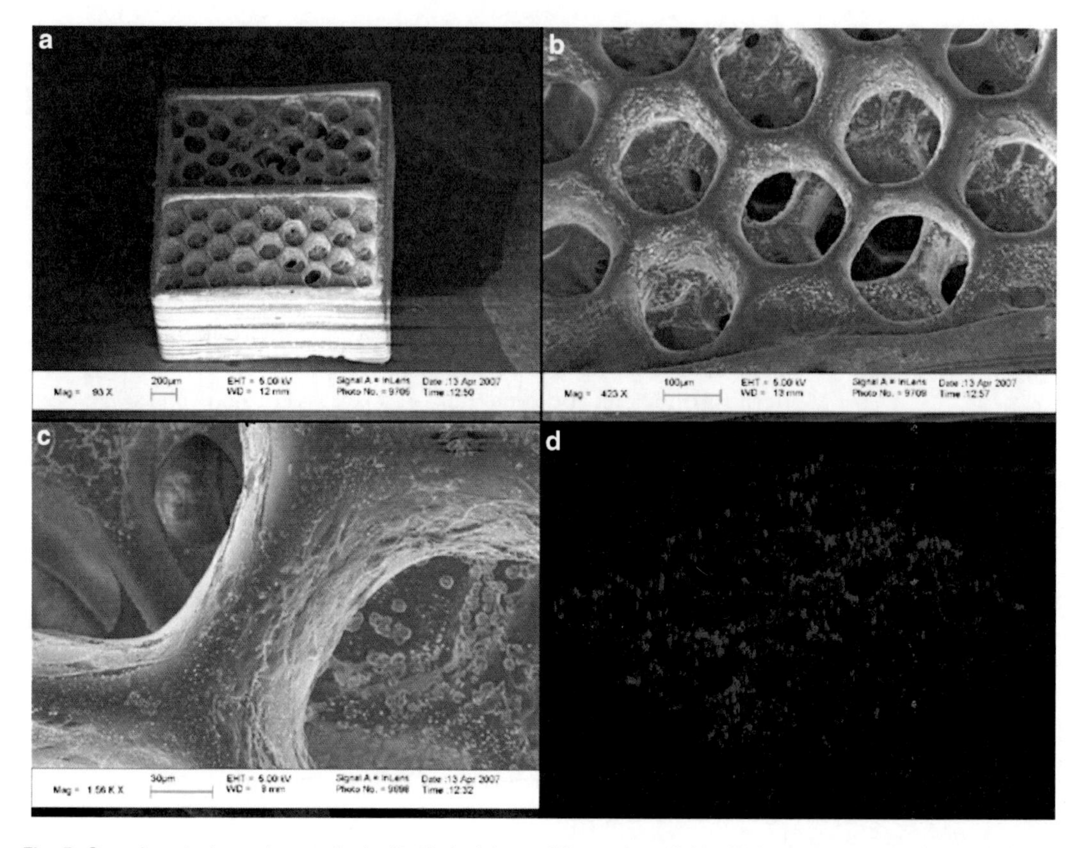

Fig. 7. Scanning electron micrographs in (**A–C**) depict a multilayered scaffold with interconnective, hexagonal-shaped porosity at increasing manification. An extracellular matrix secreted by attached D1 cells is visible in a close-up view (**C**). D1 cells were stained with a fluorescent tracer dye prior to seeding onto fibronectin-modified scaffolds. (**D**) shows a reconstructed 3D image from fluorescence confocal microscopy, indicating that cells attached efficiently to the microfabricated scaffold. (This figure is reproduced with permission of American Society of Mechanical Engineering).

layers, neighboring honeycomb layers are spaced by three rectangular rims, and the space between two honeycomb layers was measured to be 150 μm. By using UV absorber, Tinuvin 234, we were able to reduce the depth of curing to about 50 μm (see Notes).

Figure 7 micrographs show that D1 cells attach and secrete ECM onto the surfaces of the fibronectin-modified scaffold. The scaffold depicted here is composed of four layers with a wall thickness of 50 μm and hexagonal pore geometry of approximately 150 μm. See ref. 18 for cell preparation details.

4. Notes

Laser stereolithography is well established and widely employed in prototyping in a board range of industries. The capabilities of its optical and mechanical systems have been improved throughout the past few decades to meet ever-ambitious demand. For instance, multiphoton-initiated polymerization has enabled making of sub-micron 3D structures, with the critical dimension as small as 50 nm. It is not uncommon that the capability of the system exceeds that of the resin. In other words, characteristics, such as the speed of curing and the critical dimension, are often limited by the resin. It is more so in making tissue engineering scaffolds, which directly interact with cells and their biological environment. The resins of choice need to bear mechanical, chemical, and biological properties to accommodate different demands. Trade-offs may need to be made in selecting curable material, curing agent, light absorber, radical scavenger (reagent that terminates free radical), filler, bioreagent, and solvent. For example, it would be difficult to make the aforementioned PEGDA scaffolds as hard as bone structures.

Biocompatibility of photopolymers and photoinitiators has been extensively studied in the past (19). It was found that acrylates with urethane units, most dialkylacrylamide and especially trimethylolpropane triacrylate, gave outstanding biocompatibility. However, there are much fewer commercially available photoinitiators that exhibit good biocompatibility. Even they do, they are often limited at low concentration, mostly because free radicals generated from initiator are highly reactive. For instance, Irgacure 2959 is a low cytotoxic initiator, but a concentration $< 0.1\%$ (w/v) is recommended. Insufficient initiator can result in two challenges: (1) low cross-linking density and hence low mechanical strength and excessive swelling; (2) depth of curing is increased due to low absorbance. To compensate for low cross-linking density, multifunctional (tri-, tetra-, penta-) acrylates can be used as bulk material or additive. Materials that form large number of hydrogen bonds can also be stronger. It is noteworthy that some of the polymers have excellent mechanical properties in a dry state, but they may quickly lose strength due to swelling in aqueous media. To increase absorbance and reduce the depth of curing, an inert light absorber, even at low concentration, is often effective. Effect of absorber is further discussed in the context.

Some hydrogels are soft (i.e., PEGDA). They swell and deform in aqueous media. Additional supporting structure is recommended when making scaffolds that contain small features.

Oxygen is a strong radical inhibitor. Because the polymerization occurs at the top surface, oxygen is constantly replenished and terminates further polymerization. It is particularly important to shield reaction from oxygen when initiator concentration is low. Figure 8 shows that gelation time is significantly decreased in nitrogen environment. However, we found that if oxygen is not effectively removed (i.e., leaky enclosure), speed of curing could fluctuate significantly. We believe that it was because relative fluctuation of concentration of oxygen was greater than that in the ambient.

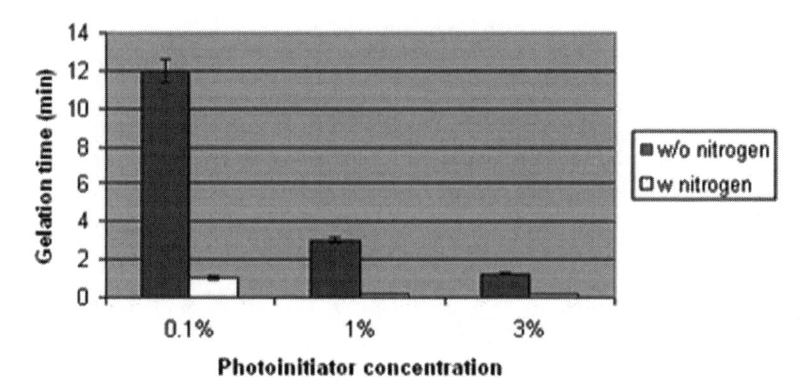

Fig. 8. Gelation time for resin containing 100% (w/v) PEG(1000)DMA with/without nitrogen protection at 80% gray scale.

Resolution of optical images is largely dependent upon the DMD chip and imaging optics. The size of a pixel in the DMD is about 10 μm. The imaging lens can further reduce pixel size, but limited by optical diffraction. Image resolution can be measured by viewing through an optical microscope or exposing a thin layer of photoresist. However, as previously mentioned, image resolution does not always determine critical dimension of a structure that can be made. Loss of feature resolution can be attributed to scattering, radical diffusion and propagation. We have observed lower patterning resolution when particles or cells are present. Additionally, radicals diffuse outside of the illuminated area and cause polymerization in the "dark" area. Light absorber is a powerful tool to improve resolution both laterally and vertically. The function of the light absorber can be understood through Beer's law (20). Assuming that a resin has an absorption constant α and a UV light intensity I_0, the UV intensity becomes I at depth z to give the following:

$$I = I_0 \exp(-\alpha \cdot z). \qquad (1)$$

Figure 9a shows the exponential decay of light intensity as z increases. I_{cure} is a threshold intensity at which the resin starts to cure. CD is the curing depth, $I(\text{CD}) = I_{\text{cure}}$. When the light absorber is added, α becomes greater (α') and CD becomes smaller (CD′); the addition of the light absorber to the resin increases the absorption coefficient α and decreases the curing depth, and the structural resolution is therefore enhanced. Figure 9b, c show the patterned polymerization of honeycomb-shaped structures with and without adding absorber. In the absence of absorber, the resin inside the pores was partially cured by undesired scattering or diffusion. With a small amount of absorber in the same resin, the structure was much better defined.

Another challenge encountered with the DMD system is the accumulation of oligomers within previously polymerized layers of

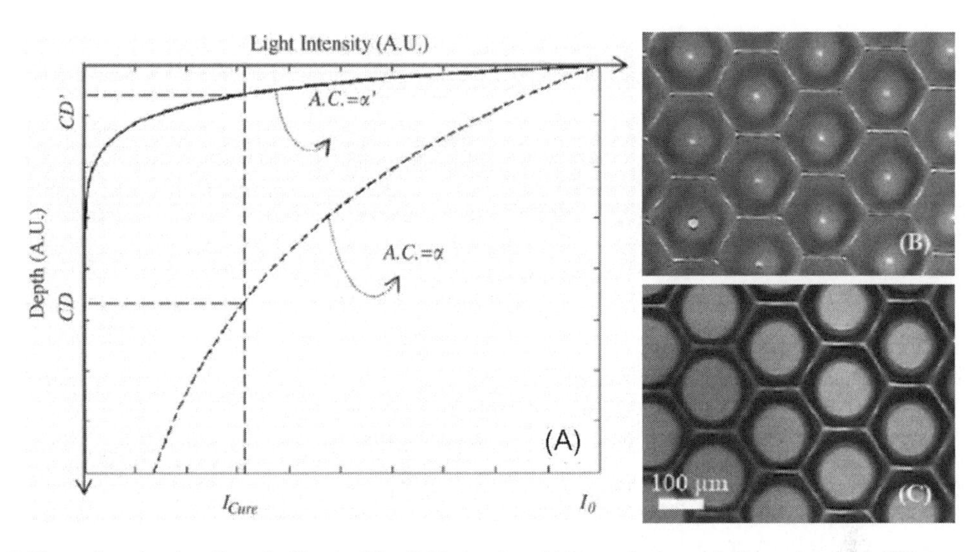

Fig. 9. The curing of resin with and without adding light absorber. (**a**) Theoretical model: The curing depth (CD) decreases when the absorption coefficient is increased after the addition of dye. (**b**) Photopatterning of PEGDA on a glass slide. The pores of a honeycomb structure are smeared by unwanted curing caused by scattered light. (**c**) Patterning the same monomer with the addition of Tinuvin 234 at a concentration of 0.2 wt%. The geometry of the honeycomb structure had improved feature resolution (this figure is reproduced with permission of American Society of Mechanical Engineering).

the multilayered construct. These oligomers are partially cured monomers, weakly cross-linked by unwanted scattered and penetrating UV light. To overcome this problem, it is necessary to purge the construct with fresh resin after each scaffold layer is made.

In experiments of cell encapsulation, we observed that cell tended to migrate to outside of the hydrogel scaffold. As a result, controlling cell density inside the scaffold became difficult. The cause is yet to be discovered. However, increasing viscosity by adding viscous additives (i.e., glycerol) may help solve the problem.

Acknowledgment

This work was supported by the National Institutes of Health (R01EB012597).

References

1. Griffith LG (2002) Emerging design principles in biomaterials and scaffolds for tissue engineering. Annu N Y Acad Sci 961:83–95
2. Orban JM, Marra KG, Hollinger JO (2002) Composition options for tissue-engineered bone. Tissue Eng 8:529–539
3. Sharma B, Elisseeff JH (2004) Engineering structurally organized cartilage and bone tissues. Ann Biomed Eng 32:148–159
4. Richardson TP, Peters MC, Ennett AB, Mooney DJ (2001) Polymeric system for dual growth factor delivery. Nat Biotechnol 19:1029–1034
5. Johnstone B, Hering TM, Caplan AI, Goldberg VM, Yoo JU (1998) In vitro chondrogenesis of bone marrow-derived mesenchymal progenitor cells. Exp Cell Res 238:265–272
6. Nuttelman CR, Tripodi MC, Anseth KS (2004) In vitro osteogenic differentiation of

human mesenchymal stem cells photoencapsulated in PEG hydrogels. J Biomed Mater Res A 68A:773–782

7. Williams CG, Kim TK, Taboas A, Malik A, Manson P, Elisseeff J (2003) In vitro chondrogenesis of bone marrow-derived mesenchymal stem cells in a photopolymerizing hydrogel. Tissue Eng 9:679–688
8. Maruo S, IK (1998) In: Proceeding of the 1998 international symposium on micromechatronics and human science, pp 115–120
9. Sachlos E, Czernuszka JT (2003) Making tissue engineering scaffolds work. Review: the application of solid free form fabrication technology to the production of tissue engineering scaffolds. Eur Cell Mater 5:29–39
10. Zhang X, Jiang XN, Sun C (1999) Micro-stereolithography of polymeric and ceramic microstructures. Sensor Actuator Phys 77: 149–156
11. Mapili G, Lu Y, Chen SC, Roy K (2005) Laser-layered microfabrication of spatially patterned functionalized tissue-engineering scaffolds. J Biomed Mater Res B Appl Biomater 75B:414–424
12. Lu Y, Mapili G, Suhali G, Chen SC, Roy K (2006) A digital micro-mirror device-based system for the microfabrication of complex, spatially patterned tissue engineering scaffolds. J Biomed Mater Res A 77A:396–405
13. Bertsch A, Lorenz H, Renaud P (1999) 3D microfabrication by combining microstereolithography and thick resist UV lithography. Sensor Actuator Phys 73:14–23
14. Bertsch A, Renaud P, Vogt C, Bernhard P (2000) Rapid prototyping of small size objects. Rapid Prototyp J 6:259–266
15. Itoga K, Yamato M, Kobayashi J, Kikuchi A, Okano T (2004) Cell micropatterning using photopolymerization with a liquid crystal device commercial projector. Biomaterials 25: 2047–2053
16. Sun C, Fang N, Wu DM, Zhang X (2005) Projection micro-stereolithography using digital micro-mirror dynamic mask. Sensor Actuator Phys 121:113–120
17. Kasturi SP, Sachaphibulkij K, Roy K (2005) Covalent conjugation of polyethyleneimine on biodegradable microparticles for delivery of plasmid DNA vaccines. Biomaterials 26:6375–6385
18. Han LH, Mapili G, Chen S, Roy K (2008) Projection microfabrication of three-dimensional scaffolds for tissue engineering. Trans ASME: J Manuf Sci Eng 130:021005-1–021005-4
19. Liu VA, Bhatia SN (2002) Three-dimensional photopatterning of hydrogels containing living cells. Biomed Microdevices 4:257–266
20. Brown B, Foote C, Iversion B (2005) Organic chemistry, 4th edn. Thomson Learning, Belmont, CA, 796

Chapter 18

Laser Sintering for the Fabrication of Tissue Engineering Scaffolds

Stefan Lohfeld and Peter E. McHugh

Abstract

Laser sintering (LS) utilises a laser to sinter powder particles. A volumetric model is sliced and processed cross section by cross section to create a physical part. In theory, all powdered materials are suitable for sintering; however, only few have been tested successfully. For tissue engineering (TE) applications of this rapid prototyping technology it is an advantage that no toxic solvents or binders are necessary. This chapter reviews the direct and indirect use of LS to fabricate scaffolds for TE from single and multiphase materials.

Key words: Laser sintering, Tissue engineering, Scaffold, Composite materials

1. General Aspects of Laser Sintering

Laser sintering (LS), also referred to as selective laser sintering (SLS), is a rapid prototyping/manufacturing technique utilising a laser to fuse or sinter powder particles to form solid objects. Two major manufacturers of these machines are 3D Systems, USA (Fig. 1), and EOS GmbH, Germany. The abbreviation SLS® is a registered trademark of 3D Systems; therefore, EOS refers to their system as a LS machine (1). LS allows the production of complex shapes like other rapid prototyping (RP) techniques, such as stereolithography, fused deposition modelling, 3D printing, etc. Closed volumes are possible; however, unsintered powder will be trapped in the volume.

As with the other rapid prototyping methods, volumetric models are divided into slices with thicknesses that are determined by the sintering process, typically in the region of 0.10–0.15 mm, and which are subsequently transferred to the scanner unit of the machine. Based on the cross sectional image of the sliced part, the laser scans selective areas of the preheated powder bed, where the powder particles melt and solidify quickly to form a

Michael A.K. Liebschner (ed.), *Computer-Aided Tissue Engineering*, Methods in Molecular Biology, vol. 868, DOI 10.1007/978-1-61779-764-4_18, © Springer Science+Business Media, LLC 2012

Fig. 1. DTM (now 3D Systems) sinterstation 2500^{plus}.

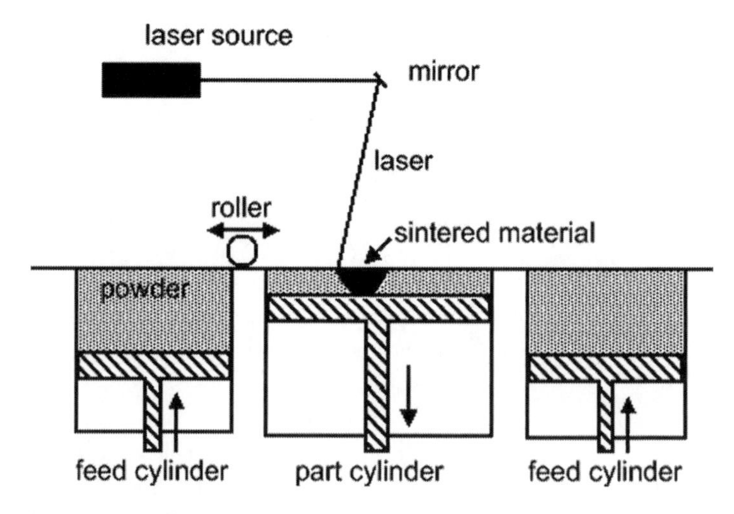

Fig. 2. Laser sintering schema.

dense part. To establish fusion with the current layer, the laser energy also partially liquefies the previous layer.

A typical machine has a part cylinder and one or more feed cylinders (Fig. 2). After each slice is scanned, the part piston is lowered by a predefined increment (slice thickness) and the feed piston is raised by the required amount to provide sufficient powder to the part bed. A roller picks up the material from the feed cylinder and distributes it over the part bed.

The sintered parts are supported by the unsintered powder surrounding them, i.e. there is no need to build support structures simultaneously as in some other RP processes.

Laser sintering machines usually are quite costly to run and a high throughput is necessary to justify the cost. The build volumes are fairly large and the laser spot diameter is focussed to 400–500 μm to allow quick sintering of large areas with sufficient fine details.

2. Laser Sintering for Tissue Engineering Applications

The minimum feature size, which directly depends on the laser spot diameter, slightly limits the usability of the machines for tissue engineering (TE) scaffolds, where in general highly porous constructs are desired, with pores bigger than the struts within the scaffold. However, a significant advantage of the SLS process over some other rapid prototyping techniques is that no toxic solvents or binders are needed.

To directly manufacture a scaffold for TE via rapid prototyping, the material not only has to be biocompatible to ensure suitability for tissue growth in vivo or ex vivo, and possibly other properties such as biodegradability, but it also has to be suitable for the chosen rapid prototyping technique. While the LS technique suggests that in principle any powder material can be processed, only a small range is commercially available. In addition, only few biocompatible materials have been used successfully to produce scaffolds via LS. A major limitation of the materials holding the most promise for TE is their suitability for LS. Polycaprolactone (PCL) is investigated most extensively for scaffolds fabricated via LS (2–6). Other materials include poly(methyl methacrylate) (PMMA), polyetheretherketone (PEEK) (7), polylactide acid PLA (8) and poly(lactic-*co*-glycolide) (PLG) (9).

Tissue engineering scaffolds generally require small features to achieve small-diameter strut networks with high porosity. Accuracy and minimum feature size are influenced by both the laser and the powder. Coarse powder particles will lead to a rough surface. The particle size should be around 50–100 μm for satisfactory surface quality on general macroscale parts. For TE scaffolds, the surface roughness is significant in relation to the overall dimensions of the part. The laser beam spot size limits the minimum feature size. With a diameter of 400–500 μm for a typical machine, in practice the minimum feature size is at least 500 μm. When using powder particles with a size of 50–100 μm, this feature, e.g. a strut of the scaffold, is just 5–10 particles thick.

Parts may be overly porous and brittle if the degree of sintering is too low. On the other hand, growth, where the feature is bigger than the actual sintered area, may become extensive if the laser adds too much energy to achieve a dense part. For tiny features as required in scaffolds it can be difficult to find settings that lead to dense sintered areas that are still within their intended dimensions. Williams et al. (5) experienced this when manufacturing PCL scaffolds with various designs and porosities, which were consistently 27% lower than those of the corresponding CAD designs. A dense cylinder, in contrast, had a designed porosity of 0%, but an actual one of 17.8%. The heat gradient in the powder surrounding the sintered area caused growth and decreased the size of the designed pores. The porosity of the cylinder, on the other hand, shows that although the laser energy caused growth, the sintering settings applied were still insufficient to achieve a dense body. As an example, Fig. 3 shows the cross section of a strut of a PCL scaffold. Clearly visible is the area that was liquefied during sintering with a laser spot diameter of 410 μm. While the sintered area still contains pores and is just over 300 μm wide, the overall thickness with all particles connected to the surface already exceeds 500 μm.

To increase the overall porosity of a scaffold to a value desirable for TE, Huang et al. (2) mixed 80 wt% of NaCl to PCL. The NaCl was leached out with water and a porosity of 89% was achieved in the scaffold with a branching and joining 3D diameter-varying flow-channel network. While PCL is mainly used for hard tissue applications, the scaffold described by Huang et al. was considered as an evolutionary step in the development of a scaffold for TE of a liver.

Fig. 3. Cross-sectional view of a PCL scaffold strut, produced at NUI, Galway on a DTM Sinterstation 2500^{plus}, powder particle size 50–110 μm, laser spot diameter 410 μm.

While in the process of LS no solvents are needed that could have negative consequences for TE applications, it has to be kept in mind that the heat of the laser in the standard LS process can influence the molecular weight of the polymer, hence degrading it. To confront this problem and to be able to expand the range of materials that can be processed via LS, surface selective laser sintering (SSLS) was introduced (8, 10). SSLS uses a diode laser with a wavelength of 0.97 μm, whose laser radiation is not absorbed by the polymer particles, which are therefore not heated. Only biocompatible carbon microparticles, which were added in a very low ratio to the actual scaffold material and which settle on the powder particle's surfaces, take up the radiation. They cause local melting, leaving most of the bulk particle cold. This technique allows the processing of thermally unstable polymers such as polylactides or polylactoglycolides and to obtain polymeric structures with bioactive proteins (8, 10). In the present case, ribonuclease was mixed to PLA, and carbon microparticles were added to absorb the laser radiation. Struts with a thickness of approximately 200 μm and a separation distance of 500 μm were sintered with a layer thickness of 200 μm. Raman spectroscopy showed no significant difference in sintered and untreated material. Degradation of the polymers was only detectable when a considerable high laser energy density of 250 W/cm was chosen for sintering.

A similar method to fabricate constructs from materials usually not suited for LS is to mix them with a binder phase. For instance, a ceramic or metallic phase with a sintering temperature too high for the LS machine is blended with a polymer to fabricate a green body. During LS, only the polymeric phase is sintered. The polymer will be removed from the green body by burning it out at an elevated temperature, while the ceramic or metallic particles will be sintered in a subsequent step at a further increased temperature. Examples of this procedure were published by Goodridge et al. (11), who added up to 15% of an acrylic binder to an apatite–mullite glass ceramic, and Lin et al. (12), where an epoxy resin/nylon blend was used as a binder for tricalciumphosphate (TCP).

3. Composites

The previously described powder mixtures can be referred to as composite powders, blended for the purpose of processing a material that cannot be laser sintered on its own. Composites scaffolds consist of two or more phases after the final manufacturing step. In this case all phases have a function beyond simple bonding of material; however, only one phase is sintered. Starting with a biocompatible polymer, a second phase may be added when the polymer is bioinert or has low mechanical properties. A bioactive phase is then added to enhance cell attachment, bone ingrowth, etc., or to increase the stiffness, especially when the scaffold is likely to be loaded after implantation. Hydroxyapatide (HA), for instance, is known for its bioactive behaviour and osteoconductivity. HA has been blended with PCL, PE, PEEK and polyvinylalcohol (PVA), and the compound powders have been processed via LS (6, 7, 13, 14).

In the 1990's, i.e. at an early stage of SLS, Lee et al. (15, 16) worked with PMMA-coated HA particles to develop bioceramic scaffolds via SLS. The drawback of the technique using coated particles was the involvement of organic solvents used in the production process of the powders. Any remnants of the solvent would cause inflammation. To avoid the necessity of solvents, Tan et al. (7) later used a physical blend of PEEK and HA powders. As PEEK is bioinert, HA was used to increase the bioactivity of the implant. Composite powders with up to 40 wt% of HA were sintered successfully.

Using a physically blended powder mixture with up to 30 wt% HA, Chua et al. (14) successfully processed PVA/HA composites via LS. As with most materials, the variation of process parameters settings for successful sintering, such as preheat temperature, laser scan speed and laser power, was very limited, and in general a considerable amount of time has to be dedicated to find the most suitable parameters when investigating the suitability of a new material for LS. Outside the established window, in ref. 14 the constructs were found to be either too brittle or suffered from burning of the material due to an excessive energy density applied.

Experiments with two different types of HA blended with PCL were conducted by Wiria et al. (6). Depending on the type of HA (either "reactor" or "sintered" powder, which differed significantly in their particulate surface areas) and depending on the amount of HA, the process parameters had to be adjusted. A range of possible setting combinations to sinter PCL with various amounts of HA was established. In vitro cell culturing showed the suitability of the processed material for cell ongrowth. A bioactivity analysis and mechanical testing of laser-sintered material complemented the investigations showing the potential of this composite for TE.

Simpson et al. (9) investigated LS of composites of 95/5 poly (L-lactide-*co*-glycolide) (PLG), HA and TCP. Solid constructs with mechanical properties similar to those of trabecular bone were successfully manufactured. However, dimensional inaccuracies occurred due to shifting of previously sintered layers when the roller passed to spread new material, creating difficulties to manufacture scaffolds with accurate shapes and dimensions. As this phenomenon may partly be caused by the nonspherical shape of the PLG particles, this problem may be rectified by an optimised powder-preparation technique.

In any case, in general it is found that the amount of the second phase is limited. If the polymer content is too low, the sintered constructs may be too brittle to handle. In their publication on PVA/HA composites, Chua et al. (14) also reported on LS of HA particles coated with PVA in a spray dry process, which unlike the blended powder mixture could not be sintered successfully. It appeared that this was rather due to the high HA content than to the nature of the composite powder, as in this case the HA content was 80 wt% and the amount of polymeric binder phase was too small to generate bonding between the particles. Maximum concentrations of HA in composites were investigated by Hao et al. (13), Savalani et al. (17) and Zhang et al. (18). Polyethylene and polyamide were used as binders, and concentrations of up to 50% of HA were achieved. Not all of the fabricated constructs were found suitable for TE due to the material choice; however, the research showed promising results in achieving mechanically stable, HA-rich composites.

In summary, a limited number of materials, including composites, have successfully been evaluated thus far; however, from the present results, LS of selected biocompatible materials for TE is promising. The build chamber size of industrial LS machines allows for the production of a reasonably large number of scaffolds at the same time. A reduction of the laser beam spot size to meet the requirements regarding minimum feature size of TE scaffolds would permit finer details and higher porosities. Machines developed especially for TE purposes, such as the SSLS machine, already take this into account. As the materials investigated are nonstandard materials for the industrial LS process, powder preparation currently is an important issue. Both size and shape of the particles are crucial for successful processing, and the preparation of the material can be a costly step in the entire process of manufacturing scaffolds for TE. In addition, when the particle size of a second phase is too small, the particles may settle on the surface of the polymeric phase and shielding can occur, also observed by Simpson et al. (9). This shielding prevents the polymeric particles establishing connection to each others and forming a mechanical stable construct. Hence a similar particle size for both phases seems favourable, so that the second phase does not severely impair the sintering behaviour of the polymeric phase. It has been proven that with careful process control the thermal degradation of the polymeric phases during LS can be kept either superficial or at least not too severe, so the biocompatible nature of the materials remains unchanged. Selection of the material depends on the task and intended site of the scaffold, and the addition of bioactive phases caters for improved performance of the scaffold in TE.

References

1. Bibb R (2006) Medical modelling. Woodhead Publishing, Cambridge, p 297
2. Huang H et al (2007) Avidin-biotin binding-based cell seeding and perfusion culture of liver-derived cells in a porous scaffold with a three-dimensional interconnected flow-channel network. Biomaterials 28(26):3815–3823
3. Lohfeld S et al (2006) Manufacturing of small featured PCL scaffolds for bone tissue engineering using selective laser sintering. J Biomech 39(Suppl 1):S216–S212
4. Smith MH et al (2007) Computed tomography-based tissue-engineered scaffolds in craniomaxillofacial surgery. Int J Med Robot Comput Assist Surg 3(3):207–216
5. Williams JM et al (2005) Bone tissue engineering using polycaprolactone scaffolds fabricated via selective laser sintering. Biomaterials 26 (23): 4817–4827
6. Wiria FE et al (2007) Poly-[epsilon]-caprolactone/hydroxyapatite for tissue engineering scaffold fabrication via selective laser sintering. Acta Biomater 3(1):1–12
7. Tan KH et al (2003) Scaffold development using selective laser sintering of polyetheretherketone–hydroxyapatite biocomposite blends. Biomaterials 24(18):3115–3123
8. Antonov EN et al (2005) Three-dimensional bioactive and biodegradable scaffolds fabricated by surface-selective laser sintering. Adv Mater 17(3):327–330
9. Simpson RL et al (2008) Development of a 95/5 poly(L-lactide-*co*-glycolide)/hydroxylapatite and beta-tricalcium phosphate scaffold as bone replacement material via selective laser sintering. J Biomed Mater Res B Appl Biomater 84B(1):17–25
10. Popov VK et al (2007) Laser technologies for fabricating individual implants and matrices for tissue engineering. J Optic Technol 74(9): 636–640
11. Goodridge RD, Dalgarno KW, Wood DJ (2006) Indirect selective laser sintering of an apatite-mullite glass-ceramic for potential use in bone replacement applications. Proc Inst Mech Eng H J Eng Med 220(1):57–68
12. Lin L et al (2007) Design and fabrication of bone tissue engineering scaffolds via rapid prototyping and CAD. J Rare Earths 25(Suppl 2): 379–383
13. Hao L et al (2006) Selective laser sintering of hydroxyapatite reinforced polyethylene composites for bioactive implants and tissue scaffold development. Proc Inst Mech Eng H J Eng Med 220(4):521–531
14. Chua CK et al (2004) Development of tissue scaffolds using selective laser sintering of polyvinyl alcohol/hydroxyapatite biocomposite for craniofacial and joint defects. J Mater Sci Mater Med 15(10):1113–1121
15. Lee G, Barlow J (1993) Selective laser sintering of bioceramic materials for implants. In: Proceedings of the solid freeform fabrication symposium, Austin, TX, 1993
16. Lee G et al (1996) Biocompatibility of SLS-formed calcium phosphate implants. In: Proceedings of the solid freeform fabrication symposium, Austin, TX, 1996
17. Savalani M et al (2007) Fabrication of porous bioactive structures using the selective laser sintering technique. Proc Inst Mech Eng H J Eng Med 221(8):873–886
18. Zhang Y et al (2008) Characterization and dynamic mechanical analysis of selective laser sintered hydroxyapatite-filled polymeric composites. J Biomed Mater Res A 86A(3): 607–616

Chapter 19

Three-Dimensional Microfabrication by Two-Photon Polymerization Technique

Aleksandr Ovsianikov and Boris N. Chichkov

Abstract

Two-photon polymerization (2PP) technique is a novel CAD/CAM-based technology allowing the fabrication of any computer-designed 3D structure from a photosensitive polymeric material with a lateral resolution down to 100 nm. The fabrication of highly reproducible scaffold structures for tissue engineering by 2PP is very important for systematic studies of cellular processes and better understanding of in vitro tissue formation. Flexibility of this technology and ability to precisely define 3D construct geometry allow the direct addressing of issues associated with vascularization and patient-specific tissue fabrication. In this chapter, we report on our recent advances in the fabrication of biomedical implants and 3D scaffolds for tissue engineering and regenerative medicine by 2PP technique.

Key words: Microfabrication, Photosensitive polymeric material, Two-photon polymerization, Scaffold, Tissue engineering

1. Introduction

Artificial fabrication of a living tissue that is able to integrate with the host tissue inside a body is a very challenging task. Natural repair of a tissue at a particular site is a result of complex biological processes, which are currently the subject of intensive research and are not yet fully understood. In order to encourage cells to form a tissue, one has to create an appropriate environment, exactly resembling that of a particular tissue type. Some cell types can preserve tissue-specific features in a 2D environment while others require a 3D environment. One of the most popular approaches in tissue engineering is based on the application of 3D scaffolds, whose function is to guide and support proliferation of cells in 3D (1). The ability to produce arbitrary 3D scaffolds is, therefore, very appealing. Few techniques, which can create 3D porous scaffolds, have been developed in the recent years (2–7). These techniques can be subdivided into passive

Michael A.K. Liebschner (ed.), *Computer-Aided Tissue Engineering*, Methods in Molecular Biology, vol. 868,
DOI 10.1007/978-1-61779-764-4_19,

and active. The passive techniques, such as phase separation, can yield porous structures with high resolution and uniform pore sizes. However, they do not allow the fabrication of exactly identical structures and provide little control over the location of individual pores. On the other hand, the active techniques, such as inkjet printing and stereolithography, provide possibility to produce any CAD-designed structure, but offer limited resolution of the order of tens of micrometers. The advantage of two-photon polymerization (2PP) technique for the fabrication of scaffolds is a combination of unprecedented resolution, high reproducibility, and the ability to fabricate true 3D structures. In particular, 2PP provides control over both external shape and internal porosity of the structure. Therefore, scaffolds fabricated by the 2PP technique enable systematic studies of the cell proliferation, acquired functionality, and tissue formation in 3D.

2. Two-Photon Polymerization Technique

2PP is a direct laser writing technique, which allows the fabrication of 3D structures by direct "recording" into the volume of the photosensitive material. Due to the threshold behavior and nonlinear nature of the 2PP process, resolution (structure size) beyond the diffraction limit of the optics used to focus the laser beam can be realized by controlling the laser pulse energy and the number of applied pulses (8, 9). Figure 1 shows scanning electron microscope images of 3D microstructures fabricated by the 2PP technique. The micro-spiders are fabricated by transferring of CAD into the real micromodel by 2PP of commercial photolithographic SU8 resin. Each structure body stands above the glass and is supported by six 1.7-μm-thick legs. The two rightmost images represent 3D photonic crystals—highly porous 3D structures exhibiting refractive index periodicity in any propagation direction. These examples allow envisioning the potential of this technology for various applications. Despite the fact that the 2PP is a relatively new technology, its application areas are rapidly expanding. Fabrication of 3D

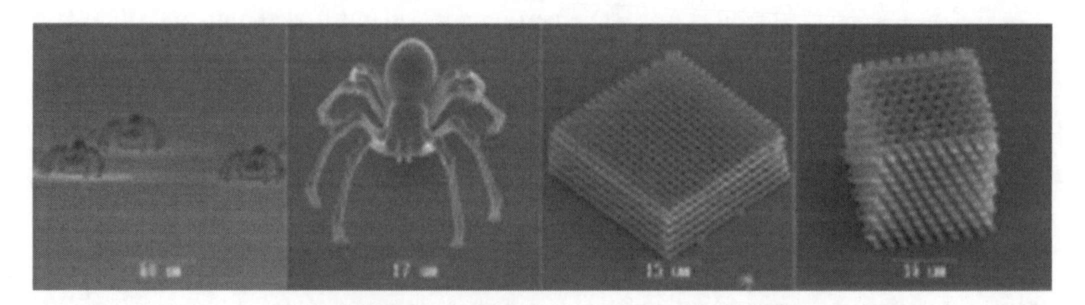

Fig. 1. Scanning electron microscope (SEM) images of 3D structures fabricated by 2PP technique.

photonic crystals by the 2PP technique has been first proposed and demonstrated by Maruo et al. (10) and by now is applied by different groups in the world. Apart from that, 2PP is also used for the fabrication of micromechanical systems (11), microfluidic devices (12), micro-optical components (13), plasmonic components (14), biomedical devices (15), scaffolds for tissue engineering (16), and cross-linking of natural proteins (17).

Similarly as in stereolithography, which is a rapid 3D prototyping technology, the 2PP is based on a photochemical process, in which light triggers a chemical reaction, leading to polymerization of a photosensitive material. Polymerization is a process in which monomers or weakly cross-linked polymers (liquid or solid) interreact to form three-dimensional network of highly cross-linked polymer (solid). Photoinitiators—molecules, which have low photo dissociation energy—are usually used to provide the material's photosensitivity. Absorption of a photon is leading to a photodissociation (bond cleavage) of a molecule and formation of highly reactive radicals, which react with the monomer and initiate radical polymerization. The reaction is terminated when two radicals react with each other.

In stereolithography, UV laser radiation, applied to scan the surface of the photosensitive material, is producing 2D patterns of polymerized material (Fig. 2a). This radiation induces photopolymerization through one-photon absorption at the surface of the material. Therefore, with stereolithography, it is only possible to fabricate 3D structures using a layer-by-layer approach. Majority of available photosensitive materials are transparent in the near-infrared and highly absorptive in the UV spectral range. One can initiate 2PP with near-infrared laser pulses within the small volume

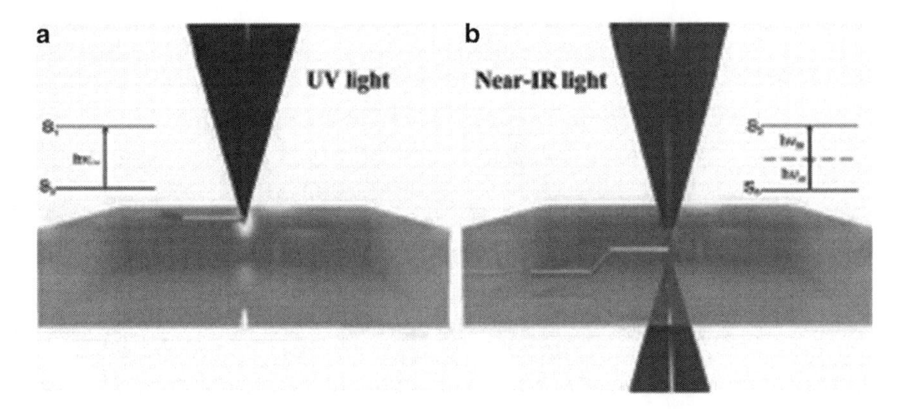

Fig. 2. Photosensitive material processing by (**a**) a single-photon absorption with UV light. Light is absorbed at the surface of the photosensitive material. 2D patterns can be produced by photopolymerization. (**b**) Two-photon absorption with near-infrared light. TPA and following chemical reactions are confined in the focal volume, and the rest of the laser radiation is passing through the material without interaction. According insets in the figures illustrate (**a**) one-photon absorption and (**b**) two-photon absorption processes (18).

of the material using tightly focused laser pulses. Figure 2 provides a simplified illustration of the difference between one-photon and two-photon-activated processing. In 2PP, a material is polymerized along the trace of the moving laser focus, thus enabling fabrication of any desired polymeric 3D pattern by direct "recording" into the volume of photosensitive material (Fig. 2b). In a subsequent processing step, the material, which was not exposed to the laser radiation, and therefore stayed unpolymerized, is removed by an appropriate solvent and the fabricated structure is revealed. The material sensitive in the UV range (λ_{UV}) can be polymerized by irradiation with the near-infrared laser light at approximately double wavelength ($\lambda_{IR} = 2\ \lambda_{UV}$), under the condition that the intensity of the radiation is high enough to initiate two-photon absorption (TPA). Since femtosecond lasers provide very high peak intensities at the moderate average laser power, they present a very suitable light source creating favorable conditions for TPA and are commonly used for 2PP technique.

The resolution of stereolithography depends on the size of the focal spot and is limited by diffraction; thus, the minimum feature size cannot be smaller than the half of the applied laser wavelength. In reality, due to the technical reasons, inherent to this technology, the best lateral resolution of stereolithography is in the range of a few micrometers (19). Since TPA is a nonlinear process displaying threshold behavior, the structural resolution beyond the diffraction limit can be realized. Structures with feature size down to 100 nm (and even better) have been demonstrated by several groups, which is almost an order of magnitude smaller than the laser wavelength (800 nm)!

Theoretical model for the multiphoton absorption (MPA) was developed in 1931 by Göppert-Meyer (20), three decades prior to its experimental observation (21). The probability of n-photon absorption is proportional to the nth power of the photon flux density; consequently, high photon flux densities are required in order to observe this phenomenon. In fact, MPA was one of the first effects demonstrated with the help of lasers, since laser can provide intensities, which are much higher than those available from other laboratory light sources. It was demonstrated that an atom can absorb two or more photons simultaneously, thus allowing electron transition to the states that cannot be reached with a single-photon absorption. Atom excitation with both one-photon absorption and TPA is compared schematically in the insets of Fig. 2a, b. TPA is mediated by a virtual state (dashed line in the inset in Fig. 2b), which has an extremely short lifetime (several femtoseconds). Thus, TPA is only possible if a second photon is absorbed before the decay of this virtual state. Note that excited energy levels S1 and S2, shown in the insets of Fig. 2a, b are not exactly the same, since the selection rules for single photon and TPA are different (22). Since the probability of the TPA is

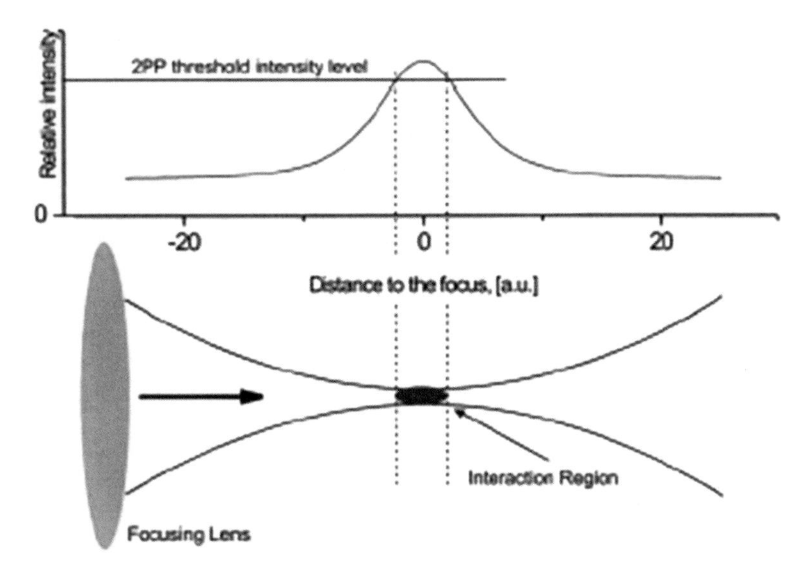

Fig. 3. Schematics of two-photon interaction region. *Upper image* illustrates laser intensity distribution along the beam propagation direction, and the relevant 2PP threshold intensity level. *Lower image* illustrates focused laser beam and according light–material interaction region. Due to the nonlinear nature of the process, intensity distribution along the propagation direction of the focused laser beam limits the light–material interaction region to the vicinity of the focal volume (18).

proportional to the square of the intensity of the laser radiation, favorable conditions for TPA in the first place are created in the focus of laser beam (see Fig. 3). Thus, the light–material interaction region is strongly confined.

3. Materials and Methods

3.1. Experimental Setup for 2PP Microfabrication

Main factors, determining the performance of the 2PP microfabrication system, are the sample positioning precision, laser system stability, and flexibility of the scanning algorithm. Schematic representation of the experimental setup is shown in Fig. 4. The femtosecond solid-state Ti:sapphire laser generates pulses with duration of 120 fs and a repetition rate of 94 MHz. The central emission wavelength of such a laser can be tuned between 720 and 950 nm. Unless otherwise noted, for all experiments described in this chapter, laser radiation at the central emission wavelength of 780 nm is applied. A small portion of the light emitted from the laser system is guided into the spectrum analyzer for continuous monitoring of the laser emission spectrum. The $\lambda/2$-plate mounted on the computer-controlled rotational stage is used to rotate polarization of the laser beam. In combination with the polarization-sensitive beam splitter, it enables continuous adjustment of the average power of the

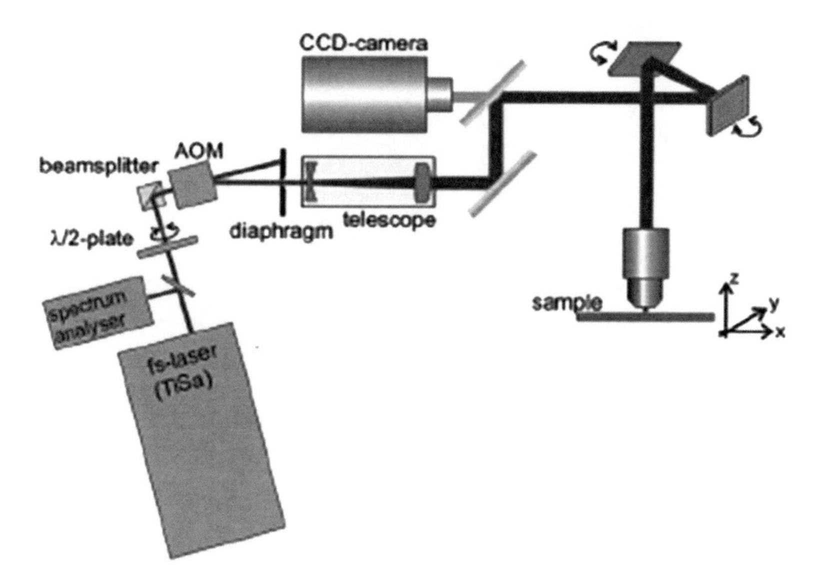

Fig. 4. Schematic representation of the experimental setup applied for 2PP photofabrication.

beam entering the acousto-optic modulator (AOM). The AOM is adjusted such that the first diffraction order of the beam can pass the diaphragm aperture while the zero order is blocked. By controlling the AOM on/off state with computer-generated TTL signal, the AOM is used as a laser shutter. In order to completely fill the aperture of a focusing optic and to achieve optimal focusing conditions, the beam is expanded to a diameter of about 10 mm by a telescope. A highly sensitive CCD camera is mounted behind the last dichroic mirror to provide online process observation. The refractive index of the polymer is slightly changed by a 2PP process, and the polymerized patterns become visible immediately after the laser exposure. The relative position of the laser focus within the sample is controlled by two galvoscanner mirrors (angular range $\pm 12.5°$, resolution 6.7 μrad) and three linear translational stages (*XYZ*, resolution 10 nm, maximal travel distance 2.5 cm). For the fabrication of structures presented in Fig. 1 and in the latter sections, unless otherwise noted, a 100× microscope objective lens (Zeiss, Plan Apochromat, NA = 1.4) was used to focus the laser beam.

3.2. Materials Used for 2PP

A great advantage for the application of the 2PP technology in microstructure fabrication is provided by the wide variety of available materials, with properties that can fit almost any end application. Most of the photosensitive materials, which are currently used for 2PP microstructuring, were originally developed for lithography and are commercially available (9). Two particular examples presented here are the ORMOCER and SU8.

ORMOCER®s are hybrid organic–inorganic polymeric materials produced from liquid precursors using sol–gel process. They include urethane- and thioether (meth)-acrylate alkoxysilanes, and contain strong covalent bonds between the organic and inorganic components. The cross-linking of inorganic and organic moieties leads to the formation of a 3D network, which provides ORMOCER®s with significant chemical and thermal stability (23). By changing the ratio in organic/inorganic network density, it is possible to design ORMOCER® materials with desired mechanical, optical, chemical, and surface properties. In this study, ORMOCOMP® (Microresist Technology GmbH, Germany), a member of the ORMOCER® family, containing 1.8% of photoinitiator Irgacure 369 (Ciba Specialty Chemicals, Switzerland), is used. This is a liquid photosensitive material, which can be locally transferred into solid phase through free radical polymerization reaction, taking place immediately after the laser exposure. After irradiation, the nonsolidified material is removed by a 1:1 solution of 4-methyl-2-penthanone and 2-propanol to reveal the generated structure.

SU8 is an epoxy-based polymer, which is widely used in lithography. From the fabrication point of view, its main difference to ORMOCER is that SU8 is initially solid. Therefore, the fabricated structure is immobilized throughout the whole fabrication time (24). SU8 is polymerized via a cationic reaction, which takes place during the postbake processing step. Refractive index of this material changes only slightly during the illumination step; therefore, there are no disturbances of the laser beam focus by the structures written beforehand. This allows very flexible patterning strategies. A standard formulation of SU8-2050 (Microresist Technology GmbH, Germany) is used to fabricate two rightmost structures shown in Fig. 1. The material is spin coated onto the 170-μm-thick glass coverslips. Processing requires prebake and postbake steps, which are performed in accordance to the protocol provided by the manufacturer (25). In order to remove the nonilluminated material, standard SU8 developer is applied.

An important issue is the biocompatibility of applied materials. Our recent studies using different cell lines demonstrated that both ORMOCOMP® and SU8 support cell growth and can be considered as biocompatible at the cellular level. Cells proliferating on these materials are able to form cell-to-cell junctions, such as gap junctions, characteristic to functional tissue. It has been demonstrated that cells can be grown of the vertical surfaces of ORMOCOMP® structures generated by 2PP technique (16).

Since virtually any photosensitive material can be structured by the 2PP technique, there are still many unexplored materials. Of particular interest are the biodegradable photosensitive materials. Scaffolds, fabricated from such materials, naturally vanish after fulfilling the function of supporting and guiding tissue regeneration. Recent efforts by polymer chemist in the area of photosensitive hydrogels indicate some promising advantages of such materials for injectable scaffolds (26) and even cell encapsulation (27).

4. Biomedical Applications

Recently, we have demonstrated several very promising biomedical applications of 2PP technique: for tissue engineering (16, 35–38), drug delivery (39–43), and medical implants (28, 44, 45), which are described in the following sections.

4.1. Scaffolds for Tissue Engineering

Advantage of 2PP for the fabrication of scaffolds for tissue engineering is a combination of unprecedented resolution, high reproducibility, and a possibility to fabricate true 3D structures.

Using the software developed at our institute, it is possible to directly transfer CAD file in the STL format (a common stereolithography data format developed by "3D Systems") to a machine code controlling the relative position of the laser focus inside the sample. Similarly to stereolithography, the structure is fabricated in a layer-by-layer approach. First, the 3D CAD data is triangulated in order to obtain a surface consisting out of triangles, i.e., in the STL format (see step A in Fig. 5). In the next step, individual contours are obtained by slicing such triangulated surface with a defined interlayer distance (see step B in Fig. 5). Later, in the fabrication step, the according structure is polymerized layer by layer by moving laser focus along the calculated trace. After the unpolymerized material, which was not irradiated with the laser, is removed by an appropriate solvent, the 3D structure is revealed (see Fig. 6a). Since 2PP is a true 3D microfabrication technique, the orientation of slices in space can be arbitrary, and not necessarily parallel to the sample surface as implied in stereolithography.

One can either polymerize only the contour of each layer or the complete layer area. In the second case, a better mechanical stability is achieved at the expense of fabrication duration. Alternatively, one

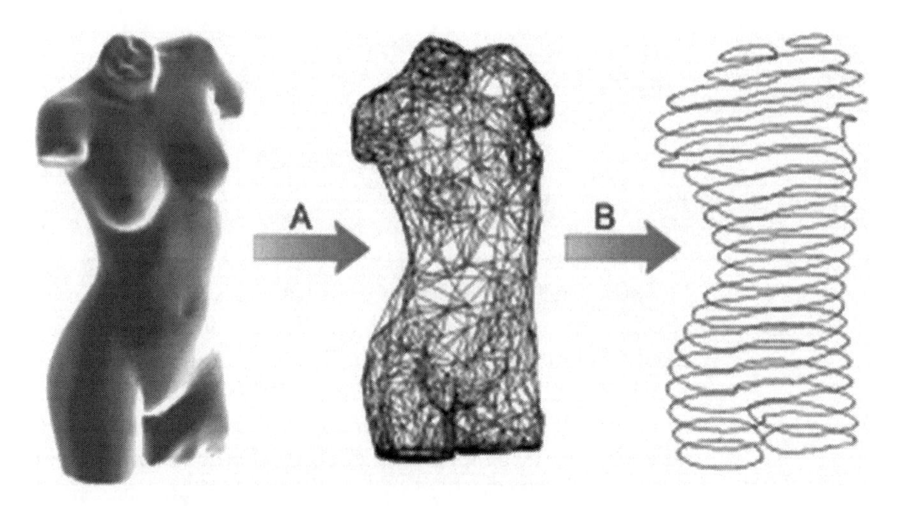

Fig. 5. CAD data processing for 3D object microfabrication.

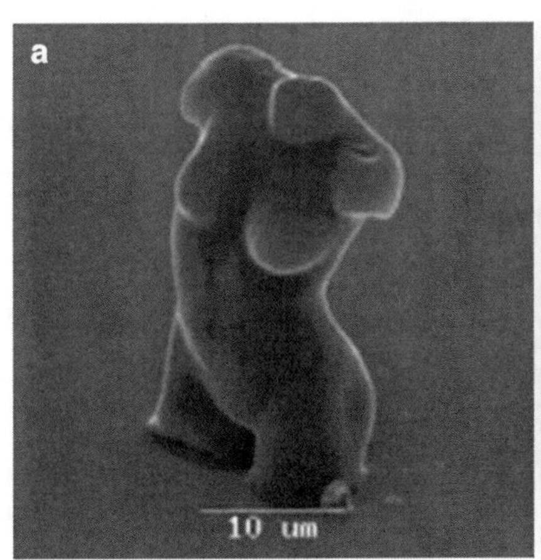

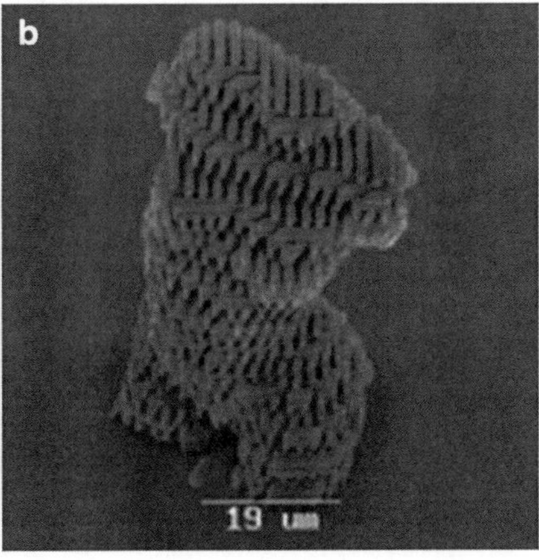

Fig. 6. SEM images of microstructures for tissue engineering fabricated by 2PP technique: (**a**) bulk microvenus model; (**b**) corresponding porous structure (42).

can polymerize only the outer shell of such structure so that it comprises the unpolymerized material inside. After the development step, the sample is subjected to the UV radiation leading to complete material cross-linking inside the structure. This approach allows substantial shortening of the fabrication time.

The particular advantage of 2PP technique for the fabrication of scaffolds for tissue engineering is the possibility to control both external shape and internal porosity of the structure (see Fig. 6). In order to produce microporous venus statue, we have polymerized each individual layer in such a way that is consists of separate parallel rods. The relative orientation of the rods in neighboring layers is orthogonal (see Fig. 6b). Such interconnected micropores, on the order of one to a few micrometers, while not accessible by cells, provide means for liquid transport inside the 3D scaffold. The size and the orientation of such pores are controlled by the settings in the microfabrication software and can be introduced into any initial CAD design. The porosity of this structure can be up to 50% and a more sophisticated pore distribution can be introduced, leading to even higher porosity of fabricated structures.

It is also possible to obtain scaffolds with a more sophisticated pore size distribution by designing them directly. Figure 7 shows a CAD design and an according 2PP-fabricated scaffold having porosity of more than 80%. The through pores, which can be accessed from each scaffold surface, are large enough to accommodate cells. A second set of pores, having diameter much smaller than the cell dimension, is introduced in order to enhance fluid transport inside the scaffold. In addition, the mechanical properties of such scaffold can be controlled by changing utilized material and adjusting porosity of the structure and the individual scaffold design. The high reproducibility of 2PP microfabrication, for the first time, allows the fabrication of large series of identical samples, and

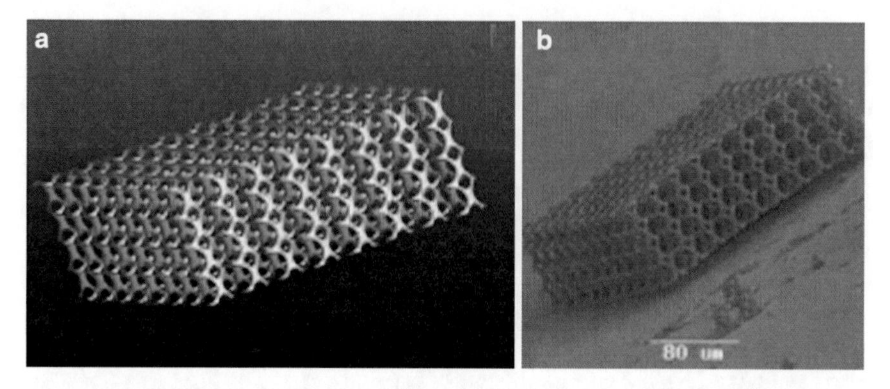

Fig. 7. Highly porous 3D scaffold with two distinct pore dimensions: (**a**) CAD concept; (**b**) SEM image of according structure fabricated by 2PP technique (18).

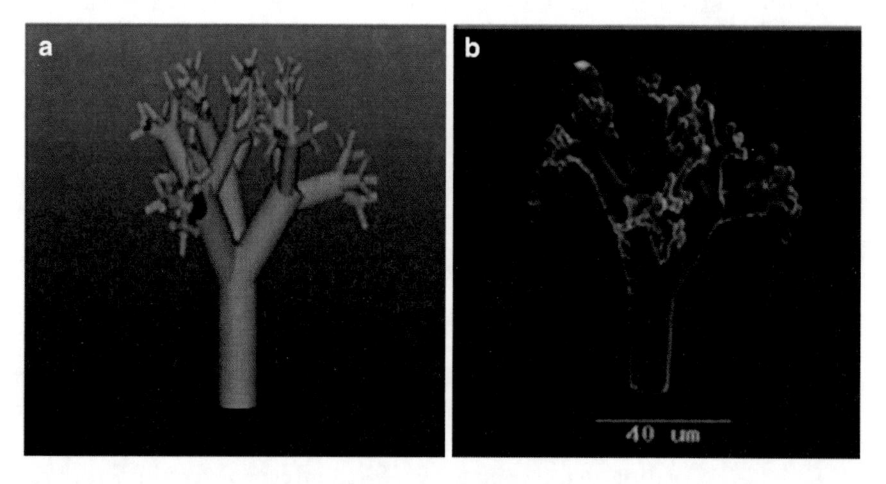

Fig. 8. An original CAD design (**a**) and a scanning electron microscope (SEM) image of a fabricated structure (**b**), which resembles a fragment of pulmonary alveoli—microcapillaries responsible for gas exchange in the mammalian lungs (16).

therefore enables systematic studies of scaffold–cell and cell–cell interactions in a 3D system.

One of the issues associated with the repair of large-volume tissue defects is vascularization. Since 2PP allows introducing pores of desired size at any location in the scaffold, it has a great potential for defining the future vascular network already in the initial scaffold fabrication step. By this way, it would be possible to integrate the newly formed vascular network, with the one already present around the defect. Figure 8 shows an original CAD design of a fragment of a branching microcapillarie network model and an according structure fabricated by 2PP of SU8 material.

High resolution of 2PP technique can also provide means for adjustable surface–cell interaction mediated by nanostructures or biomolecules deposited by photografting. Since the 2PP microstructuring takes place at room temperature, the denaturation of proteins due to high temperatures can be avoided.

4.2. Biomedical Implants

2PP technique can be also applied for the fabrication of implants and prostheses. For example, the malleus, incus, and stapes bones serve to transmit sounds from the tympanic membrane to the inner ear. Ear diseases may cause discontinuity or fixation of the ossicles, which results in a conductive hearing loss. The size of the total ossicular replacement prosthesis (TORP) is of the order of few millimeters and is varying from patient to patient. The materials used in ossicular replacement prostheses must demonstrate appropriate biological compatibility, acoustic transmission, stability, and stiffness properties. The prostheses prepared using Teflon®, titanium, Ceravital, and other conventional materials have demonstrated several problems during clinical studies, including migration, puncture of the eardrum, difficulty in shaping the prostheses, and reactivity with the surrounding tissues.

We have demonstrated the application of the 2PP technique for rapid prototyping of ORMOCER® middle-ear bone replacement prostheses (28). Figure 9 shows an original CAD design and an optical microscope image of the fabricated TORP. The 2PP technique provides several advantages over the conventional processing for scalable mass production of ossicular replacement prostheses. First, the raw materials used in this process are widely available and inexpensive. Second, 2PP can be set up in a conventional clinical environment (e.g., an operating room) that does not contain clean room facilities. It allows fabrication of patient-specific prosthesis based on the optical coherence tomography (OCT) analysis or other provided data. Moreover, the resolution required for TORPs is not very high, and therefore one can accelerate the fabrication process even further. And finally, 2PP of ossicular replacement prostheses is a straightforward single-step process, as opposed to the conventional multistep fabrication techniques. We anticipate that the number of applications of the 2PP technique for the fabrication of prosthesis will rapidly increase in the nearest future.

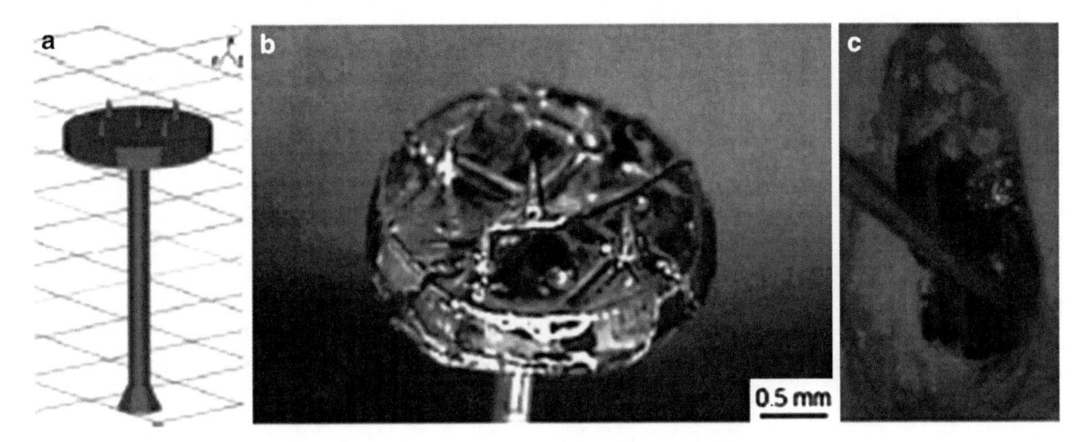

Fig. 9. Middle-ear bone replacement prosthesis: (**a**) original CAD design; (**b**) optical microscope image of a structure fabricated by 2PP in ORMOCER®; (**c**) in vitro implantation of the fabricated implant (28).

4.3. Microneedles for Drug Delivery

Transdermal drug delivery is a method that has experienced a rapid development in the past two decades, and has often shown improved efficiency over the other delivery routes (29). It avoids many issues associated with intravenous drug administration, including pain to the patient, trauma at the injection site, and difficulty in providing sustained release of pharmacologic agents. In addition, precise dosing, safety, and convenience are also addressed by transdermal drug delivery. However, only a small number of pharmacological substances are delivered in this manner today. The most commonly known example is nicotine patches. The main reason for that is the significant barrier to diffusion of substances with higher molecular weight provided by the upper layers of the skin. The top layer, called stratum corneum, is composed of dead cells surrounded by lipid. This layer provides the most significant barrier to diffusion to approximately 90% of transdermal drug applications (30, 31). A few techniques, enhancing the substance delivery through the skin, have been proposed. Two of the better-known active technologies are iontophoresis and sonophoresis. The rate of product development, involving these technologies, has been relatively slow (32, 33). This is partly conditioned by the relative complexity of the resulting systems, compared to the passive transdermal systems. One of the passive technologies is based on microneedle-enhanced drug delivery. These systems use arrays of hollow or solid microneedles to open pores in the upper layer of the skin and assist the drug transportation. The length of the needles is chosen such that they do not penetrate into the dermis, pervaded with nerve endings, and thus do not cause pain. In order to penetrate the stratum corneum, microneedles for drug delivery have to be longer than 100 μm, and are generally 300–400-μm long, since the skin exhibits thickness values that vary with age, location, and skin condition. Application of microneedles has been reported to greatly enhance (up to 100,000-fold) the permeation of macromolecules through the skin (34). The microneedles for withdrawal of blood must exceed lengths of 700–900 μm in order to penetrate the dermis, which contains blood vessels. Most importantly, microneedle devices must not fracture during penetration, use, or removal.

The flexibility and high resolution of the 2PP technique allow rapid fabrication of microneedle arrays with various geometries (Fig. 10), and study of its effect on the tissue penetration properties. Results of our studies indicate that microneedles created using the 2PP technique are suitable for *in vivo* use, and for the integration with the next-generation MEMS- and NEMS-based drug delivery devices.

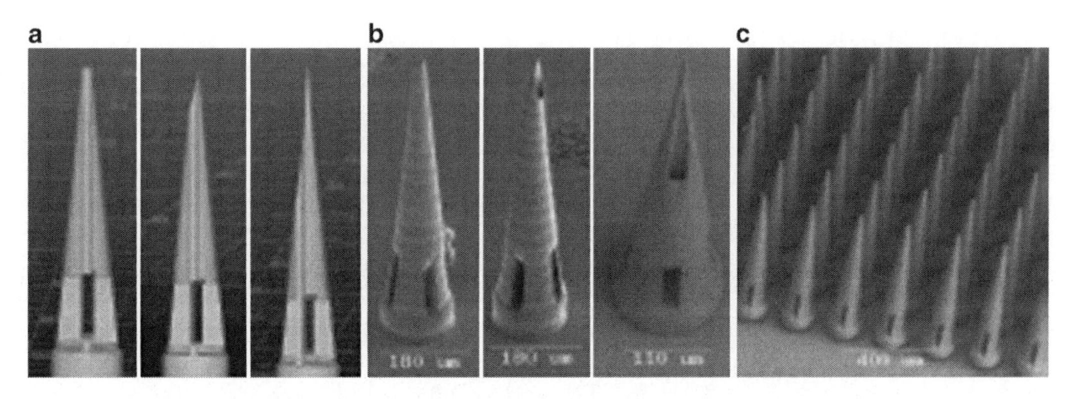

Fig. 10. Hollow microneedles for transdermal drug delivery: (**a**) cross section of original CAD design of microneedles with different channel position and tip sharpness; (**b**) SEM images of according microneedles fabricated by 2PP technique; (**c**) an array of microneedles fabricated by 2PP technique (41).

5. Summary and Outlook

The 2PP technique demonstrates great promise for biomedical applications, particularly for the fabrication of scaffolds for tissue engineering, medical implants, sensors, and transdermal drug delivery. The ability to produce arbitrary 3D structures directly from the CAD model, with sub 100-nm resolution, in combination with the right materials, provides unique opportunities for systematic studies of cell interactions in 3D geometry, guided reproduction, and regeneration of cellular microenvironment. In addition, due to the flexibility of this technology, patient-specific implants and prostheses can be fabricated.

Acknowledgments

The authors gratefully acknowledge very important contribution from their colleagues, who have been involved in different parts of this work: A. Doraiswamy, A. Haverich, R. Narayan, A. Ngezahayo, A. Ostendorf, and S. Schlie. Financial support from the DFG Excellence Cluster "ReBirth" and Transregio TR37 project is gratefully acknowledged.

References

1. Hutmacher DW (2001) Scaffold design and fabrication technologies for engineering tissues – state of the art and future perspectives. J Biomater Sci Polym Ed 12(1):107–124
2. Sachlos E, Czernuszka JT (2003) Making tissue engineering scaffolds work. Review: the application of solid freeform fabrication technology to the production of tissue engineering

scaffolds. Eur Cell Mater 5:29–39; discussion 39–40
3. Buckley CT, O'Kelly KU (2004) Regular scaffold fabrication techniques for investigations in tissue engineering. In: Prendergast PJ, McHugh PE (eds) Topics in bio-mechanical engineering. ColourBooks Ltd., Dublin, pp 147–166
4. Tsang VL, Bhatia SN (2007) Fabrication of three-dimensional tissues. Adv Biochem Eng Biotechnol 103:189–205
5. Vozzi G, Ahluwalia A (2007) Microfabrication for tissue engineering: rethinking the cells-on-a scaffold approach. J Mater Chem 17(13): 1248–1254
6. Hull C (1990) Method for production of three-dimensional objects by stereolithography. US Patent 4929402
7. Hutmacher DW, Sittinger M, Risbud MV (2004) Scaffold-based tissue engineering: rationale for computer-aided design and solid free-form fabrication systems. Trends Biotechnol 22(7):354–362
8. Sun HB, Kawata S (2004) Two-photon polymerization and 3D lithographic microfabrication. In: Fatkullin N (ed) NMR, 3D analysis, photopolymerization. Springer, pp 169–273
9. Ovsianikov A, Passinger S, Houbertz R, Chichkov BN (2006) Threedimentsional material processing with femtosecond lasers. In: Phipps CR (ed) Laser ablation and its applications. Springer series in optical sciences, pp 129–167
10. Maruo S, Nakamura O, Kawata S (1997) Three-dimensional microfabrication with two-photon-absorbed photopolymerization. Opt Lett 22(2):132–134
11. Sun HB, Kawakami T, Xu Y, Ye JY, Matuso S, Misawa H, Miwa M, Kaneko R (2000) Real three-dimensional microstructures fabricated by photopolymerization of resins through two-photon absorption. Opt Lett 25(15): 1110–1112
12. Sugioka K, Masuda M, Hongo T, Cheng Y, Shihoyama K, Midorikawa K (2004) Three-dimensional microfluidic structure embedded in photostructurable glass by femtosedonc laser for lab-on-chip applications. Appl Phys A 79:815–817
13. Gu M, Fu L (2006) Three-dimensional image formation in fiber-optical second-harmonic-generation microscopy. Opt Express 14(3): 1175–1181
14. Kiyan R, Reinhardt C, Passinger S, Stepanov AL, Hohenau A, Krenn JR, Chichkov BN (2007) Rapid prototyping of optical components for surface plasmon polaritons. Opt Express 15(7):4205–4215
15. Doraiswamy A, Jin C, Narayan RJ, Mageswaran P, Mente P, Modi R, Auyeung R, Chrisey DB, Ovsianikov A, Chichkov B (2006) Two photon induced polymerization of organic–inorganic hybrid biomaterials for microstructured medical devices. Acta Biomater 2(3): 267–275
16. Ovsianikov A, Schlie S, Ngezahayo A, Haverich A, Chichkov BN (2007) Two-photon polymerization technique for microfabrication of CAD-designed 3D scaffolds from commercially available photosensitive materials. J Tissue Eng Regen Med 1(6):443–449
17. Basu S, Wolgemuth CW, Campagnola PJ (2004) Measurement of normal and anomalous diffusion of dyes within protein structures fabricated via multiphoton excited cross-linking. Biomacromolecules 5(6):2347–2357
18. Ovsianikov A (2009) Investigation of two-photon polymerization technique for applications in photonics and biomedicine, 1. Aufl., Cuvillier, Göttingen
19. Bertsch A, Jiguet S, Bernhard P, Renaud P (2002) Microstereolithography: a review. In: Pique A, Holmes AS, Dimos DB (eds) Materials research society symposium proceedings, vol. 758, Boston, MA, 3–5 December 2002
20. Göppert-Mayer M (1931) Über Elementarakte mit zwei Quantensprüngen. Ann Phys 401: 273–294
21. Kaiser W, Garrett CGB (1961) Two-photon excitation in $CaF_2:Eu^{2+}$. Phys Rev Lett 7(6): 229–231
22. Shen YR (1984) The principles of nonlinear optics (pure and applied optics series: 1–349). Wiley, New York
23. Haas KH, Wolter H (1999) Synthesis, properties and applications of inorganic–organic copolymers (ORMOCER®s). Curr Opin Sol State Mater Sci 4:571–580
24. Serbin J, Ovsianikov A, Chichkov B (2004) Fabrication of woodpile structures by two-photon polymerization and investigation of their optical properties. Opt Express 12(21): 5221–5228
25. Microchem: http://www.microchem.com/Prod-SU8_KMPR.htm
26. Lum L, Elisseeff J (2003) Injectable hydrogels for cartilage tissue engineering. In: Ashammakhi N, Ferretti P (eds) Topics in tissue engineering 2003, eBook
27. Albrecht DR, Tsang VL, Sah RL, Bhatia SN (2005) Photo- and electropatterning of hydrogel-encapsulated living cell arrays. Lab Chip 5 (1):111–118
28. Ovsianikov A, Chichkov B, Adunka O, Pillsbury H, Doraiswamy A, Narayan RJ (2007)

Rapid prototyping of ossicular replacement prostheses. Appl Surf Sci 253(15):6603–6607

29. Chong S, Fung HL (1989) Transdermal drug delivery systems: pharmacokinetics, clinical efficacy, and tolerance development. In: Hadgraft J, Guy RH (eds) Transdermal drug delivery: developmental issues and research initiatives. Marcel Dekker, New York, NY, p 135
30. Flynn GL (1996) Cutaneous and transdermal delivery: processes and systems of delivery. In: Banker GS, Rhodes CT (eds) Modern pharmaceutics. Marcel Dekker, New York, NY, pp 239–299
31. Sivamani RK, Stoeber B, Wu GC, Zhai H, Liepmann D, Maibach H (2005) Clinical microneedle injection of methyl nicotinate: stratum corneum penetration. Skin Res Technol 11(2):152–156
32. Mitragotri S (2001) Effect of therapeutic ultrasound on partition and diffusion coefficients in human stratum corneum. J Control Release 71 (1):23–29
33. Guy RH (1998) Iontophoresis – recent developments. J Pharm Pharmacol 50(4):371–374
34. Barry BW (2001) Novel mechanisms and devices to enable successful transdermal drug delivery. Eur J Pharm Sci 14(2):101–114
35. Kiyan Y, Limbourg A, Kiyan R, Tkachuk S, Limbourg FP, Ovsianikov A, et al. (2011) Urokinase receptor associates with myocardin to control vascular smooth muscle cells phenotype in vascular disease. Arterioscler Thromb Vasc Biol 32:110–122
36. Ovsianikov A, Gruene M, Pflaum M, Koch L, Maiorana F, Wilhelmi M, et al. (2010) Laser printing of cells into 3D scaffolds. Biofabrication 2:014104
37. Ovsianikov A, Deiwick A, Van Vlierberghe S, Dubruel P, Möller L, Dräger G, et al. (2011) Laser fabrication of three-dimensional CAD scaffolds from photosensitive gelatin for applications in tissue engineering. Biomacromolecules 12:851–858
38. Ovsianikov A, Deiwick A, Van Vlierberghe S, Pflaum M, Wilhelmi M, Dubruel P, et al. (2011) Laser fabrication of 3D gelatin scaffolds for the generation of bioartificial tissues. Materials 4:288–299
39. Gittard SD, Ovsianikov A, Akar H, Chichkov B, Monteiro-Riviere NA, Stafslien S, et al. (2010) Two photon polymerization-micromolding of polyethylene glycol-gentamicin sulfate microneedles. Adv Eng Mater 12:B77–B82
40. Gittard SD, Ovsianikov A, Chichkov BN, Doraiswamy A, Narayan RJ (2010) Two-photon polymerization of microneedles for transdermal drug delivery. Expert Opin Drug Deliv 7:513–533
41. Ovsianikov A, Chichkov B, Mente P, Monteiro-Riviere NA, Doraiswamy A, Narayan RJ (2007) Two photon polymerization of polymer? Ceramic hybrid materials for transdermal drug delivery. Int J Appl Ceramic Technol 4:22–29
42. Ovsianikov A, Ostendorf A, Chichkov B (2007) Three-dimensional photofabrication with femtosecond lasers for applications in photonics and biomedicine. Appl Surf Sci 253:6599–6602
43. Doraiswamy A, Ovsianikov A, Gittard SD, Monteiro-Riviere NA, Crombez R, Montalvo E et al. (2010) Fabrication of microneedles using two photon polymerization for transdermal delivery of nanomaterials. J Nanosci Nanotechnol 10:6305–6312
44. Boland T, Ovsianikov A, Chichkov BN, Doraiswamy A, Narayan RJ, Yeong W-Y, et al. (2007) Rapid prototyping of artificial tissues and medical devices. Adv Mater Process 165:51–53
45. Gittard SD, Narayan R, Lusk J, Morel P, Stockmans F, Ramsey M, et al. (2009) Rapid prototyping of scaphoid and lunate bones. Biotechnol J 4:129–134

Chapter 20

Direct Fabrication as a Patient-Targeted Therapeutic in a Clinical Environment

Dietmar W. Hutmacher, Maria Ann Woodruff, Kevin Shakesheff, and Robert E. Guldberg

Abstract

A paradigm shift is taking place in orthopaedic and reconstructive surgery. This transition from using medical devices and tissue grafts towards the utilization of a tissue engineering approach combines biodegradable scaffolds with cells and/or biological molecules in order to repair and/or regenerate tissues. One of the potential benefits offered by solid freeform fabrication (SFF) technologies is the ability to create such biodegradable scaffolds with highly reproducible architecture and compositional variation across the entire scaffold due to their tightly controlled computer-driven fabrication. Many of these biologically activated materials can induce bone formation at ectopic and orthotopic sites, but they have not yet gained widespread use due to several continuing limitations, including poor mechanical properties, difficulties in intraoperative handling, lack of porosity suitable for cellular and vascular infiltration, and suboptimal degradation characteristics. In this chapter, we define scaffold properties and attempt to provide some broad criteria and constraints for scaffold design and fabrication in combination with growth factors for bone engineering applications. Lastly, we comment on the current and future developments in the field, such as the functionalization of novel composite scaffolds with combinations of growth factors designed to promote cell attachment, cell survival, vascular ingrowth, and osteoinduction.

Key words: Tissue engineering, Scaffolds, Rapid prototyping, Composites, Mesenchymal stem cells, Growth factors

1. Introduction

Using a hierarchical framework, tissue engineering can be subdivided into different strategies or concepts (1). One strategy is purely cell based involving the transplantation of autologous or allogeneic cell suspensions or cell sheets that are injected and/or transplanted to a defect site or an injured tissue. Another strategy utilizes biomolecules (growth factors, completely lyophilized cell fractions, peptides, polysaccharides, etc.) aimed at delivering cues

Michael A.K. Liebschner (ed.), *Computer-Aided Tissue Engineering*, Methods in Molecular Biology, vol. 868, DOI 10.1007/978-1-61779-764-4_20, © Springer Science+Business Media, LLC 2012

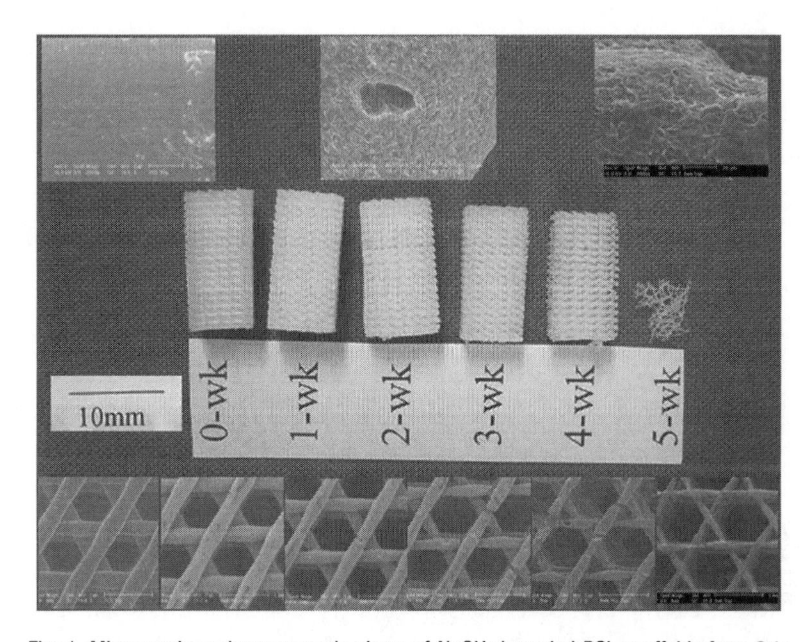

Fig. 1. Microscopic and macroscopic views of NaOH-degraded PCL scaffolds from 0 to 5 weeks. *Top inset* (**a**) shows SEM micrographs (×2,000 magnifications) of the surface texture of the scaffold filaments over time: 0, 1, and 5 weeks (*left to right*). (**b**) Macroscopic view of the degraded scaffolds. *Bottom inset* (**c**) shows SEM micrographs (×64) from 0 to 5 weeks. The scaffolds were observed to degrade via a surface erosion pathway homogenously throughout the scaffold structure, through the thinning of the filament diameters. Image reproduced from reference 2 with permission from Polymer International.

to the cells of the host tissue. Other methods are based on the use of different types of matrices (hydrogels, microspheres/beads, etc.) in combination with cells and/or biomolecules. The fourth and most frequently applied strategy focuses on seeding and culturing specific cell types in 3D environments that closely mimic natural extracellular matrix (ECM). Such 3D environments are specifically configured as cellular solids and referred to as "scaffolds" in tissue engineering nomenclature. Figure 1 depicts a series of 3D scaffolds fabricated using solid freeform fabrication (SFF), which have been degraded in NaOH over time (2).

It can be argued that the "scaffold-based tissue engineering concept" was introduced in the mid-1980s when Dr. Joseph Vacanti of the Children's Hospital approached Dr. Robert Langer of MIT with an idea to design scaffolds for cell delivery as opposed to seeding cells onto or mixing cells into naturally occurring matrices with physical and chemical properties that are difficult to manipulate (3). Today's concepts of scaffold-based tissue engineering involve the combination of a scaffold with cells and/or biomolecules that promote the repair and/or regeneration of tissues (4). These tissue-engineered constructs (TECs) are under intense investigation and various approaches and strategies are continually being developed. However, despite intense efforts, the ideal scaffold/cell or scaffold/neo-tissue construct (even for a specific tissue type) has yet to be developed. Certain minimum requirements are essential when developing TECs that must address the biochemical as well as chemical and physical properties of native tissue. Of these requirements, biocompatibility, vascularization, and chemotaxis are vital (5, 6). The scaffold must also be nonimmunogenic and free from prions involved in disease transmission and it must also possess suitable architectural qualities that are easily reproducible. There are also sterilization and delivery issues to contend with. Lastly, one must consider the temporal and spatial variations in some of these factors both in vitro and/or in vivo (7).

Despite rapid advancements in the fabrication of TECs, tissue engineering is still a long way off matching Natures' ability to grow and repair tissues and organs. Nevertheless, from an engineering perspective, the past decade has seen significant advances in scaffold design and fabrication (8). Starting with simple foams and fibers, porosity and pore interconnectivity has quickly become a central theme, with pore dimensions and material properties being vital in promoting cell seeding, migration, proliferation, and the de novo production of ECM. Significant advances have also been made in the incorporation of bioactive molecules into scaffolds, specifically in the area of bone engineering (9–12).

Native bone tissue is a major storage site for growth factors, especially for those characterized by their high affinity to heparin (heparin-binding growth factors—HBGFs), such as fibroblast growth factors (FGFs), transforming growth factor-βs (TGF-βs), and bone morphogenic proteins (BMPs). BMPs are biologically active molecules capable of inducing new bone formation and have shown significant potential for clinical use in the repair of bone defects (13, 14). Although there have been several successful studies using BMPs, there remain many unanswered questions, such as, from a more general perspective, release kinetics and timing of administration and, more specifically, patient site specificity; for example, the ideal dose of BMP in an anterior cervical spine fusion may be quite different to that of a cranial defect. Carriers for BMP play an equally important role pertaining to the optimal delivery leading to predictable and functional bone regeneration (15). In this chapter, we discuss broad criteria and constraints for SFF-based scaffold design and fabrication and then summarize the current knowledge of growth factor delivery for bone engineering applications. Lastly, we comment on the current and future developments in this domain, such as the functionalization of novel composite scaffolds with combinations of growth factors, e.g., immobilization of vascular endothelial growth factor (VEGF)/PDGF and BMP.

2. Scaffolds for Bone Engineering Fabricated by Solid Freeform Fabrication

Reviewing the current literature, it can be concluded that a conceptual shift in thinking is taking place in orthopaedic and reconstructive surgery, specifically the departure from using auto- and allografts towards a tissue engineering approach that uses biodegradable scaffolds integrated with biological cells or molecules, to regenerate tissues. This new paradigm requires scaffolds that balance temporary mechanical function with morphological properties (pore architecture, size, and interconnectivity) to aid biological delivery and tissue regeneration (16). Scaffold fabrication is a significant hurdle; complex scaffold architecture designs generated using hierarchical image-based or computer-aided design (CAD) techniques usually cannot be readily built using conventional techniques. Instead, such scaffold architectures must be built using layer-by-layer manufacturing processes known collectively as SFF. A number of review articles (2, 6) and book chapters (16, 17) exist which compare SFF scaffold fabrication methods; hence, this review focuses on how scaffolds have performed clinically in bone engineering, and discusses future directions for their use in regenerative medicine.

Melt extrusion-based SFF systems provide a powerful instrument for the generation of scaffold platforms. Recent advances in both computational scaffold design and SFF have made it possible for tissue engineers to design and fabricate a whole range of new scaffold types. One of the major benefits is the flexibility to create scaffolds with highly reproducible architecture and compositional, morphological variation across the entire matrix combined with the control of external geometry to match the boundaries of complex defects due to the computer-controlled fabrication process. The applications of SFF technologies in scaffold fabrication are wide and varied; however, only a small number have reached the clinical arena. An interdisciplinary group at the National University of Singapore has evaluated and patented the parameters necessary to process medical-grade polycaprolactone (mPCL) and mPCL composites (PCL/HA, PCL/TCP, etc.) by fused deposition modeling (FDM): a version of SFF. These so-called first-generation scaffolds have been studied for more than 5 years in a clinical setting and gained FDA approval via a 510-K application in 2006 (Osteopore Int). Schantz et al. (18) have used mPCL scaffolds as burr-hole plugs in a pilot study for cranioplasty. The clinical outcome after 12 months was positive, with all patients tolerating the implants with no adverse side effects reported. A functional and stable cranioplasty was observed in all cases. Today, more than 200 patients have received burr-hole plugs, scaffolds for orbital floor reconstruction, or other cranioplasties. Figure 2 (16) depicts schematically the application of these scaffolds in pig surgery and during human

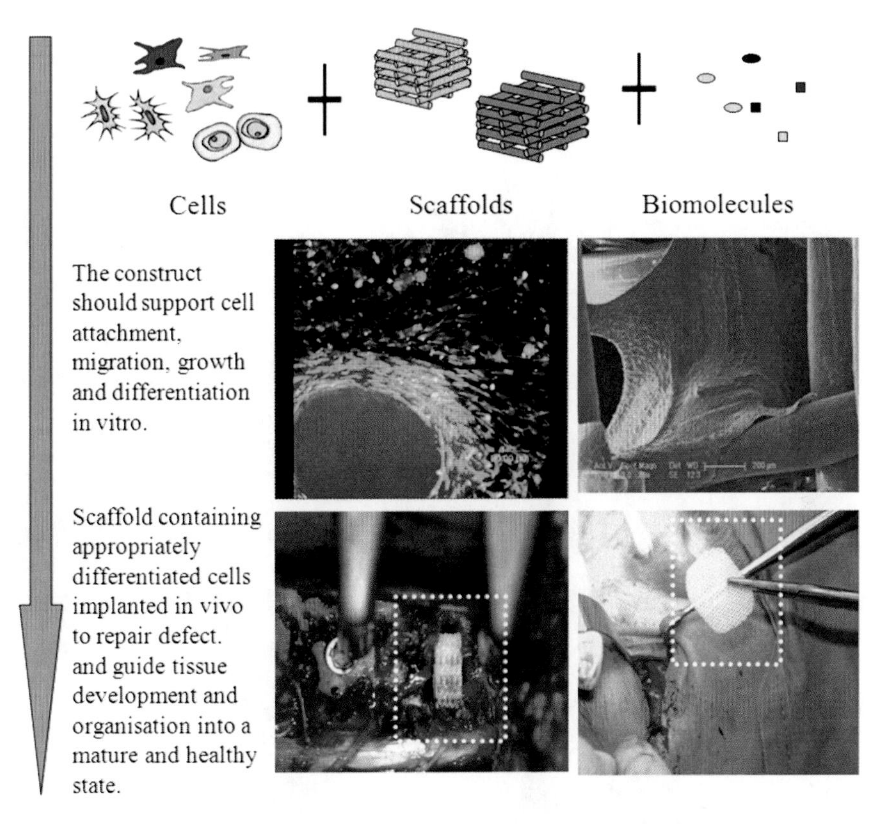

Fig. 2. Scaffold-based tissue engineering aims to promote the repair and/or regeneration of tissues through the incorporation of cells and/or biomolecules within a 3D scaffold system, which can be maintained in in vitro culture conditions until implantation. The lower two in vivo images show a medical-grade polycaprolactone–tricalcium phosphate (mPCL-TCP) composite scaffold ($14 \times 12 \times 5$ mm) being inserted during pig surgery in a spinal fusion model, and also an FDA-approved mPCL scaffold ($50 \times 50 \times 2$ mm) being utilized during orbital floor fracture repair in a patient. Depending on the size of the defect and biomechanical loading, the scaffold (Osteopore™, Singapore) can be used with or without cell and has shown in several in vitro and in vivo studies to support cell migration, growth, and differentiation and guide tissue development and organization into a mature and healthy state (16).

orbital floor fracture repair. Scaffolds can be combined with cells and/or biomolecules to produce a TEC or may be used in isolation.

The second-generation scaffolds produced by FDM for bone engineering are based on composites and have been evaluated both in vitro (19) and in vivo (7). Polymer/CaP composites confer favorable mechanical and biochemical properties, including strength via the ceramic phase and toughness and plasticity via the polymer phase; they also possess more favorable degradation and resorption kinetics, and graded mechanical stiffness. In these scaffolds, the ceramic phase is homogenously distributed in the matrix as well as exposed on the surface. An example of long-term implantation of a PCL-TCP composite scaffold, within a pig calvarial model, is shown in Fig. 3. Histological staining reveals excellent integration of the implanted scaffold, at 18 months, with new bone clearly evident through the centre of the scaffold. Contact angle

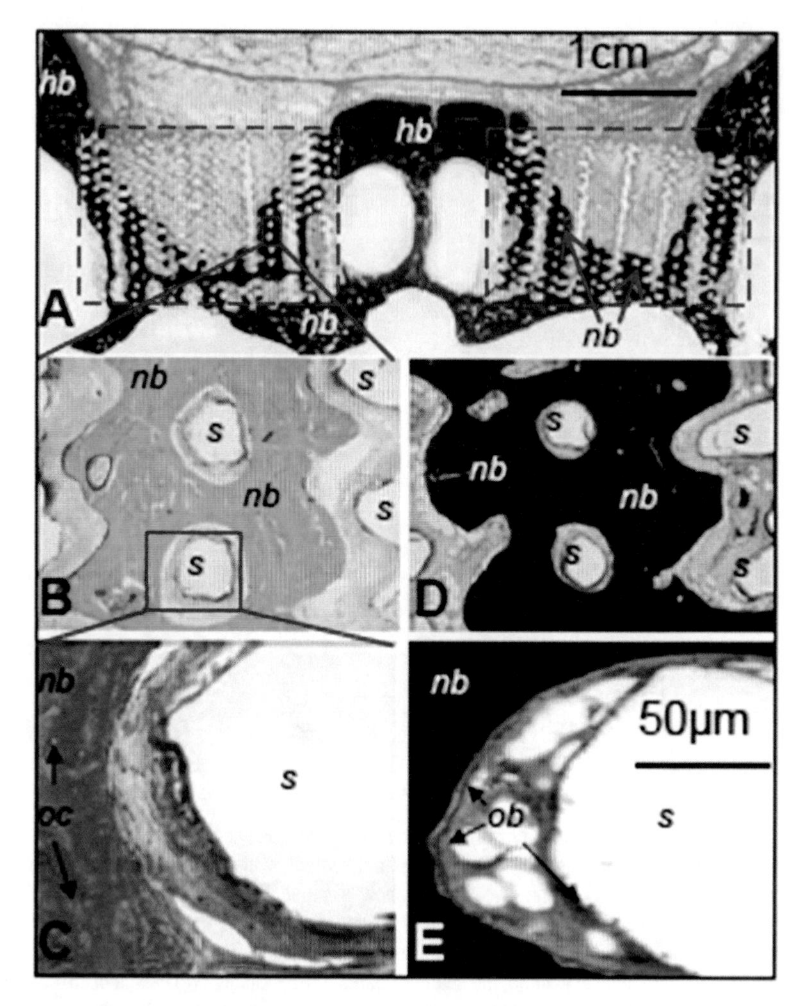

Fig. 3. Histological examination of explanted PCL-TCP scaffolds implanted for 18 months within a pig calvarial model. Von kossa (**a**, **d**, **e**) and goldners trichrome (**b**, **c**) staining reveals excellent integration of implanted scaffold with the host bone (*hb*) and a significant amount of new bone (*nb*) formation, as evidenced by osteoblasts (*ob*) and osteocytes (*oc*) within the centre of the scaffold construct. New bone penetrated the scaffold pores and osteoid can be observed around the scaffold struts (*s*).

measurements have shown that such composite surfaces were more hydrophilic and that the degradation kinetics were seen to be accelerated three to four times compared to PCL alone. Biochemical advantages include improved cell seeding, and enhanced control and/or simplification of the incorporation and immobilization of biological factors, such as BMPs (10).

2.1. Scaffolds Combined with Growth Factors and Other Biologically Active Molecules

Bone grafts are frequently used to treat conditions in load-bearing regions of the body. In the present climate of increasing life expectancy with an ensuing increase in bone-related injuries, bone grafting is gradually being replaced by bone engineering, where a scaffold is implanted to provide adequate load bearing and enhance tissue regeneration. However, scaffolds in combination with internal or external fixation are, in many cases, not sufficient to regenerate a critical-sized bone defect. Analysis of tissue engineering literature indicates that future generations of engineered scaffolds will not be successful by simply integrating drug delivery systems within the scaffolds. Instead, using knowledge from the fields of drug delivery within biomaterial science, multifunctional scaffolds – where the 3D template itself acts as a biomimetic programmable and multidrug delivery device – should be designed.

2.2. Accelerated Bone Repair with Growth Factors

Bone tissue is a major storage site for growth factors, especially for those characterized by their high affinity for polysaccharides, heparin, and other proteoglycans, such as the PDGFs, the TGF-βs, and the BMPs (20). BMPs are biologically active molecules capable of inducing new bone formation, and show potential for clinical use in bone defect repair. Three types of BMP-based bone tissue engineering therapies have been tried to date: cell therapy, gene therapy, and cytokine therapy. Cell therapy involves transplantation of autogenous bone marrow mesenchymal cells differentiated by BMP. Gene therapy involves transduction of genes encoding BMPs into cells at the repair site. Cytokine therapy involves recombinant BMP and carriers that retain and release BMP as needed, and is considered the most promising approach for practical use (21).

Many of the growth factors currently being investigated for fracture healing are HBGFs that are produced by the osteoblasts and stored within the matrix of bone. Of these, FGF-1 and FGF-2 have been extensively investigated as candidates for fracture healing. FGF-1 has been shown to aid in the bridging of a parietal bone critical defect (22) and to increase the bone–implant interface when used in combination with titanium-based scaffolds. FGF-2 has been more extensively studied for its use in fracture healing, and it has been shown that daily intravenous injections of FGF-2 for up to 2 weeks can enhance bone formation in rats (23). Members of the TGF-β superfamily have also been shown to be successful in enhancing osteogenesis. Both TGF-β1 and -β2 are known to increase differentiation of committed osteoprogenitor cells, and are potent stimulators of bone repair in calvarial and long-bone defects when administered alone or in combination with either insulin-like growth factor-I (IGF-I) or IGF-I/growth hormone. In addition, growth factors can directly regulate the availability of other growth factors; FGFs have been shown to regulate the expression of VEGF and HGF, factors that are also mitogenic for osteoprogenitor cells (24).

Likewise, combined local delivery of IGF-I in combination with TGF-β1 in a hydrogel scaffold has been shown to significantly enhance bone formation in a rat tibial segmental defect over IGF-I alone (25) suggesting that perhaps the mode of delivery also affects the potential for growth factor-mediated bone healing. It can be readily observed that much discussion in the growth-factor-delivery literature emphasizes the great importance of the delivery system for therapeutic effectiveness. The delivery system must fulfill clinical and wound-healing functions. These fundamental, but undefined, general performance targets are well known to surgeons and scientists designing and developing growth factor delivery systems.

Regrettably, specific details for performance targets have not been sufficiently defined. Consequently, rather than having a programmable delivery system available for PDGF and BMP, for example, contemporary scaffolds are nonprogrammable and arbitrary. Details on performance design targets from a translational research point of view are often lacking. Table 1 defines some broad scaffold criteria for growth factor delivery, and Table 2 details important considerations, which affect scaffold design and growth factor loading (26).

Unless the scaffold delivery system specifics are clearly defined, optimal performance targets for the delivery system and growth factor therapeutics will not be achieved. For example, in the clinic, a growth factor such as BMP may be released too rapidly from the delivery system after the surgeon places it at the clinical target.

Table 1
Role of scaffolds in growth factor delivery

Provide a convenient vehicle for the surgeon to place the growth factor at the clinical target
Be biocompatible prior to biodegradation, and the biodegradation products should be biocompatible
Localize, protect, and release the required dose of the growth factor at the appropriate time or times for therapeutic effect
Enable bone growth (osteoconduction), prevent soft tissue prolapse, and biodegrade in register with new bone formation

Table 2
Delivery system specifics

The period after injury, the delivery system should release the growth factor that is based on the wound healing kinetics (pharmacokinetics)
The temporal delivery, for example pulsatile and bolus
The dose of the growth factor to be delivered
The rate at which the scaffold/matrix system must be removed from the clinical target
How fast will bone form at a designated clinical target (long bones versus spine versus maxillofacial bone and cortical versus cancellous) and, therefore, at what rate must the scaffold/matrix system biodegrade?

Therefore, a therapeutic (i.e., physiological) dose may not be realized. Consequently, to compensate for this mismatch, a supraphysiologic dose (milligram versus nanograms) will be required at high cost. This problem occurs at present when milligram doses of BMP are needed for spinal fusions and nonunions (24).

The dilemma of the mismatch between delivery system properties and clinical target requirements has been a serious issue that merits significant attention (27). Recent reviews on delivery systems for BMPs and other growth factors have discussed poly(demineralized bone matrix, hyaluronan, gelatin), a hyaluronan–polyethylene glycol combination, calcium phosphates (tricalciumphosphate and apatitic hydroxyapatites), and collagen (REF). Collagen, for example, has been selected as a delivery system/scaffold for rhBMP-2 for the *InFuse* product, produced by Medtronic, and also for rhBMP-7 (also known as *OP-1*), produced by Stryker. The logic for using collagen as a delivery vehicle relates to type I bovine collagen having FDA approval, and collagen carriers typically have a biphasic release profile characterized by an initial burst release of BMP with a small sustained release thereafter. However, it may not be optimal to have a rapid bolus release from collagen during the destructive phase of osseous wound injury repair. In the destructive phase, the environment is acidotic, hypoxic, and lytic, and constructive cells may be inadequate to respond to BMP. Moreover, BMP may be squandered by lytic degradation. Therefore, to achieve a clinically effective BMP dose, potentially supraphysiologic loading (in milligrams) of BMP is required. Hence, one might argue from a clinical point of view that this poorly designed delivery system substantially escalates both the cost and risk to the patient. There, thus, exists a compelling clinical need to design and develop a programmable delivery system scaffold that allows multiple growth factor release during the constructive phase of wound healing (28).

To our knowledge, no multiple-growth-factor-releasing scaffold systems of high porosity (>80%) are currently clinically available for the treatment of medium-to-high load-bearing bone defects. To address this therapeutic challenge, groups are now looking to marry leading-edge scaffold technologies; for example, biomechanically loadable composite scaffolds (produced by CAD and rapid prototyping) and microparticle delivery systems, incorporating important bone-related growth factors which possess controllable release kinetics, as depicted schematically in Fig. 4. Groups are presently working on combining a well-established scaffold technology with controlled-release technology (29–34) to provide a leading-edge solution to the therapeutic challenge of critical-sized bone defect regeneration.

In a collaborative effort between Hutmacher's group (QUT, Australia), Shakesheff's group (University of Nottingham, England), and Guldberg's group (Georgia Tech, USA), it has been hypothesized that a composite scaffold (already successfully

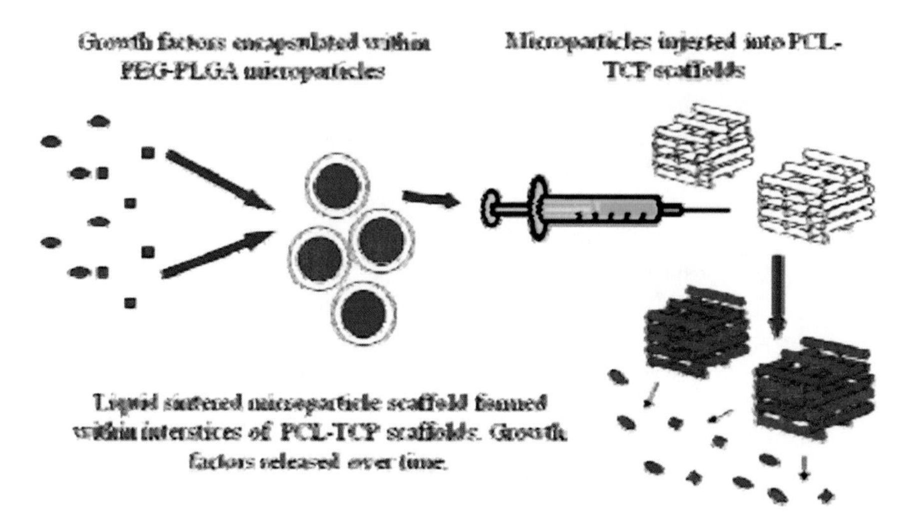

Fig. 4. A microparticle system allows controllable release of multiple GFs by use of biodegradable plasticizers comprising triblock copolymer PEG-PLGA-PEG. The aqueous suspension of particles becomes a microporous liquid-sintered scaffold at body temperature (upon injection into the high-strength PCL-TCP scaffolds), thus filling the 500–1,000-μm PCLTCP scaffold pores with a foam-like microporous structure. GFs are then released over time from the foam-encapsulated microparticles.

utilized in low load-bearing bone defects) can be biomechanically optimized and combined with controlled delivery of angiogenic (PDGF/VEGF) and osteoinductive (BMP) molecules producing a biologically active engineered bone graft system with mechanical properties suitable for load-bearing applications, in combination with internal or external fixation (35, 36) as shown in Fig. 5.

The first aim of this engineered bone graft system would be to stimulate vascularization; thus, one must consider that all vessel formation results from a multistep sequential cascade in which multiple factors temporally interact. For example, angiogenesis starts with the activation and migration of endothelial cells into the surrounding matrix, and the formation of an immature and unstable vessel network (37). After a series of remodeling and pruning processes, a stable and mature vessel network is formed by the recruitment of smooth muscle cells and pericytes. During this process, VEGF and angiopoietin-2 act synergistically to destabilize preexisting blood vessels, induce degradation of basement membrane, and promote the proliferation and migration of endothelial cells. Other factors, such as angiopoietin-1 and PDGF-BB (human recombinant PDGF), act at later stages to stabilize the newly formed vessels. Chemoattraction of pericytes and smooth muscle cells helps to stabilize capillary network formation by a PDGF-BB-controlled process. Recently, it was shown that VEGF is able to stimulate angiogenesis and acts synergistically with BMP2 in bone repair in a rat skull defect model (35). Hence, it

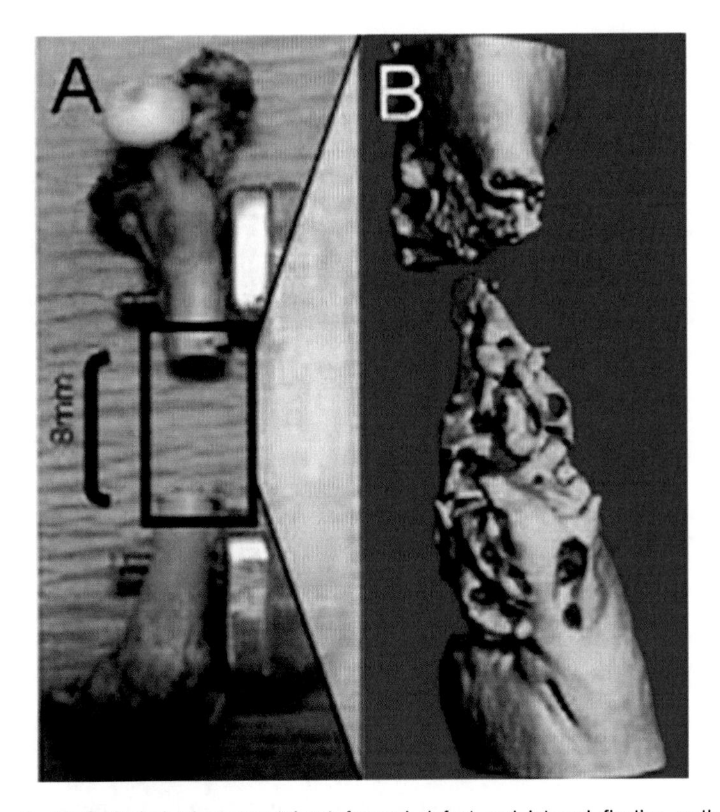

Fig. 5. (**a**) Critical-sized segmental rat femoral defect and internal fixation method. (**b**) Incomplete bone bridging 12 weeks following implantation of a PCL scaffold without cells and/or growth factors.

may be hypothesized that sustainable and time-controlled release of VEGF165 (human recombinant VEGF), PDGF-BB, and BMP2 provides an accelerated bone-regenerative environment within a critical defect site. In order to facilitate the formation and stabilization of endothelial tubules in constructs during initial healing, they are loaded with microspheres, which contain VEGF165 on the surface and PDGF-BB in the bulk material. A second slow-degrading microsphere system delivers BMP2 to stimulate the osteogenic differentiation of progentitor cells and preosteoblasts moving into the defect area 7–10 days post-surgery; this timescale of release is depicted schematically in Fig. 6.

Methodologies are, thus, underway to engineer scaffolding devices that sustain tissue growth and maturation in the presence of functionalized growth factor combinations. Therefore, successful bone tissue engineering could be achieved through the delivery of bioactive growth factors in bioresorbable scaffolding matrices functionalized with VEGF/PDGF and BMP. This method is particularly attractive as it can boost local osteoprogenitor cell recruitment, proliferation, and differentiation.

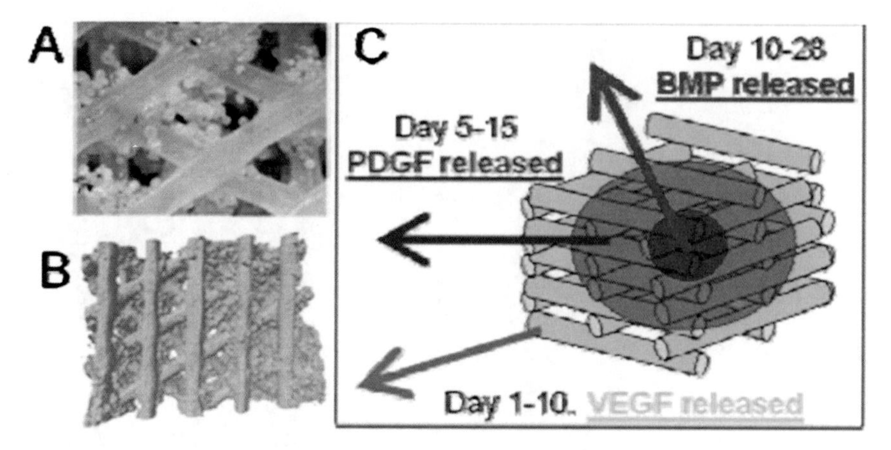

Fig. 6. (**a**) Preliminary data showing microparticles (containing growth factors) sintered to a PCL-TCP scaffold. (**b**) Microparticle distribution throughout the scaffold was verified by microCT. (**c**) Schematic of zonal and temporal release of growth factors from scaffolds. Microparticles comprising different percentage of PEG-PLGA-PEG degrade at different rates releasing growth factors based on the targeted release criteria.

3. Conclusions

Solid freeform fabrication techniques are experiencing increasing application in many biomedical fields, including regenerative medicine and tissue engineering. In general, scaffold-based tissue engineering requires a well-defined internal structure with interconnected porosity and SSF techniques allow us to design and fabricate such scaffolds. From a biological point of view, the designed scaffold should serve several functions, including (1) the ability to act as an immobilization site for transplanted cells, (2) the formation of a protective space to prevent unwanted tissue growth into the wound bed and allow healing with differentiated tissue, (3) directing the migration or growth of cells via surface properties of the scaffold, and (4) directing migration or growth of cells via release of soluble molecules, such as growth factors, hormones, and/or cytokines. Future work must provide further compelling evidence that growth factor-loaded scaffolds offer the right balance of capability and practicality to be suitable for fabrication of materials in sufficient quantity and quality to move holistic tissue engineering technology platforms into a clinical application.

In addition to the considerations of scaffold performance-based tissue engineering strategies, practical considerations pertaining to manufacture also arise. From a clinical point of view, it must be possible to manufacture scaffolds under good manufacturing practice (GMP) conditions in a reproducible and quality-controlled fashion at an economic cost and speed. To move the current *off-the-shelf* tissue engineering practices to the next frontier, manufacturing processes will be required to accommodate the incorporation of multiple growth factors with different release kinetics. These approaches are working towards the creation of a customized, highly reproducible construct possessing a controlled spatial distribution of growth factors. Constructs should feature the additional versatility of controlled microstructure, degradability, and mechanical properties inherent to the chosen scaffold material. This would facilitate the ultimate regeneration of soft or hard tissue defects within a specific anatomical site.

References

1. Hutmacher DW, Cool S (2007) Concepts of scaffold-based tissue engineering – the rationale to use solid free-form fabrication techniques. J Cell Mol Med 11(4):654–669
2. Lam CXF, Teoh SH, Hutmacher DW (2007) Comparison of degradation of PCL and PCL-TCP scaffolds in alkaline medium. Polym Int 6:718–728
3. Vacanti CA (2006) The history of tissue engineering. J Cell Mol Med 10:569–576
4. Hutmacher DW, Sittinger M, Risbud MV (2004) Scaffold-based tissue engineering: rationale for computer-aided design and solid free-form fabrication systems. Trends Biotechnol 22:354–362
5. Bach AD, Arkudas A, Tjiawi J, Polykandriotis E, Kneser U, Horch RE, Beier JP (2006) A new approach to tissue engineering of vascularized skeletal muscle. J Cell Mol Med 10:716–726
6. Kneser U, Stangenberg L, Ohnolz J, Buettner O, Stern-Straeter J, Mobest D, Horch RE, Stark GB, Schaefer DJ (2006) Evaluation of processed bovine cancellous bone matrix seeded with syngenic osteoblasts in a critical size calvarial defect rat model. J Cell Mol Med 10:695–670
7. Zhou Y, Chen F, Ho ST, Woodruff MA, Lim TM, Hutmacher DW (2007) Combined marrow stromal cell-sheet techniques and high-strength biodegradable composite scaffolds for engineered functional bone grafts. Biomaterials 28:814–824
8. Hollister SJ (2005) Porous scaffold design for tissue engineering. Nat Mater 4:518–524
9. Oest ME, Dupont KM, Kong H-J, Mooney DJ, Guldberg RE (2005) Quantitative assessment of scaffold and growth factor-mediated repair of critically sized bone defects. J Orthop Res 17415756 (P,S,E,B,D)
10. Rai B, Teoh SH, Hutmacher DW, Cao T, Ho KH (2005) Novel PCL-based honeycomb scaffolds as drug delivery systems for rhBMP-2. Biomaterials 26:3739–3748
11. Suciati T, Howard D, Barry J et al (2006) Zonal release of proteins within tissue engineering scaffolds J Mater Sci Mater Med 17 (11):1049–1056
12. Kanczler JM, Barry J, Ginty P et al (2007) Supercritical carbon dioxide generated vascular endothelial growth factor encapsulated poly (DL-lactic acid) scaffolds induce angiogenesis in vitro. Biochem Biophys Res Commun 352:135–141
13. Kain MS, Einhorn TA (2005) Recombinant human bone morphogenetic proteins in the treatment of fractures. Foot Ankle Clin 10:639–650, viii
14. Govender S, Csimma C, Genant HK, Valentin-Opran A, Amit Y, Arbel R, Aro H, Atar D, Bishay M, Borner MG, Chiron P, Choong P, Cinats J, Courtenay B, Feibel R, Geulette B, Gravel C, Haas N, Raschke M, Hammacher E, van der Velde D, Hardy P, Holt M, Josten C, Ketterl RL, Lindeque B, Lob G, Mathevon H, McCoy G, Marsh D, Miller R, Munting E, Oevre S, Nordsletten L, Patel A, Pohl A, Rennie W, Reynders P, Rommens PM, Rondia J, Rossouw WC, Daneel PJ, Ruff S, Ruter A, Santavirta S, Schildhauer TA, Gekle C, Schnettler R, Segal D, Seiler H, Snowdowne RB, Stapert J, Taglang G, Verdonk R, Vogels L, Weckbach A, Wentzensen A,Wisniewski T (2002) Recombinant human bone morphogenetic protein-2 for treatment of open tibial fractures: a prospective, controlled, randomized study of four hundred and fifty patients. J Bone Joint Surg Am 84-A:2123–2134
15. De Long WG, Jr., Einhorn TA, Koval K, McKee M, Smith W, Sanders R, Watson T (2007) Bone grafts and bone graft substitutes in orthopaedic trauma surgery. A critical analysis. J Bone Joint Surg Am 89:649–658
16. Hutmacher DW, Woodruff MA (2008) Design, fabrication and characterisation of scaffolds via solid free form fabrication techniques. In: Chu PK, Liu X (eds) Handbook of fabrication and processing of biomaterials. CRC Press/Taylor and Francis Group, pp 45–68

17. Hutmacher, Dietmar W, Woodruff, Maria A (2007) Composite scaffolds for bone engineering. In: Fakirov S, Bhattacharyya D (eds) Handbook of engineering biopolymers: homopolymers, blends and composites. Hanser Gardner, Germany, Munich, pp 773–798
18. Schantz JT, Lim TC, Ning C, Teoh SH, Tan KC, Wang SC, Hutmacher DW (2006) Cranioplasty after trephination using a novel biodegradable burr hole cover: technical case report. Neurosurgery 58:ONS-E176; discussion ONS-E176
19. Zhou YF, Sae-Lim V, Chou AM, Hutmacher DW, Lim TM (2006) Does seeding density affect in vitro mineral nodules formation in novel composite scaffolds? J Biomed Mater Res A 78:183–193
20. Mohan S, Baylink DJ (1991) Bone growth factors. Clin Orthop Relat Res 263:30–48
21. Hutmacher DW, Garcia AJ (2005) Scaffold-based bone engineering but using genetically modified cells. Gene 347:1–10
22. Cuevas P, de Paz V, Cuevas B, Marin-Martinez J, Picon-Molina M, Fernandez-Pereira A, Gimenez-Gallego G (1997) Osteopromotion for cranioplasty: an experimental study in rats using acidic fibroblast growth factor. Surg Neurol 47:242–246
23. Tabata Y (2008) Current status of regenerative medical therapy based on drug delivery technology. Reprod Biomed Online 16(1):70–80
24. Gruber R, Koch H, Doll BA, Tegtmeier F, Einhorn TA, Hollinger JO (2006) Fracture healing in the elderly patient. Exp Gerontol 41(11):1080–1093
25. SSIB, AREJ (2004) Bone defect repair in rat tibia by TGF-beta1 and IGF-1 released from hydrogel scaffold. Cell Tissue Bank 5:223–230.136
26. Giannoudis PV, Einhorn TA, Marsh D (2007) Fracture healing: the diamond concept. Injury 38(Suppl 4):S3–S6
27. S, Ryder J, Shemilt I, Mugford M, Harvey I, Song F (2007) Clinical effectiveness and cost-effectiveness of bone morphogenetic proteins in the non-healing of fractures and spinal fusion: a systematic review. Health Technol Assess 11(30):1–150, iii–iv
28. Robinson Y, Heyde CE, Tschöke SK, Mont MA, Seyler TM, Ulrich SD (2008) Evidence supporting the use of bone morphogenetic proteins for spinal fusion surgery. Expert Rev Med Devices 5(1):75–84
29. Shakesheff KM, France RM, Quirk RA Porous matrix, inventors. International Patent WO2004084968-A1, GB2415142-A, EP1605984-A1, Regentec LTD
30. Pratoomsoot C et al (2008) A thermoreversible hydrogel as a biosynthetic bandage for corneal wound repair. Biomaterials 29(3): 272–281
31. Howdle SM, Watson MS, Whitaker MJ, Popov VK, Davies MC, Mandel FS, Don WJ, Shakesheff KM et al (2001) Supercritical fluid mixing: preparation of thermally sensitive polymer composites containing bioactive materials. Chem Commun 1:109–110
32. Whitaker MJ et al (2005) The production of protein-loaded microparticles by supercritical fluid enhanced mixing and spraying. J Control Release 101(1–3):85–92
33. Suciati T, Daniel H, Barry J, Everitt N, Shakesheff K, Rose F (2006) Zonal release of proteins within tissue engineering scaffolds. J Mater Sci Mater Med 17:1049
34. Kanczler JM et al (2007) Supercritical carbon dioxide generated vascular endothelial growth factor encapsulated poly(DL-lactic acid) scaffolds induce angiogenesis in vitro. Biochem Biophys Res Commun 352(1):135–141
35. Oest ME, Dupont KM, Kong HJ, Mooney DJ, Guldberg RE (2007) Quantitative assessment of scaffold and growth factor mediated repair of critically-sized bone defects. J Orthop Res 25(7):941–950
36. Rai B, Oest ME, Dupont KM, Ho KH, Teoh SH, Guldberg RE (2007) Combination of platelet rich plasma with polycaprolactone-tricalcium phosphate scaffolds for segmental bone defect repair. J Biomed Mater Res 81 (4):888–899
37. Peng H et al (2005) Synergistic enhancement of bone formation and healing by stem cell-expressed VEGF and bone morphogenetic protein-4. J Bone Miner Res 20:2017

Chapter 21

Microstereolithography-Based Computer-Aided Manufacturing for Tissue Engineering

Dong-Woo Cho and Hyun-Wook Kang

Abstract

Various solid freeform fabrication technologies have been introduced for constructing three-dimensional (3-D) freeform structures. Of these, microstereolithography (MSTL) technology performs the best in 3-D space because it not only has high resolution, but also fast fabrication speed. Using this technology, 3-D structures with mesoscale size and microscale resolution are achievable. Many researchers have been trying to apply this technology to tissue engineering to construct medically applicable scaffolds, which require a 3-D shape that fits a defect with a mesoscale size and microscale inner architecture for efficient regeneration of artificial tissue. This chapter introduces the principles of MSTL technology and representative systems. It includes fabrication and computer-aided design/computer-aided manufacturing (CAD/CAM) processes to show the automation process by which measurements from medical images are used to fabricate the required 3-D shape. Then, various tissue engineering applications based on MSTL are summarized.

Key words: Microstereolithography, Scaffold, Tissue engineering, Computer-aided design/computer-aided manufacturing

1. Microstereo lithography Technology

1.1. Background

Microstereolithography (MSTL) technology is a novel micromanufacturing process that builds true three-dimensional (3-D) microstructures. It has a higher resolution than other solid freeform fabrication (SFF) technologies and is attractive for 3-D complex microstructures. Since it was first introduced in the early 1990s by Ikuta, many research groups have developed various systems to enhance the accuracy and widen the fields of application. Among other applications, the technology has been used for a lab-on-a-chip and microfluidics, as well as prototype fabrication. Recently, approaches have been developed to apply the technology to tissue engineering.

Michael A.K. Liebschner (ed.), *Computer-Aided Tissue Engineering*, Methods in Molecular Biology, vol. 868, DOI 10.1007/978-1-61779-764-4_21,

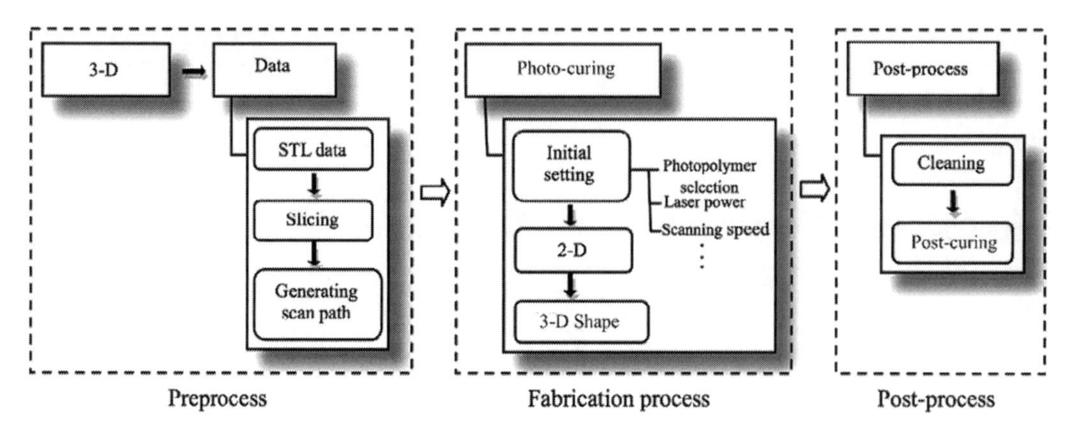

Fig. 1. Process diagram of MSTL.

1.2. 3-D Micro-structure Fabrication Process

As shown in Fig. 1, the MSTL process used to construct 3-D freeform shapes can be divided into three steps: preprocessing, fabrication, and post-processing. The following paragraph introduces each step in detail.

1.2.1. Fabrication

Figure 2a shows an MSTL system based on the scanning method widely used in conventional stereolithography (SL) (1). Each layer is fabricated using a filling process followed by a cross-sectional fabrication process. The filling process forms a new liquid layer of photopolymer on the solidified surface. In the cross-sectional fabrication process, a UV laser beam focused to a diameter of a few microns is irradiated on the open surface of a UV-curable liquid photopolymer. Photons existing in the UV light make a photoinitiator to produce free radicals and reactive cations. They subsequently polymerize monomers in the photopolymer, and the photopolymer becomes solidified. Various theoretical models for simulating the shape of the solidified polymer are proposed (2, 3). Figure 2b shows the theoretical shape of a polymer solidified by a focused laser beam (2). The focal point is set at the surface of the polymer and the intensity of the focused beam decreases with the beam's absorption into the polymer. The depth and width of the solidified region exponentially increase as the scanning speed decreases. To enhance the accuracy of a fabricated microstructure, it is necessary to verify the conditions required to obtain the desired solidified depth and width. Based on the solidification phenomena, a two-dimensional (2-D) cross-sectional shape is fabricated along the laser scanning path. A new layer of liquid photopolymer is then added. Consequently, a complex 3-D microstructure can be obtained by stacking up each cross-sectional shape. The 3-D shape illustrated in Fig. 3a was fabricated by stacking up the different cross sections shown in Fig. 3b; Fig. 3c shows the final result.

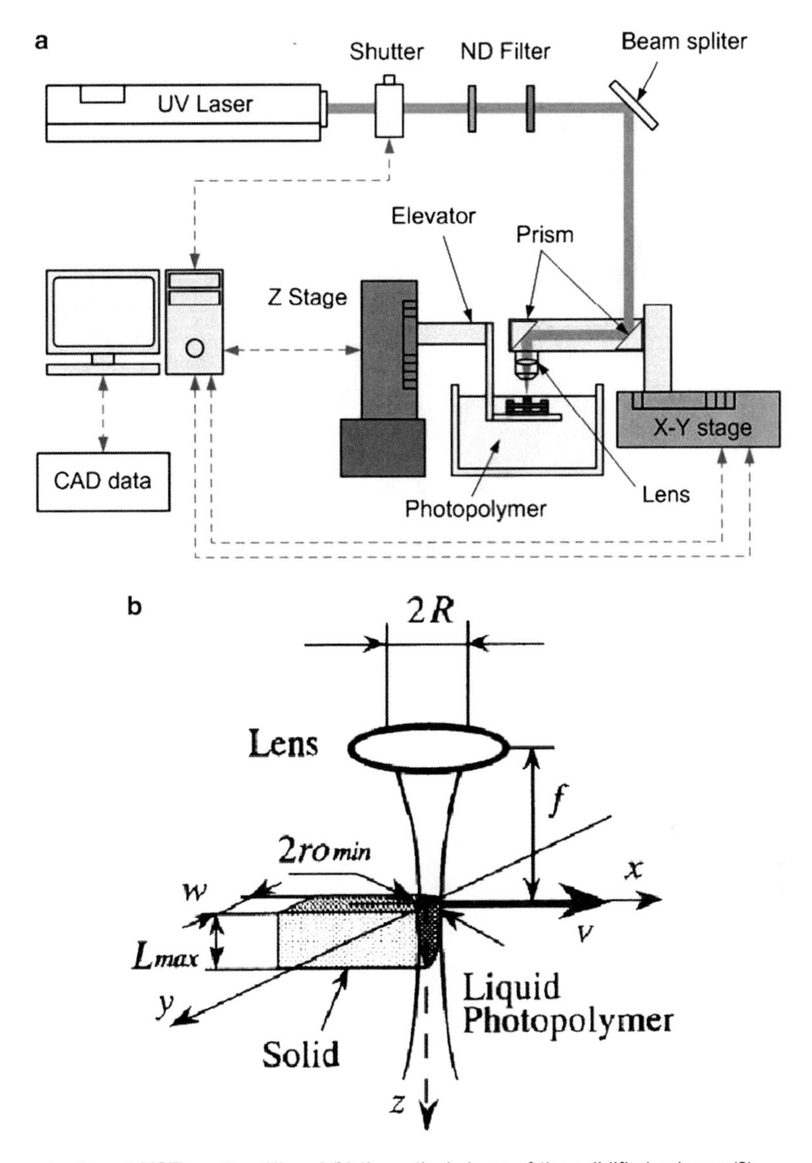

Fig. 2. (**a**) A scanning-based MSTL system (1) and (**b**) theoretical shape of the solidified polymer (3).

1.2.2. Post-processing

MSTL can be used to create a designed structure inside a photocurable resin reservoir. Unsolidified resin remains around the fabricated structure. Various cleaning solvents, such as isopropyl alcohol or ethanol, can be used to dissolve the unsolidified resin (5). Immersing the structure in a solvent can cause swelling, distorting the structure. Therefore, a solvent must be selected by considering not only the dissolution performance, but also the resulting distortion to the fabricated structure. The structure is irradiated again by a high-power UV lamp to terminate the polymerization of the residual resin that did not react perfectly in the fabricated structure.

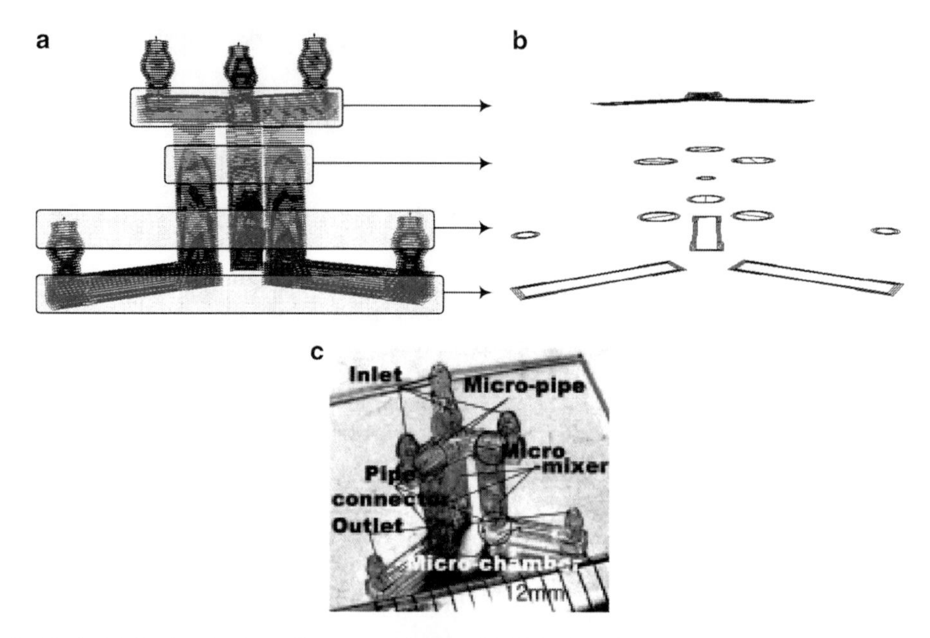

Fig. 3. (**a**) Stacked 3-D microstructure, (**b**) cross sections, and (**c**) fabricated 3-D microstructure (4).

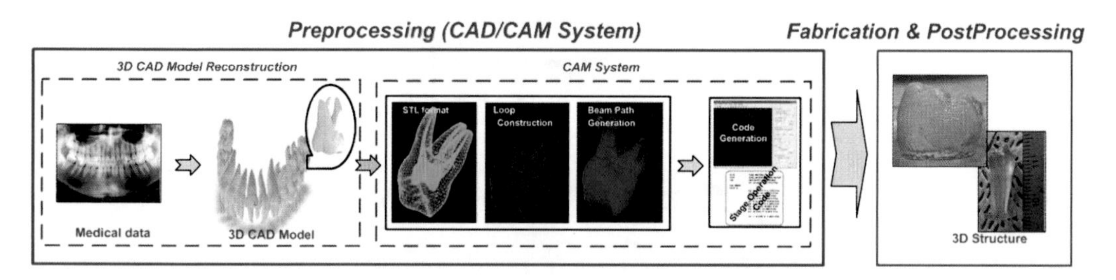

Fig. 4. Preprocessing procedure.

1.2.3. Preprocessing

Computer-aided design/computer-aided manufacturing (CAD/CAM) system preprocessing is used for automatic code programming. The code is loaded into the control computer of the MSTL system to fabricate a 3-D microstructure. This process can be divided into 3-D CAD model reconstruction and CAM processes for medical applications of MSTL (Fig. 4). The reconstruction is used to obtain a 3-D CAD model from voxel-based volumetric image data obtained from medical imaging equipment, such as computer tomography (CT) and magnetic resonance imaging (MRI). This procedure can be realized by applying reverse engineering technology based on a nonuniform rational B-spline (NURBS) surface or a general B-spline surface (6). Various software packages have been developed and commercialized for medical applications of CAD technology.

The CAM process is used to obtain the final code for fabricating 3-D structures using the reconstructed 3-D CAD model.

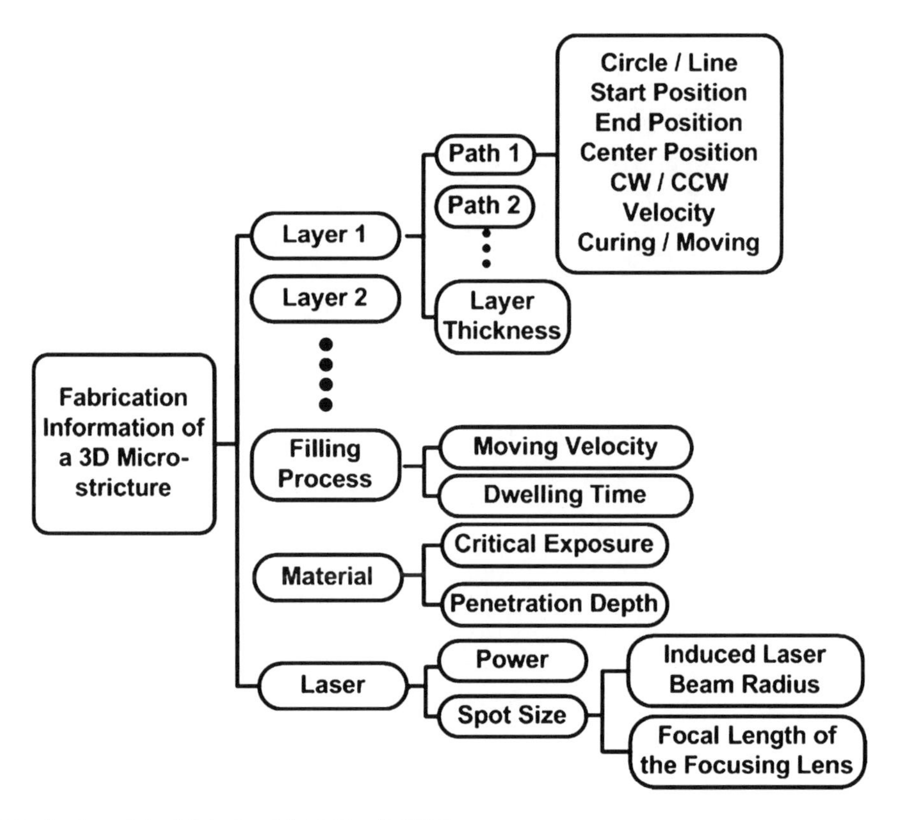

Fig. 5. Data structure of the fabrication information for MSTL.

Fabrication information, such as the beam path, layer thickness, and scanning speed, is required to generate the code. Figure 5 shows the data structure to represent detail fabrication process (4). The data include the fabrication information for each layer, since MSTL technology is a layer-by-layer process. These are obtained by storing the laser beam path and layer thickness in the data structure. The data also contain information about the filling process, material, and laser and optics settings for the MSTL fabrication process. In the data structure, the beam path data can be obtained by slicing a 3-D CAD model using computer technology. Usually, the STL CAD format is used to generate the sliced shape of a 3-D CAD model. This format, proposed by 3D Systems, Inc., is for a 3-D surface model composed of triangular facets. Each triangular facet is defined by three points and one normal vector (left image of Fig. 6). A 2-D sliced shape of a 3-D CAD model is generated by calculating the geometric intersection points between the slicing plane and the three lines in a facet (right image of Fig. 6). The required beam paths can be obtained by applying a beam path generation strategy with the sliced shape. For example, when a zigzag strategy is applied, the beam path information can be constructed by calculating intersection points between the 2-D sliced shape and intersecting lines defined by the user (Fig. 7).

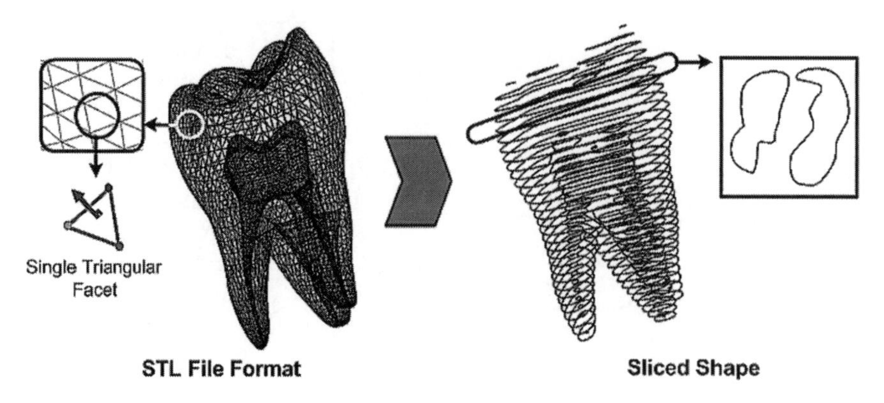

Fig. 6. STL format and sliced result of a molar CAD file.

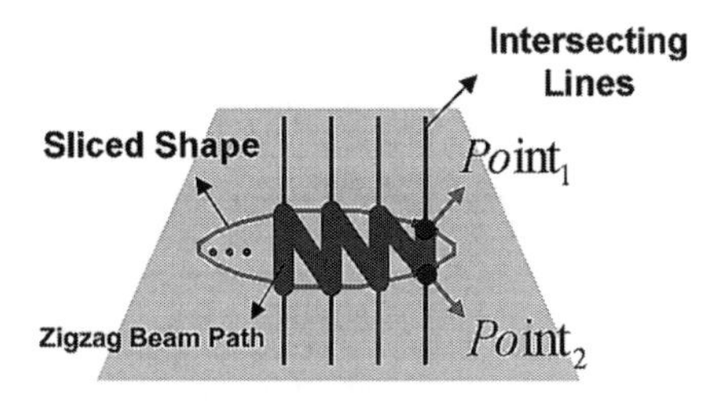

Fig. 7. Zigzag path generation by point calculations between the sliced shape and intersecting lines.

Various path generation strategies have been studied to fabricate precise 3-D structures (7, 8). The obtained beam path information is transferred into a code in the CAM system that will be loaded into the MSTL system to fabricate the 3-D structure.

1.3. Developed MSTL Systems

Ikuta first developed a series of integrated hardened (IH) polymer stereolithography processes for MSTL systems (9–12). He proposed a mass-IH process (Fig. 8a) to provide for both simultaneous scanning and uniform accuracy over the entire area. An array of single-mode optical fibers is utilized in this process. The total number of optical fibers can be infinite in principle. Ikuta also proposed a super-IH process, which can solidify a liquid polymer from a pinpoint position in 3-D space by focusing a laser beam inside the polymer due to the critical exposure below which the polymer does not begin to solidify (12). The resulting resolution is better than 1 μm in 3-D space (Fig. 9). Although the fabrication process based on the scanning method has a high resolution, it limits the yield rate for mass production because the fabrication of a 3-D structure generally takes a few hours. To avoid this disadvantage, Bertsch proposed a projection-based MSTL system using a liquid crystal display (LCD) or a digital micromirror device (DMD) as a dynamic pattern generator (13, 14). Because the dynamic mask was capable of fabricating a layer after just one irradiation, the

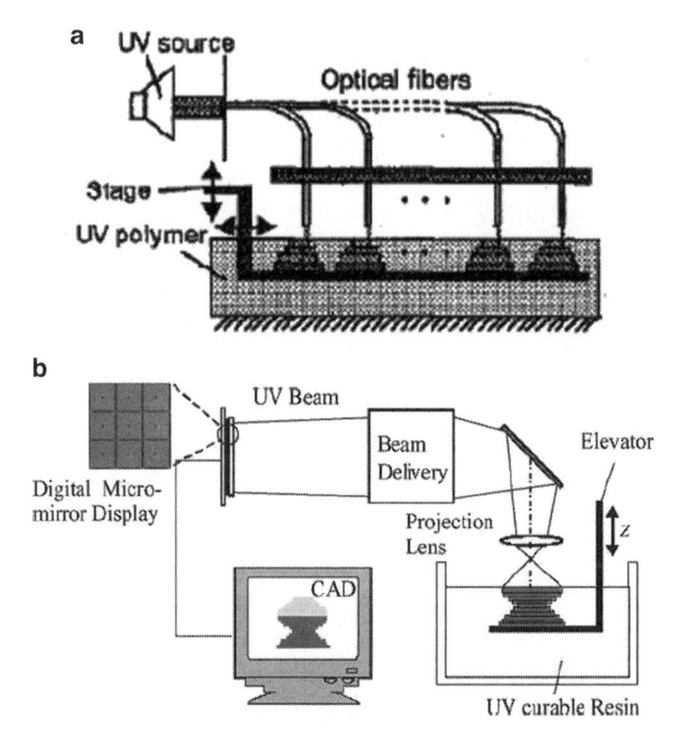

Fig. 8. (**a**) Mass-IH process (12) and (**b**) super MSTL apparatus using a pattern generator (15).

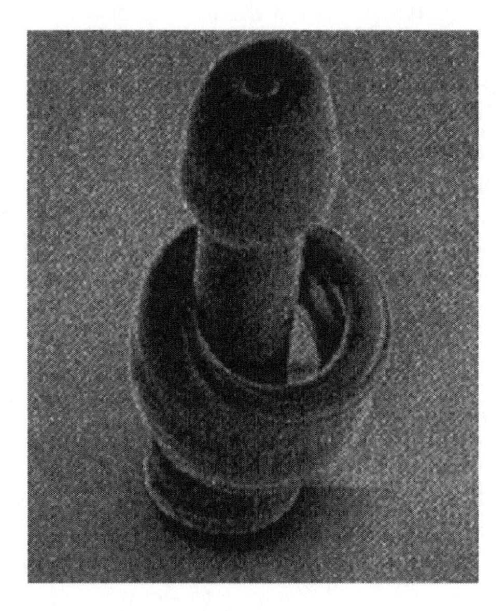

Fig. 9. Micro-quoits fabricated by the super-IH process (12).

process time was reduced dramatically. The LCD technique has some intrinsic drawbacks, such as large pixel sizes, low switching speed (~20 ms), low optical density of the refractive elements, and UV light absorption. Using a DMD as the pattern generator prevents these problems because its mirror elements have high reflexibility for UV light. The DMD modulates the light by collectively controlling the micromirror arrays to switch the light on and off for each individual pixel. Zhang fabricated 3-D microstructures, such as a matrix and microspring array, with 0.6-μm smallest features using MSTL technology with a DMD (Fig. 8b) (15). Roy enhanced the resolution by reducing the curing depth of a similar MSTL system by adding an absorbing dye (TINUVIN 234, 0.2 wt.%) into poly(ethylene glycol)diacrylate (PEGDA) (16). Only a small amount of monomer solution was required during fabrication because heavy liquid-like perfluorohexane acted as a "filler" material below the PEGDA.

2. Tissue Engineering Applications of MSTL

MSTL is useful for fabricating 3-D structures. Especially for scaffolds, the direct fabrication method performs well and is a simple process compared to indirect fabrication methods, which is introduced in Subheading 2.2. However, this method requires photocurable materials to fabricate a structure. Such a material limitation is a large barrier for scaffold fabrication. Many researchers are trying to overcome this difficulty by developing new photocurable biomaterials. In this section, applications of the direct fabrication method to tissue engineering are classified and described according to the biodegradability of the materials.

2.1. Direct Fabrication Technologies

2.1.1. MSTL Technology Using Biodegradable Materials

In tissue engineering, an implanted scaffold should eventually degrade itself in the body, whereas a variety of nonbiodegradable materials are used as supports for fixing bone defects, dental implant materials, or stents for extending blood vessels. However, if tissue regeneration is to be accomplished by tissue engineering technology, the implanted materials should be replaced by regenerated tissue. Biodegradable materials have been developed to satisfy such an eventual objective.

Biodegradable Photopolymer

MSTL is based on changing a liquid monomer or oligomer into a polymer by photopolymerization. Polypropylene fumarate (PPF), trimethylene carbonate (TMC)-based materials, and gelatin-based materials have been developed as a result of efforts to find adequate biodegradable materials.

PPF was first synthesized by Sanderson et al. through transesterification of diethyl fumarate and propylene glycol with a para-toluene sulfonic acid catalyst in 1988 (17). After that, various synthetic methods were developed by different researchers. The mechanical properties of synthetic PPF are similar to those of trabecular bone. This polymer can be polymerized by UV or heat,

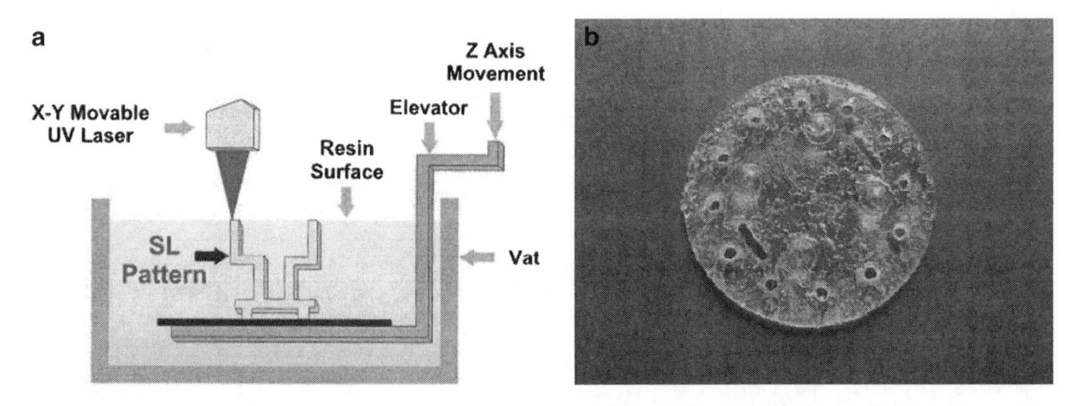

Fig. 10. (**a**) Laser-curing device (3D Systems 250/40 SLA) and (**b**) SLA-manufactured PPF/DEF prototype (18).

but solidification using UV is preferred in the field of scaffold fabrication because MSTL technology can make more precise 3-D structures. Since the first trial of UV polymerization using a conventional SL system by Mikos (18), several similar attempts have been made to use this technique for tissue engineering (Fig. 10).

And, Cho has presented a series of research results regarding scaffold fabrication using a high-resolution MSTL system and PPF/DEF (19–21). A high-accuracy scaffold that retained a pore size error of several tens of microns was fabricated (19) and the mechanical properties and characteristics of the cell adhesion were analyzed (Fig. 11) (19, 20). Surface modifications to the scaffolds for improving the cell adhesion and proliferation were made using biomimetic and RGD coatings. On the other hand, Yaszemski introduced new formula to improve mechanical properties of PPF (22).

TMC is polymerized by ring-opening polymerization, and various TMC-based photopolymers have been introduced by combining initiator materials, such as trimethylolpropane (TMP) and polyethylene glycol (PEG). Diverse TMC-based photopolymers have been synthesized by mixing TMC and TMP or TMC and PEG of various molecular weights. The photopolymerization characteristics and cell proliferation ability of the polymers were studied by Matsuda (23). In a recent application of this material to scaffold fabrication, TMC/TMP scaffold (Fig. 12) fabricated by MSTL was applied into cartilage tissue regeneration (1). Acrylate-based monomers consisting of a biodegradable basis monomer derived from gelatin and different reactive diluents were also synthesized (Fig. 13) (24) and the mechanical properties and biocompatibility of the photopolymers were investigated.

Composite Materials

Most polymers used for MSTL do not possess sufficient mechanical properties for bone tissue engineering. A variety of polymers and

Fig. 11. SEM images of a 3-D high-accuracy scaffold taken at various angles (19). Scaffold with 0.8mm line pitch is depicted in (**a & b**) while a scaffold with 0.5 mm line pitch is shown in (**c–f**)

inorganic compounds have been blended in attempts to enhance the mechanical properties. For example, incorporating hydroxyapatite into the PPF/DEF materials enhances the mechanical properties and cell affinity (25).

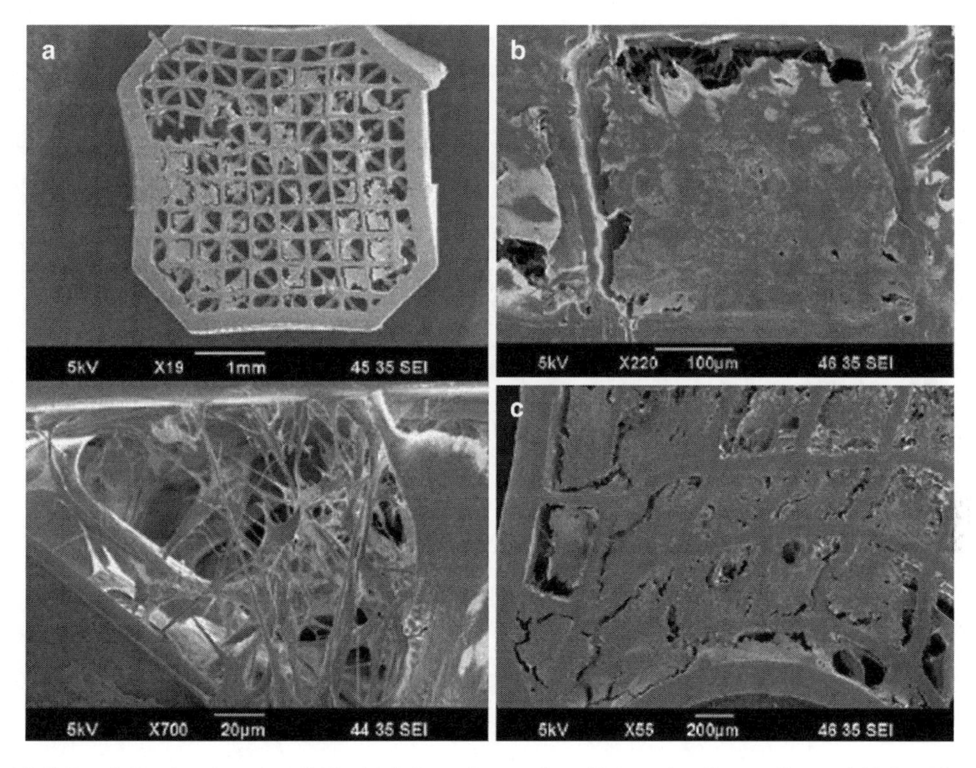

Fig. 12. Cultured chondrocytes of scaffolds: (**a**) 3 days after seeding, (**b**) 3 weeks after seeding, and (**c**) 2 weeks after seeding (1).

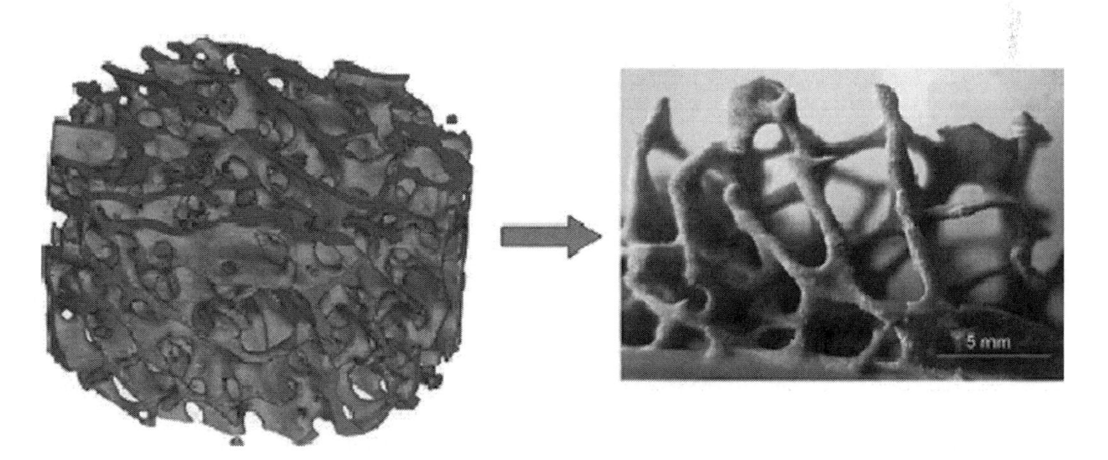

Fig. 13. Sponge-like structure prepared by SL (24).

2.1.2. MSTL Technology Using Nonbiodegradable Materials

Because no need exists to consider the biodegradability of a photopolymer, it is easier to develop a nonbiodegradable material with a high process resolution and strength. Structures can be fabricated through MSTL technology using PEGDMA as a biocompatible polymer, and research has been directed toward fabricating a 3-D scaffold structure and examining its cell proliferation effect (25). Other researchers have investigated the possibility of enhancing the

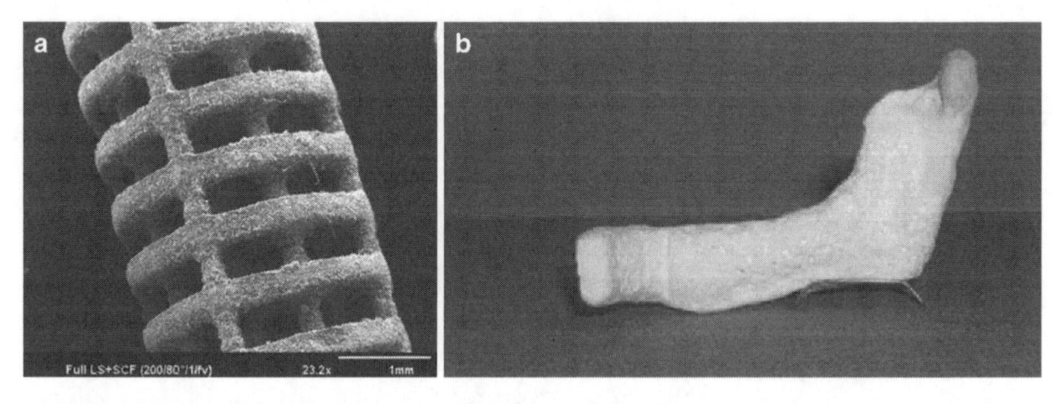

Fig. 14. (**a**) Composite sample of HA/OCM-2™ and (**b**) composite implant for a mandibular part (28).

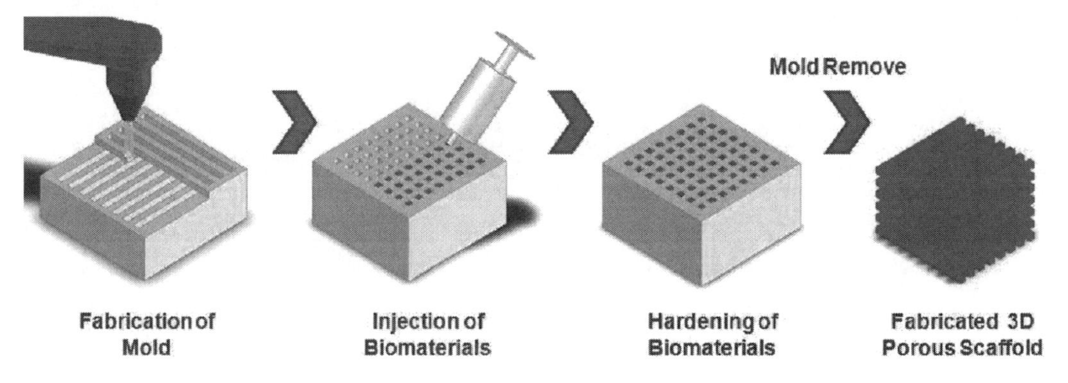

Fig. 15. Fundamental process of indirect SFF technology.

cell adhesion by combining RGD with scaffolds, and have fabricated scaffolds including growth factors by incorporating heparin (26). Commercial polymers, which have low cell toxicity, have also been used to fabricate scaffolds (Fig. 14) (27, 28).

2.2. Indirect Solid Freeform Fabrication Technologies

Indirect SFF technology based on MSTL is used to fabricate 3-D biocompatible scaffolds with a combined molding process. This technology was proposed to allow a wide variety of biomaterials to be used for tissue engineering applications. Figure 15 shows the fundamental process of indirect SFF technology based on MSTL. The biomaterial is injected into the high-resolution 3-D mold fabricated by MSTL. A 3-D porous scaffold can be fabricated after selectively removing the mold. This process is known as lost-mold shape forming (29). Additional molding and a mold removal process are necessary, unlike direct fabrication methods. This results in a more complex process for scaffold fabrication, but when one considers that only a small number of biomaterials have been approved for clinical treatment, the ability to choose from more biomaterials is one of the most important factors in medical applications of tissue engineering. Several indirect SFF technologies developed for tissue engineering are introduced in the following sections.

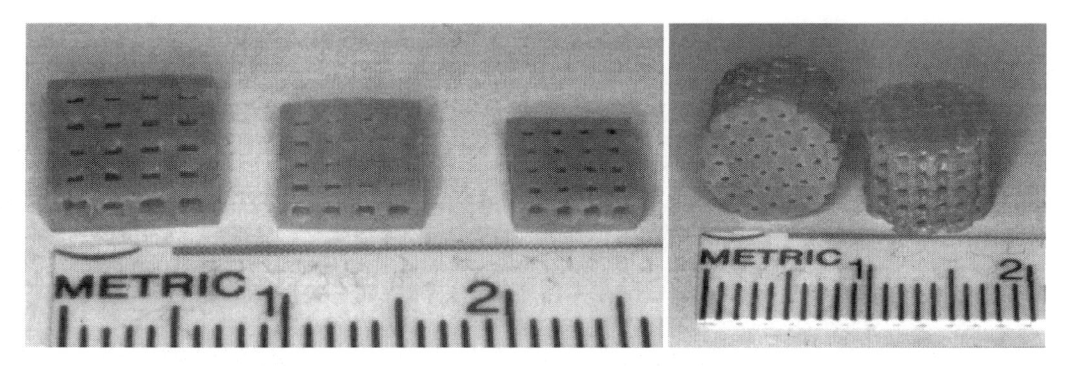

Fig. 16. Hydroxyapatite scaffold (24).

The mold removal process must not affect the injected biomaterials. Since the photopolymer material used for MSTL can be removed at temperatures above 700–800°C, highly heat-resistant ceramic material was first tried for scaffold fabrication. High-strength ceramic scaffolds were fabricated using a sintering process. Feinberg fabricated a negative mold using commercialized epoxy resin (29). After injecting a hydroxyapatite slurry, the negative mold was removed during a sintering process at 1,300°C, resulting in a high-strength hydroxyapatite structure. He demonstrated the usability of the proposed process by fabricating various 3-D structures (Fig. 16). Woesz used sintering to fabricate calcium phosphate scaffolds for bone tissue regeneration (30). Preosteoblasts, MC3T3-E1, were seeded in the 3-D porous scaffolds and the resulting cell proliferation and osteoblastic differentiation were examined.

Dichen created 3-D chitosan/gelatin scaffolds using MSTL (31). High-resolution MSTL technology was used to fabricate a mater-patterned 2-D PDMS mold. Chitosan/gelatin material was cast and freeze dried on the mold, resulting in a 2-D shape composed of natural polymer. A 3-D structure was created by assembling the 2-D layers. Cell culture results using hepatocytes were also presented. Figure 17 shows a fabricated chitosan/gelatin porous structure.

The studies described above used the existing sintering and 2-D molding techniques for the mold removal process. On the other hand, Liska developed a new soluble photopolymer for SL that had a high photoreactivity and mechanical strength (32). The soluble photopolymer could be removed by an alkali solution. A biocompatible copolymer was molded using the developed resin. And, Vallet-Regi used the etching properties of a commercialized photopolymer (33, 34). A 3-D mold was fabricated using Accrura SI10 from 3D Systems, Inc., and biomaterial was injected. The mold was then etched by a 2 M NaOH solution. Using this process, a scaffold composed of agarose, hydroxyapatite, and β-tricalcium phosphate was fabricated for bone tissue regeneration. Cho studied the molding process for various biomaterials, such as polycaprolactone, bone cement, and poly(lactic-co-glycolic) acid, for tissue engineering (35). The soluble photopolymer proposed by

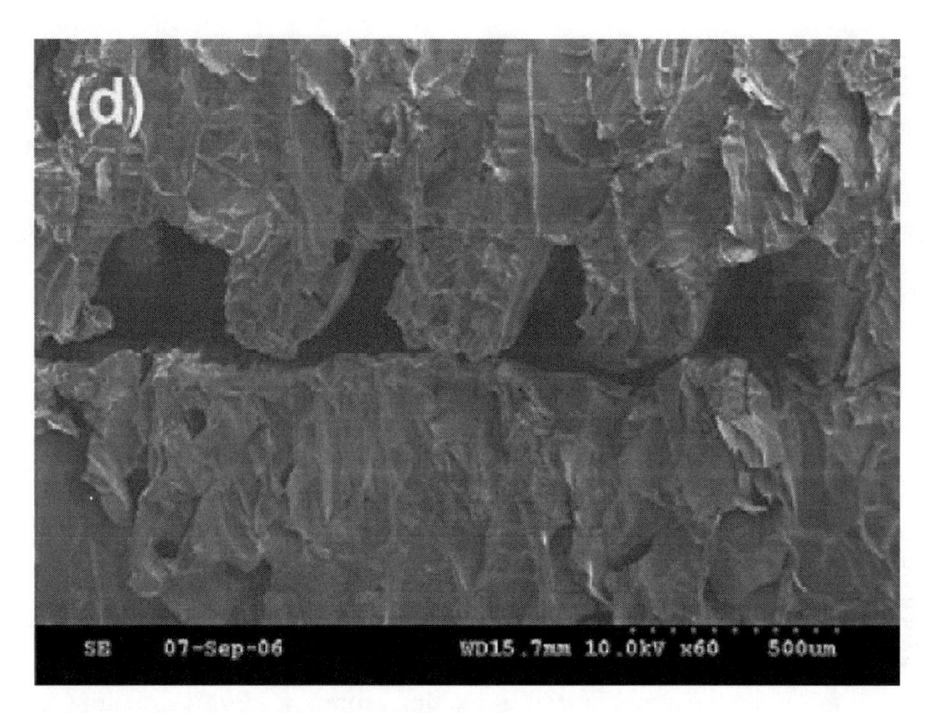

Fig. 17. SEM image of a fabricated chitosan/gelatin structure (31).

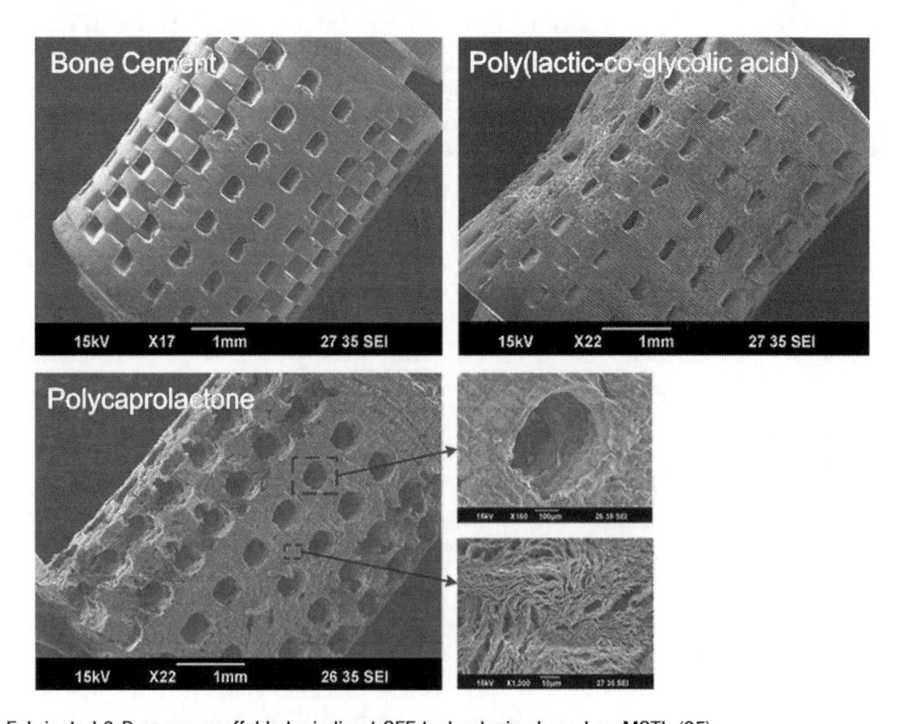

Fig. 18. Fabricated 3-D porous scaffolds by indirect SFF technologies based on MSTL (35).

Liska (32) and high-resolution MSTL technology (5) were used to fabricate 3-D scaffolds, and the lost-mold shape forming process was exploited using various biomaterials. Figure 18 shows several fabricated porous scaffolds composed of biomaterial.

3. Summary

MSTL is the best system for fabricating sophisticated 3-D micro-shapes among micro-SFF technologies. Various research groups have worked to increase the resolution of MSTL, and a resolution below 1 μm has been obtained, the highest reported resolution among all micro-SFF technologies. The technology can be automated by a CAD/CAM system based on medical image data from CT or MRI for medical applications.

Many researchers have attempted to apply this technology to tissue engineering. Some groups are actively carrying out research to develop new biocompatible/biodegradable photopolymers that will make it possible to apply MSTL to tissue regeneration. Lost-mold shaping processes based on MSTL are also being studied to diversify the selectivity of biomaterials.

MSTL provides high flexibility to the 3-D porous scaffold design for tissue regeneration compared to other micro-SFF technologies. This advantage will lead to 3-D porous scaffolds with superior abilities for tissue regeneration.

References

1. Lee S-J, Kang H-W, Park JK, Rhie J-W, Hahn SK, Cho D-W (2008) Application of microstereolithography in the development of three-dimensional cartilage regeneration scaffolds. Biomed Microdevices 10(2):233–241
2. Nakamoto T, Yamaguchi K, Abraha PA, Mishima K (1996) Manufacturing of three-dimensional micro-parts by UV laser induced polymerization. J Micromech Microeng 6(2):240–253
3. Schaeffer P, Bertsch A, Corbel S, Jézéquel JY, André JC (1997) Relations between light flux and polymerized depth in laser stereophotolithography. J Photochem Photobiol A: Chem 107:283–290
4. Kang H-W, Lee IH, Cho D-W (2004) Development of an assembly-free process based on virtual environment for fabricating 3D microfluidic systems using micro-stereolithography technology. Trans ASME: J Manuf Sci Eng 126(4):766–771
5. Jacobs PF (1992) Rapid prototyping & manufacturing: fundamentals of stereolithography. Society of Manufacturing Engineers in cooperation with the Computer and Automated Systems Association of SME
6. Sun W, Starly B, Nam J, Darling A (2005) Bio-CAD modeling and its applications in computer-aided tissue engineering. Computer-Aided Design 37:1097–1114
7. Lee IH, Cho D-W (2003) Micro-stereolithography photopolymer solidification patterns for various laser beam exposure conditions. Int J Adv Manuf Technol 22(5–6):410–416
8. Huang Y-M, Jeng J-Y, Jiang C-P (2003) Increased accuracy by using dynamic finite element method in the constrain-surface stereolithography system. J Mater Process Technol 140:191–196
9. Ikuta K, Hirowatari K (1993) Real three dimensional micro fabrication using stereo lithography and metal molding. Proc IEEE MEMS: 42–47
10. Ikuta K, Hirowatari K, Ogata T (1994) Three dimensional micro integrated fluid systems (MIFS) fabricated by stereo lithography. Proc IEEE MEMS: 1–6
11. Ikuta K, Ogata T, Tsubio M, Kojima S (1996) Development of mass productive micro stereolithography. IEEE: 301–306
12. Ikuta K, Maruo S, Kojima S (1998) New micro stereo lithography for freely movable 3D micro structure. Proc IEEE MEMS: 290–295
13. Bertsch A, Bernhard P, Vogt C, Renaud P (2000) Rapid prototyping of small size objects. Rapid Prototyping J 6(4):259–266
14. Bertsch A, Bernhard P, Renaud P (2001) Microstereolithography: concepts and

applications. In: Proceedings of the 8th IEEE international conference, vol 2. pp 289–298
15. Sun C, Fung N, Wu DM, Zhang X (2005) Projection micro-stereolithography using digital micro-mirror dynamic mask. Sen Actuators A 121:113–120
16. Han L-H, Mapili G, Chen S, Roy K (2008) Projection micro-printing of three-dimensional scaffolds for tissue engineering. Trans ASME, vol 130. pp 021005-1-4
17. Sanderson JE (1998) Bone replacement and repair putty material from unsaturated polyester resin and vinyl pyrrolidone. United States Patent 4722948, pp 1–14
18. Cooke MN, Fisher JP, Dean D, Rimnac C, Mikos AG (2002) Use of stereolithography to manufacture critical-sized 3D biodegradable scaffolds for bone ingrowth. J Biomed Mater Res 64B(2):65–69
19. Lee JW, Lan PX, Kim B, Lim GB, Cho D-W (2008) Fabrication and characteristic analysis of a poly(propylene fumarate) scaffold using micro-stereolithography technology. J Biomed Mater Res B Appl Biomater 87(1):1–9
20. Lee JW, Lan PX, Kim B, Lim GB, Cho D-W (2007) 3D scaffold fabrication with PPF/DEF using micro-stereolithography. Microelectron Eng 84(5–8):1702–1705
21. Lee JW, Anh NT, Kang KS, Seol Y-J, Cho D-W (2008) Development of a growth factor-embedded scaffold with controllable pore size and distribution using micro-stereolithography. TERMIS-EU chapter meeting, Porto, Portugal, 22–26 June 2008
22. Lee K-W, Wang S, Fox BC, Ritman EL, Yaszemski MJ, Lu L (2007) Poly(propylene fumarate) bone tissue engineering scaffold fabrication using stereolithography: effect of resin formulations and laser parameters. Biomacromolecules 8(4):1077–1084
23. Kwon IK, Matsuda T (2005) Photo-polymerized microarchitectureal constructs prepared by microstereolithography using liquid acrylated-end-capped trimethylene carbonate-based prepolymers. Biomaterials 26:1675–1684
24. Schuster M, Turecek C, Varga F, Lichtenegger H, Stampfl J, Liska R (2007) 3D-shaping of biodegradable photopolymers for hard tissue replacement. Appl Surf Sci 254:1131–1134
25. Lee JW, Lan PX, Seol YJ, Cho D-W (2007) PPF/DEF-HA composite scaffold using micro-stereolithography. TERMIS-EU chapter meeting, London, UK, 4–7 Sept 2007, pp 1745–1746
26. Mapili G, Lu Y, Chen S, Roy K (2005) Laser-layered microfabrication of spatially patterned functionalized tissue-engineering scaffolds. J Biomed Mater Res B Appl Biomater 75 (2):414–424
27. Engelmayr GC, Papworth GD, Watkins SC, Mayer JE, Sacks MS (2006) Guidance of engineered tissue collagen orientation by large-scale scaffold microstructures. J Biomech 39:1819–1831
28. Popov VK, Ivanov AL, Roginski VV, Volozhin AI, Howdle SM (2004) Laser stereolithography and supercritical fluid processing for custom-designed implant fabrication. J Mater Sci Mater Med 15:123–128
29. Chu T-MG, Halloran JW, Hollister SJ, Feinberg SE (2001) Hydroxyapatite implants with designed internal architecture. J Mater Sci Mater Med 12:471–478
30. Woesz A, Rumpler M, Stampfl J, Varga F, Fratzl-Zelman N, Roschger P, Klauschofer K, Fratzl P (2005) Towards bone replacement materials from calcium phosphates via rapid prototyping and ceramic gelcasting. Mater Sci Eng C 25:181–186
31. Jiankang H, Dichen L, Yaxiong L, Bo Y, Bingheng L, Qin L (2004) Fabrication and characterization of chitosan/gelatin porous scaffolds with predefined internal microstructures. J Mater Sci Mater Med 15:123–128
32. Schuster M, Infuhr R, Turecek C, Stampfl J, Varga F, Liska R (2006) Photopolymers for rapid prototyping of soluble mold materials and molding of cellular biomaterials. Monatsh Chem 137:843–853
33. Cabanas MV, Pena J, Roman J, Vallet-Regi M (2006) Room temperature synthesis of agarose/sol–gel glass pieces with tailored interconnected porosity. J Biomed Mater Res A 78A(3):508–514
34. Sanchez-Salcedo S, Nieto A, Vallet-Rgi M (2008) Hydroxyapatite/β-tricalcium phosphate/agarose macroporous scaffolds for bone tissue engineering. Chem Eng J 137:62–71
35. Kang H-W, Cho D-W (2007) Indirect solid freeform fabrication (SFF) using microstereolithography technology. TERMIS-AP, Tokyo, Japan, 3–5 Dec 2007

INDEX

Michael A.K. Liebschner (ed.), *Computer-Aided Tissue Engineering,* Methods in Molecular Biology, vol. 868,
DOI 10.1007/978-1-61779-764-4,

N

O

P

Q

R

S

T

V

W

Printed by Publishers' Graphics LLC